Environmental Engineering

Series Editors: U. Förstner, R.J. Murphy, W.H. Rulkens

Springer-Verlag Berlin Heidelberg GmbH

R.K. Jain · Y. Aurelle ·
C. Cabassud · M. Roustan · S.P. Shelton (Eds.)

Environmental Technologies and Trends

International and Policy Perspectives

With 202 Figures

 Springer

Series Editors

Prof. Dr. U. Förster

Arbeitsbereich Umweltschutztechnik
Technische Universität Hamburg-Harburg
Eißendorfer Straße 40
D-21073 Hamburg, Germany

Prof. Robert J. Murphy

Dept. of Civil Engineering and Mechanics
College of Engineering
University of South Florida
4202 East Fowler Avenue, ENG 118
Tampa, FL 33620-5350, USA

Prof. Dr. ir. W. H. Rulkens

Wageningen Agricultural University
Dept. of Environmental Technology
Bomenweg 2, P. O. Box 8129
NL-6700 EV Wageningen, The Netherlands

Editors

Prof. Dr. Ravi K. Jain   University of Cincinnati, OH, USA
Prof. Dr. Yves Aurelle   INSA, Toulouse, France
Dr. Corinne Cabassud   INSA, Toulouse, France
Dr. Michel Roustan   INSA, Toulouse, France
Prof. Dr. Stephen P. Shelton   University of New Mexico, USA

ISBN 978-3-642-63913-5

Cataloging-in-Publication Data applied for

Die Deutsche Bibliothek – Cip-Einheitsaufnahme
Environmental technologies and trends: international and policy perspectives /
R. K. Jain ... - Berlin ; Heidelberg ; New York ; Barcelona ; Budapest ; Hong Kong ; London ;
Milan ; Paris ; Santa Clara ; Singapore ; Tokyo : Springer, 1996
  (Environmental engineering)
  ISBN 978-3-642-63913-5   ISBN 978-3-642-59235-5 (eBook)
  DOI 10.1007/978-3-642-59235-5
NE: Jain, Ravi K.

Production: ProduServ GmbH Verlagsservice, Berlin
Typesetting: Data-conversion by Fotosatz-Service Köhler oHG, Würzburg
SPIN: 10514881   61/3020 – 5 4 3 2 1 0 – Printed on acid-free paper

Preface

This manuscript was made possible by the exceptional support provided by INSA (Institut National des Sciences Appliquées) Toulouse, the University of New Mexico and the University of Cincinnati College of Engineering.

The authors, as listed in this book, took the time to prepare excellent manuscripts focusing on scientific and technical areas relevant to emerging environmental issues. These manuscripts were rigorously reviewed and refereed by scientists and engineers before inclusion in this book. An introductory chapter was prepared to summarize and integrate technical issues covered and the last chapter was written to present policy perspectives.

The editors are most grateful to the contributors, sponsor organizations, and many colleagues who were kind enough to assist us in making this manuscript possible. Background information about the editors, principal authors and other contributors to this manuscript follows.

Editors

Professor Dr. Ravi K. Jain
Associate Dean for Research and International Engineering
College of Engineering
University of Cincinnati
Mail Location 0018
Cincinnati OH 45221-0018
U.S.A.

Ravi K. Jain received B.S. and M.S. degrees in Civil Engineering from California State University and a Ph.D. in Civil Engineering from Texas Tech. He studied Public Administration and Public Policy at Harvard, earning an M.P.A. degree. Dr. Jain has directed a staff of over 200 engineers and scientists conducting interdisciplinary research for the US Army, and was the Founding Director of the Army Environmental Policy Institute (AEPI). He has held research and faculty appointments at the University of Illinois (Urbana-Champaign) and Massachusetts Institute of Technology. He has been a Littauer Fellow at Harvard University and a Fellow, Churchill College, Cambridge University. He has been a consultant to federal agencies, international organizations, and private industry. Dr. Jain has served on numerous National Task Forces and Advisory Councils for the Department of Defense, NSF, Navy, Army, EPA and NAS. He has published ten books and numerous journal articles and technical reports. His two most recent books include: Environmental Assessment (McGraw Hill, 1993); and Management of Research and Development Organizations-Managing the Unmanageable (Wiley, 1991).

Professor Dr. Yves Aurelle
INSA Toulouse
Directeur du Departement G.P.I
Complexe Scientifique de Rangueil
31077 Toulouse Cédex
France

Dr. Aurelle is a Professor of Chemical Engineering at the National Institute of Applied Sciences of Toulouse, France, and Head of the Department of Industrial Process Engineering. He has been an engineer in Industrial Chemistry (1967), Engineer Doctor (1974) and State Doctor (1980). Over his career, he has performed research in chemistry at the Societe National des Petroles d'Aquitaine, Research Center of Lacq and later in water and wastewater treatments at the National Institute of Applied Sciences. He is currently Research Team Manager in physical chemical treatment of water in the Laboratoire d'Ingénierie des Procédés de l'Environment (L. I. P. E.). Specializing for more than twenty years in the treatment of hydrocarbon pollution, he has developed with ELF Company many new treatment processes. He has published over 150 reports and journal articles in environmental engineering literature. He received the Chevalier grade of Academic Palms in 1994.

Dr. Corinne Cabassud
INSA Toulouse
Departement Génie des Procédés Industriels
Complexe Scientifique de Rangueil
31 077 Toulouse Cédex
France

Dr. Cabassud is an Associate Professor in Chemical Engineering in the Department of Génie des Procédés Industriels at the Engineer School INSA Toulouse, France. She graduated with a degree in Chemical Engineering from the Engineer School ENSIC Nancy, France (1983) and received a Ph. D. in Chemical Engineering in 1986 after a research study concerning membrane biological reactors applied to water denitrification. From 1986 to 1993 she was Research Engineer and then Research Team Leader in the Membrane Laboratory of the Lyonnaise des Eauz Company. Specializing in membrane processes, she participated in research and development of new ultrafiltration modules and processes for drinking water production. Dr. Cabassud joined INSA in 1993 and she is now in charge of research on membrane separations for environmental applications.

Professor Dr. Michel Roustan
INSA Toulouse
Department Génie des Procédés Industriels
Complexe Scientifique de Rangueil
31 077 Toulouse Cédex
France

Michel Roustan, Ph.D., is a Professor of Chemical Engineering in the Department of Industrial Engineering Processes ("Génie des Procédés Industriels" – GPI) at the Engineer School INSA Toulouse (France). He received his engineer diploma in Chemical Engineering (INSA). He obtained his Ph.D. in 1978 after a research concerning mixing in the gas-liquid reactors. From 1986–1993 he was Head of the Department of GPI-INSA and currently he is Director of the Research Laboratory "Laboratoire d'Ingénierie des Procédés de l'Environnement" (LIPE EA 833) at INSA and in charge of the Doctoral Studies Programme. Dr. Roustan works in the field of multiphasic reactors (gas-liquid-solid) applied to the wastewater or drinking water processes. He contributed to the conception and development of several new processes operating now in full scale. He is the author of nearly 180 papers in chemical engineering (journal articles, oral communications and book chapters) and has directed 27 Ph.D. Dissertations.

Professor Dr. Stephen P. Shelton
Department of Civil Engeneering
University of New Mexico
Albuquerque NM 87131
U.S.A.

Stephen P. Shelton, Ph.D., is a Professor of Civil Engineering at the University of New Mexico, Albuquerque, New Mexico USA. During his tenure at UNM, he held numerous administrative positions including Department Chair, Technical Lead for Research for the New Mexico Hazardous Waste Consortium, and Interim Director of the Army Environmental Policy Institute. He formally has been on faculties at the University of South Carolina and the University of Texas. A graduate of the University of Tennessee, Dr. Shelton is registered as a professional engineer in six states and a Diplomat in the American Academy of Environmental Engineers. Over his career, he has performed research for government and industry in many areas related to environmental engineering and policy. He has published over 100 reports and journal articles in the environmental science, engineering and policy literature.

Principal Authors

Gérard Antonini, Professor, Université de Technologie de Compiègne, Division Génie des Transferts et Energétique, Département Génie Chimique, B. P. 64206 Compiègne Cédex – France

Philippe Aptel, Université Paul Sabatier, Laboratoire de Génie Chimique et electrochimie – URA CNRS – 192, 118, Route de Narbonne, 31062 Toulouse Cédex – France

Michel Astruc, Professor, Université dé Pau et Pays de l'Adour, Laboratoire de Chimie Analytique, Avenue de l'Université, 64000 Pau – France

Jean-Marc Audic, Senior Research Engineer, Centre International de Recherche sur l'Bau et l'Environnement (Research Center of Lyonnaise des Eaux) 38, Rue du Président Wilson, 78230 LePecq – France

Dominique Bastoul, Associate Professor, Institut National des Sciences Appliquées de Toulouse (INSA), Laboratoire d'Ingéniérie des Procédés de l'Environnement, Complexe Scientifique de Rangueil, 31077 Toulouse Cédex – France

Nouredine Bel Hadj Tahar, Ph.D. Student, Groupe Génie des Procédés de Séparation et Membranes, Laboratoire de Génie Chimique et Electrochimie – URA CNRS – 192, Université Paul Sabatier, 118, Route de Narbonne, 31062 Toulouse Cédex – France

Vincent Boisdon, Research Engineer (Research Center of Compagnie Generale des Eaux), Anjou Recherche, Chemin de la Digue, BP 76, 78603 Maisons Laffitte – France

Christophe Bonnin, Group Manager, Anjou Recherche (Research Center of Compagnie Générale des Eaux), Chemin de la Digue, BP 76, 78603 Maisons Laffitte – France

Richard C. Brenner, Environmental Engineer, U.S. Environmental Protection Agency, 26 W. Martin Luther King Drive, Mail Stop 420, Cincinnati, OH 45268, U.S.A.

Bernard Capdeville, Professor, Institut National des Sciences Appliquées de Toulouse (INSA), Laboratoire d'Ingeniérie des Procédés de l'Environnement, Complexe Scientifique de Rangueil, 31077 Toulouse Cédex – France

Michael Clifton, Senior Scientist, Groupe Génie des Procédés de Séparation et Membranes, Laboratoire de Génie Chimique et Electrochimie – URA CNRS – 192, Université Paul Sabatier, 118, Route de Narbonne, 31062 Toulouse Cédex – France

Hubert Debellefontaine, Professor, Institut National des Sciences Appliquées de Toulouse (INSA), Laboratoire d'Ingéniérie des Procédés de l'Environnement, Complexe Scientifique de Rangueil, 31077 Toulouse Cédex – France

Géraldine Deiber, Ph.D. Student, Institut National des Sciences Appliquées de Toulouse (INSA), Laboratoire d'Ingéniérie des Procédés de l'Environnement, Complexe Scientifique de Rangueil, 31077 Toulouse Cédex – France

Bernard Delanghe, Associate Professor, Université de Pau et Pays de l'Adour, Laboratoire de Chimie Analytique, Avenue de l'Université, 64000 Pau – France

John D. Ditmars, Manager, Environmental Sciences and Engineering Group, Environmental Assessment Division, Argonne National Laboratory, 9700 South Cass Ave. – EAD/900, Argonne, IL, 60439, U.S.A.

Christian Fonade, Professor, Institut National des Sciences Appliquées de Toulouse (INSA), Département GBA – URA CNRS 544, Complexe Scientifique de Rangueil, 31077 Toulouse Cédex – France

Jean-Noël Foussard, Associate Professor, Institut National des Sciences Appliquées de Toulouse (INSA), Laboratoire d'Ingeniérie des Procédés de l'Environnement, Complexe Scientifique de Rangueil, 31077 Toulouse Cédex – France

Donald J. Freeman, P.E., Environmental Engineer, U.S. Army Armament Research Development and Engineering Center (ARDEC), Picatinny Arsenal, New Jersey 07806–5000, U.S.A.

Dimitre Hadjiev, Associate Professor, Institut Universitaire de Technologie, Université de Caen, 14000 Caen Cédex – France

Robert Hausler, Professor, Université du Québec à Montreal, Case postale 8888, succursale Centre-Ville, Montreal (Québec), Canada, H3C 3P8

Edmond Julien, Associate Professor, Institut National des Sciences Appliquées de Toulouse (INSA), Laboratoire d'Ingéniérie des Procédés de l'Environnement, Complexe Scientifique de Rangueil, 31077 Toulouse Cédex – France

Germain Lacoste, Professor, ENSIGC, 18, Chemin de la Loge, 31078 Toulouse Cédex – France

Alain Laplanche, Professor, Ecole National Supérieure de Chimie de Rennes, Laboratoire Chimie des Nuisances et Génie de l'Environnement, Avenue du Général Leclere, 35700 Rennes – France

Pierre Le Cloirec, Professor, Ecole National Supérieure des Techniques Industrielles et des Mines d'Alés, 6, Avenue de Claviéres, 30319 Alés Cédex – France

Xavier Lefebvre, Research Engineer, Institut National des Sciences Appliquées de Toulouse (INSA), Laboratoire d'Ingéniérie des Procédés de l'Environnement, Complexe Scientifique de Rangueil, 31077 Toulouse Cédex – France

Marie-Hélène Manero, Associate Professor, I.U.T. génie Chimique, Chemin de la Loge, 31078 Toulouse Cédex – France and Institut National des Sciences Appliquées de Toulouse (INSA), Laboratoire d'Ingeniérie des Procédés de l'Environnement, Complexe Scientifique de Rangueil, 31077 Toulouse Cédex – France

Hugo Matamoros, Ph. D. Student, Institut National des Sciences Appliquées de Toulouse (INSA), Laboratoire d'Ingéniérie des Procédés de l'Environnement, Complexe Scientifique de Rangueil, 31077 Toulouse Cédex – France

Michel Mauret, Research Engineer, Institut National des Sciences Appliquées de Toulouse (INSA), Laboratoire d'Ingéniére des Procédés de l'Environnement, Complexe Scientifique de Rangueil, 31077 Toulouse Cédex – France

Pierre Mocho, Ph.D. Student, Ecole National Supérieure des Techniques Industrielles et des Mines d'Alés, 6, Avenue de Claviéres, 30319 Alés Cédex – France

Philippe Moulin, Ph. D., Groupe Génie des Procédés de Séparation et Membranes, Laboratoire de Génie Chimique et Electrochimie – URA CNRS – 192, Université Paul Sabatier, 118, Route de Narbonne, 31062 Toulouse Cédex – France

Etienne Paul, Lecturer, Institut National des Sciences Appliquées de Toulouse (INSA), Laboratoire d'Ingéniérie des Procédés de l'Environnement, Complexe Scientifique de Rangueil, 31077 Toulouse Cédex – France

Jane E. Rael, Ph. D. Candidate, Department of Civil Engineering, University of New Mexico, Albuquerque, NM 87131, U.S.A.

Jean-Christophe Rouch, Research Engineer, Groupe Génie des Procédés de Séparation et Membranes, Laboratoire de Génie Chimique et Electrochimie – URA CNRS – 192, Université Paul Sabatier, 118, Route de Narbonne, 31062 Toulouse Cédex – France

André Savall, Professor, Université Paul Sabatier, Laboratoire de Génie Chimique et Electrochimie – URA CNRS – 192, 118, Route de Narbonne, 31002 Toulouse Cédex – France

Christophe Serra, Ph.D. Student, Groupe Génie des Procédés de Séparation et Membranes, Laboratoire de Génie Chimique et Eletrochimie – URA CNRS – 192, Université Paul Sabatier, 118, Route de Narbonne, 31062 Toulouse Cédex – France

Ed Dean Smith, Chief, Environmental Engineering Division, U.S. Army Construction Engineering Research Laboratory, 2902 Newmark Drive, Champaign, IL 61821, U.S.A.

Francis L. Smith, Graduate Assistant, University of Cincinnati, Civil & Environmental Engineering, P. O. Box 210071, Cincinnati, OH 45221–0071, U.S.A.

Ian Smyth, Senior Research Fellow, Department of Mechanical Engineering, University of Southampton (UK)

George A. Sorial, Senior Research Associate, University of Cincinnati, Civil & Environmental Engineering, P. O. Box 210071, Cincinnati, OH 45221-0071, U.S.A.

Makram Suidan, Schneider Professor and Department Head, Civil and Environmental Engineering, University of Cincinnati, P. O. Box 210071, Cincinnati, OH 45221–0071, U.S.A.

Martin Thew, Professor of Fluid Technology, Mechanical Engineering Department, University of Southampton (UK)

Hany Zaghloul, Environmental Engineer, U.S. Army Construction Engineering Research Laboratory (USACERL), P. O. Box 9005, Champaign, IL 61826–9005, U.S.A.

Other Contributors

Susan Bill, Coordinator, Professional and Executive Development, Research and International Engineering, University of Cincinnati, P.O. Box 210018, Cincinnati, OH 45221–0018, U.S.A.

Todd Bragdon, Graduate Student, Civil and Environmental Engineering, University of Cincinnati, P. O. Box 210071, Cincinnati, OH 45221–0071, U.S.A.

Shafiq Islam, Assistant Professor, Civil and Environmental Engineering, University of Cincinnati, P. O. Box 210071, Cincinnati, OH 45221–0071, U.S.A.

Jean-Pierre Soula, Professor, Centre de Communication et de Gestion, INSA Toulouse, Complexe Scientifique de Rangueil, 31077 Toulouse Cédex, France, U.S.A.

Donna Vitucci, Consulting Editor, Research and International Engineering, University of Cincinnati, P.O. Box 210018, Cincinnati, OH 45221–0018, U.S.A.

Françoise Voillot, Directeur, Service des Relations Exterieeures et Internationales, INSA Toulouse, Complexe Scientifique de Rangueil, 31077 Toulouse Cédex, France

Contents

1 Introduction

Introduction

Costs associated with pollution control and related environmental requirements have been increasing steadily. In the United States of America alone, these costs in 1994 climbed to over $ 115 billion. By the year 2000, these costs are expected to top $ 200 billion per year (EPA environmental investments, EPA-230-11-90-083, 1990). Total annualized costs, by funding source, of U. S. pollution control for 1972, 1980, 1987 and projections for 2000 in 1986 dollars are provided in Fig. 1. These costs include operating costs, interest (at 7 percent), and depreciation. For 1990 these costs were estimated at $ 100 billion (1986 dollars) or $ 115 billion (1990 dollars) or 2.1 percent of the U. S. GNP (CEQ, 1990, p. 50). With the considerable investment nations are making in the environment area, **environmental technologies** and related **trends** and policy considerations should be of great interest to the technical and public policy community.

Pollution control costs in all industrialized countries, much like the United States, are increasing. Developing countries are also in a difficult pollution control environment. Pollution control costs are an extra burden to evolving industries strapped for resources; however, failure to adequately control environmental pollution only defers an even greater cost to future generations. Success in managing the paradox of emerging industry with concomitant pollution control has met with mixed success in emerging industrial economies.

This book seeks to provide the reader with a diverse international perspective on emerging pollution control technologies. The ability of a government to implement policies that support sound environmental management is too often compromised by a limited knowledge of emerging technology. The emerging technologies described herein provide a reference to assist policy makers broaden their perspective on the possibilities for environmental management. Major topics covered include:

- Drinking Water
- Air Pollution
- Wastewater Treatment
- Hazardous Waste Management
- Soil and Groundwater Contamination

Articles provided in these topics address specific technologies as well as summarize recent technological innovations. In addition to these technical issues, the final chapter discusses ways to integrate environmental technology and environmental policy. Integration of fundamentally sound science and engineering in the environmental policy making process is needed. It is hoped that the papers and discussions contained herein can in some way contribute to catalyzing this needed change.

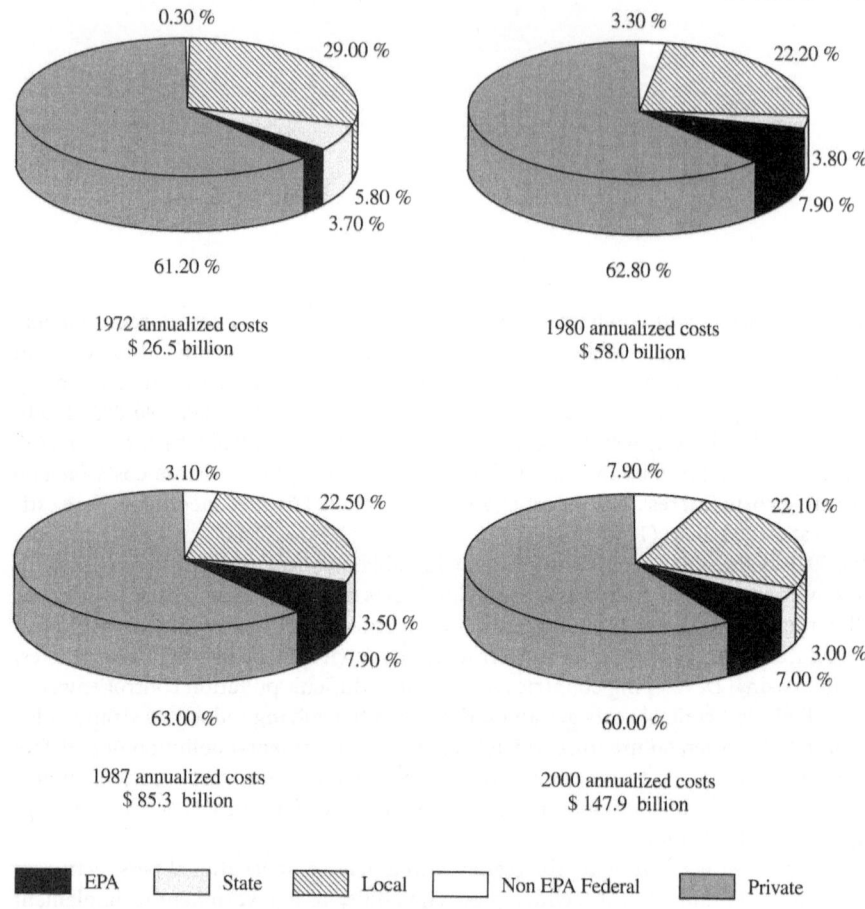

Fig. 1. Total annualized costs of U.S. pollution control by funding source for 1972, 1980, 1987 and projections for 2000, with present implementation in 1986 dollars

Chapter 2 on Drinking Water focuses on contemporary technologies, existing regulations, and future trends in drinking water supply and treatment.

Roustan and Cabassud present a comparative evaluation of environmental regulations in Europe and the United States. They also discuss the consequences of regulations on quality and treatment of drinking water. Two new technologies, membrane filtration and ozonation, are proposed and their utility and efficiency are discussed in detail.

Delanghe et al. report results on the adsorption of organic micropollutants on different types of activated carbon, those being grains, powders, and fibers. They show that performance of fiber activated carbon is significantly better than that of granular activated carbon, and is quite similar to that of powder activated carbon. The breakthrough curves obtained with fiber activated carbon are particularly steep, suggesting a smaller mass transfer resistance than for the granular activated carbon.

Laplanche et al. discuss results on the modeling of pesticide removal from drinking water by ozonation processes. Comparisons between bubble towers and static mixers are reported based on their hydrodynamic characteristics and chemical reactivity.

Julien and Aurelle evaluate the performance of pervaporation as an alternative technique of micropollutants removal from drinking water. It was found that compared to air-stripping, pervaporation and stripping on microporous membranes need smaller column volume with a comparable operating cost. It is suggested that further research is necessary to lower the membrane cost.

Chapter 3 on Air Pollution presents various methodologies to remove volatile organic compounds and odors from the air stream. This chapter also discusses induction heating to regenerate activated carbon for recycling organic compounds.

Manero discusses the advantages and disadvantages of five volatile organic compound removal techniques. Adsorption, scrubbing, and condensation systems remove pollutants with a recovery of solvents, whereas biological and oxidizing systems remove pollutants by destroying them. He argues that there are no single processes to remove all types of contaminants and choice of removal technique must be made on a case by case basis.

Suidan et al. discuss biofilters as an alternative to the treatment of gas streams contaminated with low to moderate concentrations of volatile organic carbons. They evaluate performance of three biofilter media and used toluene as the model volatile organic carbon. The pelletized media appear to be the most effective trickle bed air biofilter because of its high treatment efficiency at high organic loads and ease of biomass control.

Bonnin et al. discuss the importance of odor control in wastewater treatment plants located near sensitive, densely populated area. Their study investigates reliable deodorizing methods to achieve better control of industrial plant sizing and operating parameters. In addition, models prepared for their study provide a computerized tool capable of scaling a chemical washing deodorization plant, predicting the pollutants remaining in the treated air, and the reagents required permanently to ensure an absence of olfactory pollution.

Mocho and Le Cloirec evaluate induction heating to regenerate activated carbon for recycling volatile organic compounds. Their study demonstrates that efficiency of induction regeneration mainly depends on the granulometery of activated carbon and of susceptors.

Chapter 4 on Wastewater Treatment addresses the issues related to biological and physico-chemical treatment techniques of wastewater treatment.

Paul et al. discuss the effect of grease solubilization and optimal process monitoring on grease aerobic digestion. Grease residues contain a high oxygen demand and present several difficulties in wastewater treatment. Their study demonstrates that saponification and neutralization of raw grease led to its partial solubilization and emulsification by fatty acids, and therefore enhanced the substrate degradation rate.

Aptel et al. discuss membrane gas liquid contactors in water and wastewater treatment. Their study was based on a comparison between straight modules and coiled modules. Their results indicate that a simple change in the design of membrane hollow fiber contactors can largely improve mass transfer: a coiled configuration is 200 to 300 % more efficient compared to a straight configuration.

Fonade analyzes the role of hydrodynamic effects on the performance of biological treatment of wastewaters. It is argued that, in addition to the microorganism physico-chemical environment, hydrodynamic complexity and the characteristics of the reactor itself (whether cylindrical reactor or lagoons) would play an important role in the overall treatment efficiency.

Bastoul et al. describe different multiphasic reactors that could be applied to the aerobic biological treatment of urban wastewater. The hydrodynamics, mass transfer performance, and removal efficiency data are discussed for four different multiphasic reactors.

Aurelle reviews the current status of the physico-chemical processes for wastewater treatment. It is suggested that future treatment innovations must focus on increased efficiency, smaller dimension, soundproofing, and minimizing olfactory pollution from the effluent.

Hadjiev et al. present a new design of hydrocyclonic separator capable of effectively treating a three component (solid-liquid-liquid) system. Laboratory tests and field data related to treatment of complex oily wastewater are discussed. The potential integration of hydrocyclones with other wastewater treatment technologies, specifically membranes, is considered in cases where more rigorous treatment is required.

Cabassud et al. give an overview of the conventional treatment methods and presents more recent procedures based on membrane procedures for treating oily wastewater from industrial effluents. This study concludes that future innovative techniques for oily wastewater treatment may involve improvement of ultrafiltration membrane effectiveness and the use of combinations of different membrane techniques. Specifically, combining the ultrafiltration process with a partial chemical destabilization process may be effective.

Savall and Bel Hadj Tahar examine the adequacy of electrochemical oxidation of phenol for wastewater treatment. Lead dioxide was used as an electrochemical catalyst for the oxidation of aromatics in aqueous acidic solutions. The study shows that electrochemical oxidation of phenol and its derivatives need higher electric charge on a lead dioxide anode than on a tin dioxide anode. The performance of these electrodes could be improved by using either a separator or a graphite cathode.

Foussard et al. give a detailed overview of the wet air oxidation process for treating aqueous organic wastes. The process is well suited for fairly concentrated, nonbio-degradable, toxic wastes. However, its use is rather limited because of the associated capital costs. This study demonstrates that by introducing small amounts of hydrogen peroxide and ferrous salts continuously, and by using heterogeneous catalysts oxidation, kinetics may be significantly enhanced.

Chapter 5 focuses on recent developments in **hazardous waste treatment and management.**

Antonini provides a brief state-of-the-art review of the hazardous waste treatment methodologies. Relative advantages and disadvantages of various physico-chemical and thermal treatment methods are discussed. A case study concerning a mobile incineration unit under oxygen enriched conditions is also presented.

Falcon et al. present wet peroxide oxidation process results from laboratory as well as pilot plant studies. Their study indicates that the addition of metal ion catalysts (Fe, Cu, and Mn) increases process efficiency compared to the conventional wet peroxide oxidation process. In addition, economical data for the pilot plant are assessed.

Lacoste reports results of heavy metal recovery by electrolytic techniques. A computer system was developed to control the evolution of the metal deposits. The computer system is capable of changing hydrodynamic and electrical parameters based on the response of the system.

Smith presents a state-of-the-art review of the plasma arc technology for the destruction and vitrification of military hazardous wastes. Adequacy of plasma arc technology for treating heavy metal contaminated wastes are discussed.

Rael and Shelton discuss results of laboratory scale studies for in-situ removal of hexavalent chromium and divalent cadmium from contaminated groundwater. Based on the batch studies, it is shown that treatment efficiency as high as 97% for chromium and 98.7% for cadmium was achieved. Results of the subsequent column investigations indicate that the batch study sorption projections were low.

Chapter 6 presents assessments of the cleanup process and strategies, electrokinetic decontamination, and mulitdisciplinary approaches all related to **soil and groundwater contamination.**

Ditmars reviews of the recent assessments of the cleanup process and application of strategies and technologies to enhance decision-making for cleanup. The solution to such problems involves both regulatory change to allow more flexibility in decision-making and the introduction of technology to improve decision-making. This study provides examples of emerging decision approaches and technologies which speed up site characterization and remediation.

Astruc et al. report results from an electrokinetic soil decontamination study. Their study focuses on the transport of nitrate and heavy metals from kaolinite and sewage sludges. Their study indicates that heavy metal decontamination rates in kaolinite and sludges were predominantly controlled by liquid-solid interactions. Effective removal of cadmium, zinc, and nickel were obtained.

Hausler et al. present a multidisciplinary approach to groundwater management by invoking the total quality theory and applying it to environmental engineering.

To facilitate the interactions between scientists from Europe and the United States of America, at the end of each paper several recent abstracts of related research reprinted in the international journals are appended. This information should be a useful resource and pave the way for increased level of crosscontinental environmental research.

Chapter 7, Environmental Trends and Related Policy Perspectives, provides a discussion of ways that emerging environmental technologies could be integrated into environmental policy formulation. Understanding the complex scientific and engineering issues related to emerging environmental technologies and future trends is a critical part of the environmental policy mosaic. The focus of the discussion in this

Chapter seeks to describe broad environmental trends and policy perspectives within the context of the intersect among science, engineering, and policy.

References

Council on Environmental Quality (CEQ). 1990. Environmental Quality. U.S. Government Printing Office, Washington, D.C.
Environmental Protection Agency (EPA). 1990. Environmental Investments. EPA-230-11-90-083, pp. 2–5

2 Drinking Water

2.1
Drinking Water Production: Processes and Emerging Technologies

MICHEL ROUSTAN · CORINNE CABASSUD

Abstract

In the field of drinking water production it is necessary to establish new regulations and guidelines, due to the decrease of the resource water quality. The paper presents the regulations in Europe and in the US, and the consequences on the evolution of the water treatment plants, depending on the water quality objectives and the quest for quality. Then two new technologies are proposed, concerning the using of membranes and ozone. The membrane processes are described and their application at some full scale plants are presented. The using of ozone as oxidation and disinfection agent is presented with the effect of the hydraulics on the quality of the water. The advantages of these two technologies are also described.

2.1.1
The Regulations

In the field of drinking water production, the development of even more specific and efficient analytical methods and apparatus, the increasing quest for quality, and the decrease in adequate resource water have pushed towards the establishment of new regulations and guidelines in Europe and in the US.

European Regulation

To characterize the required quality for a drinking water 62 parameters have been identified, which can be classified as organoleptic, physico-chemical, toxic and microbiological parameters and pesticides.

For each of these 62 parameters the European regulation sets the maximum level and defines the analytical method and sampling frequency that must be used.

- The turbidity guide line is 2 NTU.
- The maximum content of total pesticides is fixed to 0.5 µg/l, with a maximum of 0.1 µg/l per pesticide. More stringent values concern some specific pesticides like aldrine, dieldrine and hexachlorobenzene (respectively 0.03, 0.03 and 0.01 µg/l).

Concerning the microbiological parameters, sampling volumes are regulated (Table 1) and European regulation does not consider, at this time, contaminants like Giardia Cysts.

Table 1. European Regulations on microorganisms in drinking water

	Unit	Regulation level
Total coliforms	N/100 ml	0 for 95% of the measurements
Thermotolerant coliforms	N/100 ml	0
Fecal Streptococcus	N/100 ml	0
Clostridium	N/100 ml	5
Pathogenic Staphylococcus	N/100 ml	0
Salmonella	N/5 l	0
Entero Viruses	N/10 l	0

US Regulation

The US Environmental Protection Agency (USEPA) is currently developing regulations, including enforceable maximum contaminant levels (MCLs), for numerous drinking water contaminants. Certain waterborne contaminants in drinking water may be limited by a maximum contaminant level (MCL) or by a treatment technique set by USEPA. For each MCL promulgated, USEPA must specify best available technologies (BATs) which are feasible and effective for removal of the contaminant.

The Surface Water Treatment Rule (SWTR) requires all systems using surface water sources or underground waters under the influence of surface waters, to disinfect providing a multi-barrier treatment including both filtration and disinfection. The aim is to achieve a minimum reduction of 99.9 percent (3 logs) in Giardia Cysts and 99.99 percent (4 logs) in viruses. Regardless of the technology employed, the treatment must meet turbidity performance criteria. The filtered water turbidity must be < 0.5 NTU in 95 percent of the measurements collected each month with no sample having a turbidity > 5 NTU.

The USEPA is developing a regulation for disinfectants and their by-products. The final rule is expected by June 1995. Presently, the only disinfection by-products regulated are the trihalomethanes (THM). The current THM regulation is 100 µ/l, as the sum of all THM species present.

The California SWTR has listed technologies approved to treat surface water. For example, four filtration processes are enabled: conventional filtration, direct filtration, diatomaceous earth filtration and sand filtration.

Other or new technologies can only be used if they have been specified BAT by USEPA; that means they must meet the following stringent criteria:

- have the removal efficiencies required
- be demonstrated at field scale
- be compatible with other treatment technologies
- have a reasonable lifetime
- be commercially available
- be affordable to large utilities

Concerning the underground waters, USEPA is developing a Ground-Water Treatment Rule (GWTR) that will be promulgated in June 1995. Some ground-water treatment plants will be required to install disinfection and a 4 logs removal or inactivation of viruses may be used as a criterion, as for the surface-waters.

2.1.2
The Theory of Evolution in the Field of Water Treatment

The demand for quality, the pressure of the regulations, the decrease in adequate water resources, the development of even more specific analytical apparatus and the growing knowledge in physico-chemical phenomena and in chemical engineering have induced a significant evolution in the schematic conception of a drinking water treatment plant (Fig. 1).

From 1930 to 1992 the plants have been compacted and automated, the unit operations have been optimized and the quality of the produced water has also been improved. Prechlorination has been proven dangerous to humans due to possible formation of chlorinated organic compounds, and has been reduced and often eliminated.

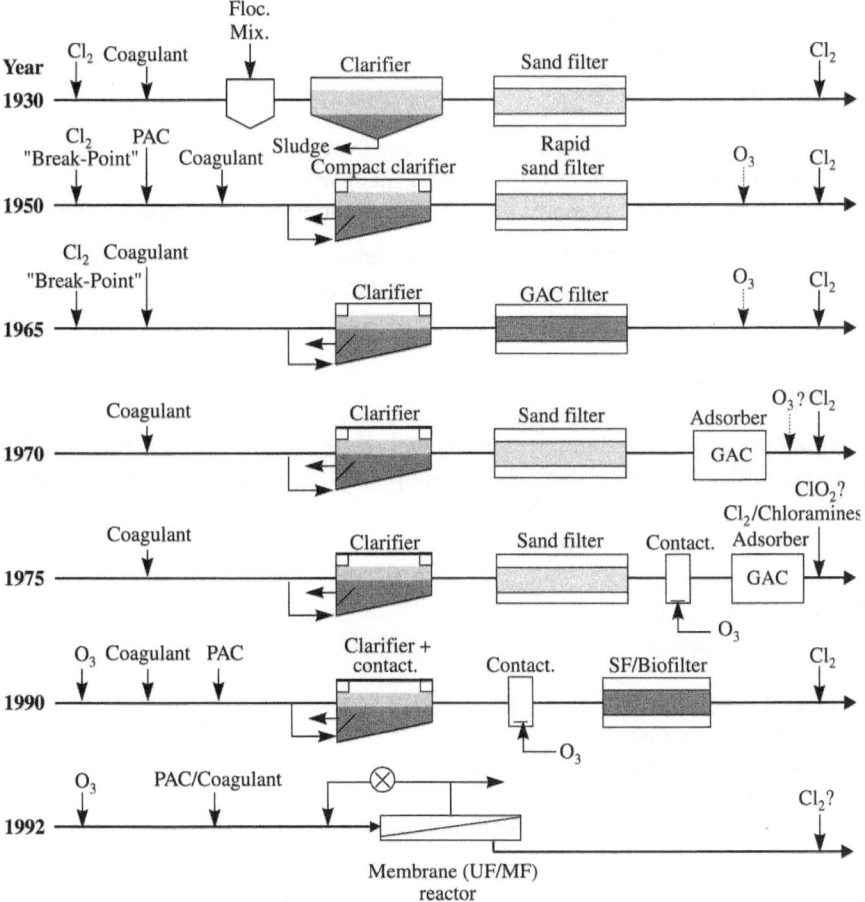

Fig. 1. The theory of evolution in the field of water treatment (example of surface water)

Since 1965 the fight against toxic organic compounds and especially pesticides has been the starting point for the development and use of Granulated Activated Carbon (GAC) filters and then GAC adsorbers.

Moreover we can distinguish two recent main evolutions:

- the use of ozone, first to reduce the chlorination doses and thus the disinfection by-products since 1950, and then to simultaneously improve the efficiency of GAC adsorption.
- the use of membrane technologies in water treatment. They now replace the process clarification + sand filtration, the latter process being nearly 2000 years old !

The Problematic of Raw Water Treatment

The difficulty in drinking water treatment is linked on one side to the numerous compounds being in the raw water and, on the other side, to the great variations of each compound concentration with time and seasons. Surface or underground waters may contain:

- particles (sand, clays, silt…)
- colloidal material (clays, hydroxides, organic matters…)
- microorganisms (bacteria, viruses, parasites) and algae
- soluble mineral compounds (nitrates, iron, manganese)
- organic compounds (estimated by COT, UV measurements, pesticides, THMFP).

Table 2 suggests here a classification of some selected usual waters. We distinguish 5 classes of underground waters and a surface water. For each class a specific treatment plant schematic can be established, as will be seen later.

The Conventional Plants

The basic principles used in the conventional treatment operations include gravity separation (flotation or settling), chemico-physical or biological reactions and filtration.

The unit operations include:

- grit removal
- coagulation, filtration
- aeration, oxidation
- settling
- sand filtration
- oxidation, ozonation
- Granulated Activated Carbon adsorption (GAC filters)
- ion exchange
- biological treatments (for nitrates or phosphorus removal for example)
- chlorination

Additional treatments can be used to modify the ionic quality of the water, to correct hardness or to prevent corroding, for example.

Table 2. Classification of the waters used for drinking water production

				Surface waters
I	II	III	IV	VI
Occasional turbidity	(I) + iron	(I) + pesticides	(I) + nitrates	Turbidity
suspended solids	and/or Manganese		V	Suspended solids
SiO_2			(IV) + pesticides	Organic compounds:
Bacteria				• COT
Viruses				• UV_{254}
				• pesticides
				• THMFP
				Bacteria
				Viruses
				Algae

For the six previously identified water classes, the schematic conventional treatment plant is presented in the Figs. 2 to 6.

Most of these processes suffer at least from one of the following limitations:

- Addition of a chemical reagent.
  One of the objectives of potable water utilities is the production of water with constant high quality. To achieve this, the treatment conditions must be adjusted continuously in accordance with the raw water quality, especially by fixing the right dose of coagulating agent and by adjusting the flocculation pH. Automation of the plant operation is a possible way to manage this problem, however, it is impossible to avoid absolutely any under or overdosage of reagents. As a consequence the resulting water quality is not always optimal.
- Formation of undesirable chemical by-products due to the chlorination of organic compounds mostly during the pre-oxidation or final disinfection steps.
- If the coagulation or precipitation step is not properly handled, the treated water may contain high residuals of reagents, thereby marring the final quality of the distributed water.
- Generation of sludges which contain a concentrated amount of *man-made* products in addition to the species coming from the resource. Among them are aluminum and iron hydroxides, and/or organic polymers used as coagulants and flocculants. These non natural compounds are no longer allowed to be discharged into rivers.

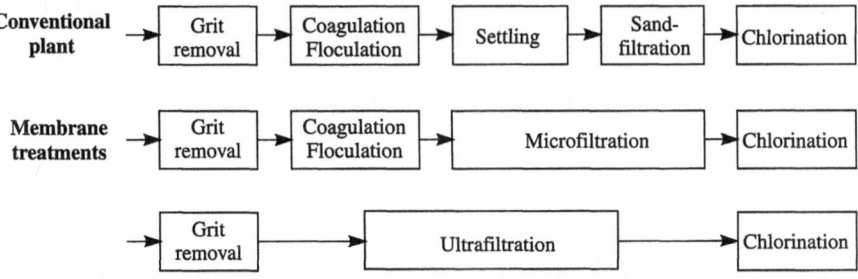

Fig. 2. The alternative treatment plants for a basic ground water (Water n°I)

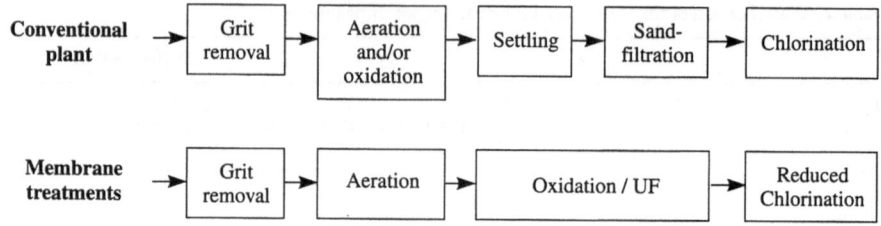

Fig. 3. The alternative treatment plants for a ground water with iron and/or manganese (Water n°II)

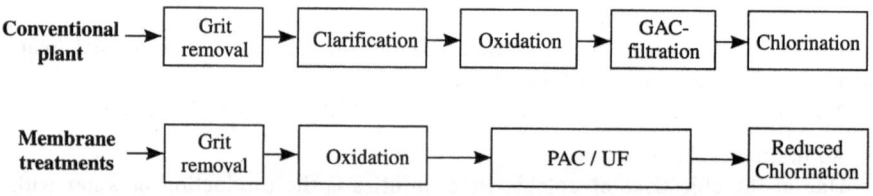

Fig. 4. The alternative treatment plants for a ground water with pesticides (Water n°III)

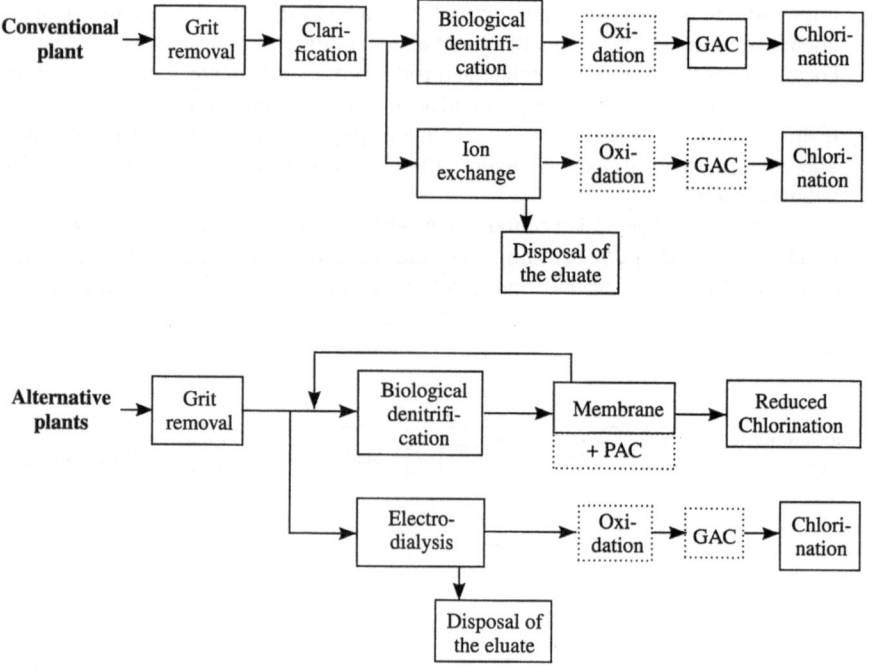

Fig. 5. The alternative treatment plants for an Underground water with nitrates (Water n°IV) – with nitrates and with pesticides (Water n°V)

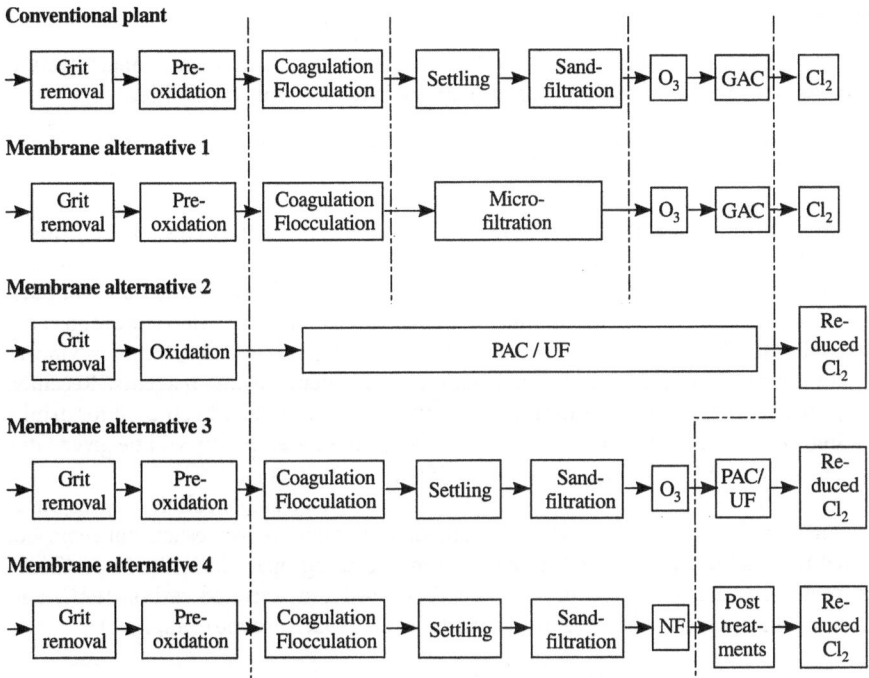

Fig. 6. The alternative treatment plants for a surface water (Water n°VI)

- GAC filters have a possibility of breaking down when they are saturated, thus inducing high and non foreseeable pesticides peaks in the produced water. This could be avoided by using GAC regenerations at least every six months.
- Final quality and reliability, which depend on the contaminant load in the raw water.

To propose suitable technical solutions to solve a part of the major drawbacks of the conventional processes, the public and industrial research has focused for nearly ten years on two promising technologies:

- membranes
- ozonation,

which now define the new tendencies for drinking water production.

2.1.3
New Tendency I: Membranes

Membrane filtration is a pressure driven separation process. A membrane is an absolute filter and, provided that its pore-size or cut-off is property chosen, it will reliably separate selected components from water.

Membrane Selectivity

Figure 7 shows a comparison of main membrane processes and conventional filtration processes (sand filtration) in terms of their particle-size removal abilities.

- Sand filtration operates as a depth filter. It can partially reject the species up to 1 μm, like algae, sand, cysts and the biggest clays. But it is not an absolute filter and it is an excellent support for microorganisms and algae growing.
- Microfiltration membranes are theoretically able to reject particles up to 0.1 μm to 1 μm, like most of the suspended solids and the bacteria. The viruses and salts can freely cross the membrane. To assure a good retention of small species and colloids, microfiltration has been sometimes coupled with a pre-flocculation (Bourdon et al., 1988; Moulin et al., 1991; Vickers et al., 1994). This process keeps the previously described drawbacks of the conventional treatments using reagents. Recently, processes using microfiltration without pre-flocculation have been used for drinking water production. Some results concerning the water quality will be given later (Yoo et al, 1994; Jacangelo and Adham, 1994; Tazi-Pain et al., 1994).
- Ultrafiltration membranes permit removal of suspended solids, colloids, algaes, parasites, bacteria and viruses. The ions and small organic molecules are not removed. Ultrafiltration can replace the flocculation + settling operations. When pesticides removal is needed a new process coupling powdered activated carbon (PAC) and ultrafiltration has been developed (Baudin et al., 1992; Anselme et al., 1991).

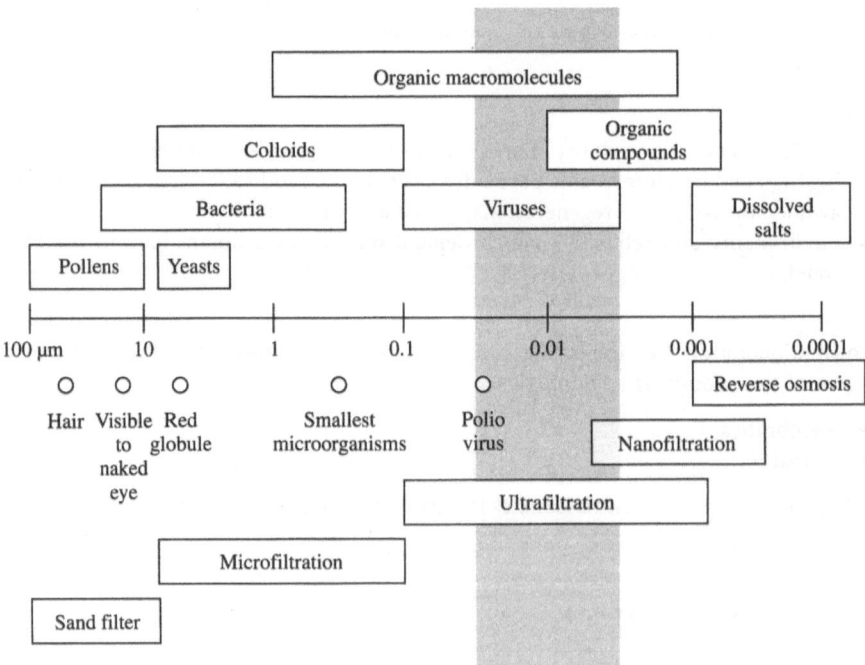

Fig. 7. Comparison of membrane processes in terms of their removal abilities

- Nanofiltration membranes (Moulin et al., 1993) remove some of the dissolved salts (essentially the divalent species) and some of the small dissolved organic compounds. These membranes are very sensitive to particle fouling and must often be used as a refining treatment after a complete water clarification.

Fluid Dynamics and Membrane Process

Membranes are offered in four different shapes (plate and frame, spiral wound, tubular, hollow fibers). For water treatment, the criteria to be taken into consideration for a system are the following:

- pretreatment requirements,
- ease to clean by backwashing,
- low energy consumption,
- high ratio filtration area/volume, i.e. compactness.

Hollow fiber systems offer the best arrangement and this geometry has been chosen for drinking water production.

Table 3 described the two most usual membrane processes for drinking water production.

Process A has been developed since 1986 and the first plant has been under operation since 1988 in Amoncourt – France. It uses cellulosic ultrafiltration hollow fibers, having a molecular weight cutoff of 100 kDa (Aptel and Vial, 1986; Bersillon et al., 1991; Thébault et al., 1992). The fibers are assembled in 100 mm bundles. Seven bundles are assembled to constitute a large module, which has been especially designed for drinking water production and which is, at present, the largest commercially available module in the world, offering a 50 m² membrane surface area. This module has been patented.

The hollow fibers have a dense inner skin and are operated in an inside-out flow configuration: the raw water to be treated flows inside the fibers and the permeate is

Table 3. Specifications of the two most used membrane processes for drinking water production

		PROCESS A	PROCESS B
Membranes	process	Ultrafiltration	Microfiltration
	material	Cellulose Acetate	Polypropylene
	cut-off	100 kDa	0.2 μm
Module Configuration		Hollow fiber	Hollow fiber
Flow configuration		Inside-Out	Outside-in
Internal fiber tube diameter		0.93 mm	0.31 mm
External fiber tube diameter		1.67 mm	0.65 mm
Effective membrane surface area per module		50 m²	10 m²
Recommended operating conditions			
• pressure		5–20 psi	3–15 psi
• flux		60–75 gfd	70 gfd
• flow velocity		0–1.5 m/s	0 m/s
Cleaning procedure specification		Liquid backwash	Air backwash
• recommended effective backwash pressure		32 psi	100 psi

obtained at the outer side of the fibers. The main flow is tangential to the fiber axis and the recommended flow velocity can vary from 0, corresponding to dead-end filtration, to less than 1 m/s, depending on the raw water quality. In the case of a coupling with powdered activated carbon, the velocity can be increased to 1.5 m/s. As the inner fiber diameter is small, the flow is laminar in any case, thus inducing low energy consumptions.

The process B (developed by Memcor-Australia) uses hydrophobic polypropylene microfiltration hollow fibers, having a 0.2 μm mean pore size. Fibers are assembled in a bundle to constitute a 10 m² module (Olivieri et al., 1991; Moulin et al., 1993).

The hollow fibers are outer skinned fibers and the flow configuration is outside-in. The raw water flows at the outside of the fibers in a dead-end mode, i.e. with 100% yield.

Membrane Fouling and Fouling Remediations

For drinking water production, the flux through the membranes can be limited by the following phenomena, very often considered all-together and called "fouling":

- adsorption
- deposition of particles and/or organic matters as a surface cake, or inside the membranes pores and structure, in the case of microfiltration membranes.
- precipitation of soluble species on or inside the membrane.

The latter phenomenon is very often difficult to distinguish from cake deposition and is currently considered and measured as a deposit.

The flux through ultrafiltration or microfiltration membranes is then given by:

$$J = \frac{\Delta P}{\mu \, (R_m + R_a + R_d)} \tag{1}$$

where R_m = membrane resistance,
R_a = adsorption resistance,
R_d = total deposit resistance,
ΔP = transmembrane pressure.

From equation (1), if $R_m > (R_a + R_d)$ the mass transfer is controlled by the membrane. If $(R_a + R_d) > R_m$ we have fouling resistance control and the flow of permeate is rapidly independent of the nature of the membrane.

Adsorption is a long-term phenomenon and is hydraulically irreversible. Chemical cleanings have to be operated to recover the initial mass transfer properties of the membrane. As many contaminants or foulants being in the raw water are capable of hydrophobic interactions, hydrophobic membranes are more likely to adsorb these materials (Fane 1994). The frequency and the kind of the necessary chemical cleaning depend on the raw water quality and on the membrane used. For example, using the process A for the treatment of karstic underground-water necessitate a regeneration procedure with some specific soap, less often than once a year.

The deposit formation and mechanism are linked to the pore size and pore size distribution, to the surface charge carried by the membrane and to the operating conditions.

Microfiltration membranes can retain particles and other species on their surface but also inside the pores, and then fine particles collecting may arise inside the cake. The term R_d may include four different mechanisms.

Ultrafiltration membranes collect the particles and suspended solids on their surface. The membrane is regenerated (deposits of particles are removed) using backflush procedures.

For the process A, the backflush (described on Fig. 8) consists in placing the permeate at the outside of the fibers under a pressure greater than the feed pressure. The mean backflush transmembrane pressure varies from 2 to 2.5 bars. The direction of the flow through the wall of the hollow fiber is then reversed and the cake can be detached and then transported out of the fiber by the circulating flow. The effect of this backflush permits a 100% recovery of the initial flux for the cellulosic hollow fiber, developed by Lyonnaise des Eaux – France, but the recovery was less or even null for all the other tested hollow fiber materials (Cabassud et al., 1992).

For the process B using microfiltration outside-in hydrophobic hollow fibers, the recommended backwash is completely different: clean compressed air at a pressure of 6 bars is forced into the center and through the walls of the fibers, thus lifting the deposit from the outer surface.

Then many rinsing and flushing procedures are applied to remove the dislodged cake from the module. The backwash procedure has been patented by Memcor (Jacangelo, 1994).

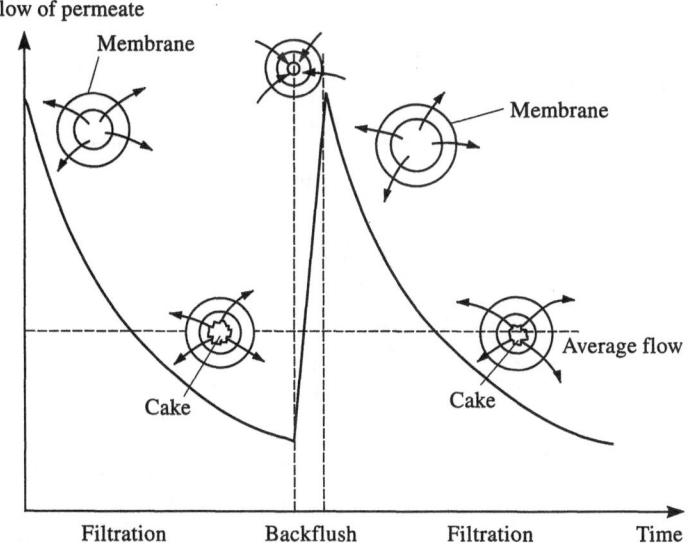

Fig. 8. Principle of backflush for the UF process

2.1.4
Quality of Output Water

Ultrafiltration

As loaded as the resource water may be, the turbidity of the ultrafiltered water is constant and remains less than 0.2 NTU. Figure 9 shows the great improvement of the distributed water quality in Douchy – France after the start-up of the ultrafiltration plant (Bersillon et al., 1991; Cabassud et al., 1991).

The 100 kDa ultrafiltration membrane behaves as an absolute filter so UF provides the ability to reject micro-organisms, bacteria, viruses and parasites like Giardia cysts.

Numerous studies have been focused on the disinfection ability of this UF membrane (Jacangelo et al., 1991). Table 4 presents results obtained on the removal of Giardia, Cryptosporidium, Escherichia Coli, Pseudomonas Aeruginosa and MS2 viruses for synthesized waters containing high contaminant contents (Jacangelo and Adham, 1994).

For surface-water treatment, membranes have been proved to remove cysts of Giardia and phages of Shigella and Escherichia Coli and the removal is higher than the one obtained with conventional treatments including preozonation, settling, sand filtration, ozonation and GAC filtration (see Tables 5a and 5b). The use of those membranes can then permit to reduce the chlorine doses necessary for the final disinfection.

The dissolved organic substances are partially eliminated by the UF membrane. The reduction of the permanganate demand in the permeate varies between 10 and 30%, the TOC removal between 10 and 15% (Cabassud et al., 1991) and the trihalomethane precursors (THMFP) removal is about 20%. The UF membranes do not remove micropollutants smaller than 10 nm in size and 100 kDa in molecular weight.

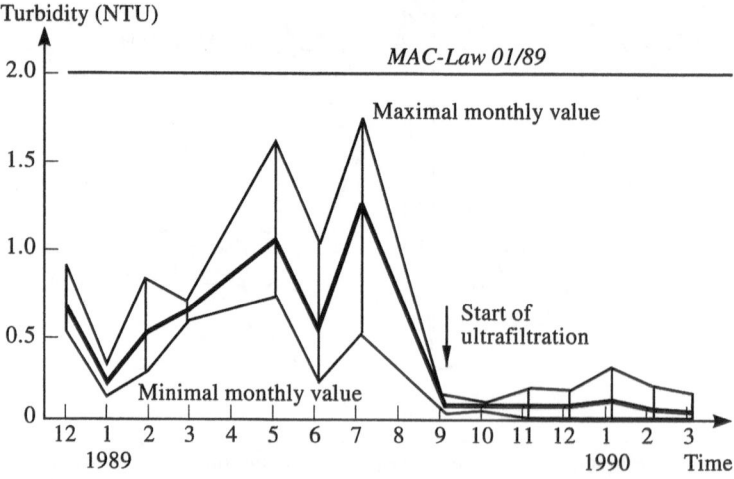

Fig. 9. Turbidity of the distributed water before and after the starting up of the ultrafiltration plant

Table 4. Removal of microorganisms by ultrafiltration (Process A), and microfiltration (Process B)

	Ref.	PROCESS A			PROCESS B		
		Feed	Permeate	Log Removal	Feed	Permeate	Log Removal
Giardia Muris	J94	5.4×10^4/l	<1/l	>4.7	5.4×10^4/l	<1/l	>4.7
	J94	1.5×10^5/l	<1/l	>5.2	1.5×10^5/l	<1/l	>5.2
	C94				2.7×10^4/l	<1/l	>4.4
Crystosporidium parvum	J94	2.6×10^4/l	<1/l	>4.4	2.6×10^4/l	<1/l	>4.4
	J94	8.3×10^4/l	<1/l	>4.9	8.2×10^4/l	<1/l	>4.9
Escherichia Coli	J94	6.6×10^7/100 ml	<1/100 ml	>7.8	6.6×10^7/100 ml	<1/100 ml	>7.8
	J94				9.8×10^7/ml	<1 ml	>6.0
Pseudomonas Aeruginosa	J94	5.3×10^8/100 ml	<1/100 ml	>8.7	1.5×10^8/100 ml	<1/100 ml	>8.2
MS2 Virus	J94			>6	1.6×10^7/ml	<1/100 ml	<0.5

J = Jacangelo; C = Coffey.

For underground-waters containing micropollutants or for surface-waters having high TOC and sometimes pesticides contents, a coupling of powdered activated carbon (PAC) with UF membranes has been developed (Anselme et al., 1991; Anselme et al., 1992). For an added PAC dose of 40 mg/l, the TOC removal in Suresnes-France on Seine river water is average 40 % in winter and 60 to 70 % in summer. Pesticides and PAH are very efficiently adsorbed on PAC. The results obtained on the river Seine, downstream from Paris (Anselme et al., 1991), are presented on Table 6 and show that the European regulation values are respected. Furthermore, preoxidation before PAC + UF treatment allows the increase of the UV adsorbance removal and to solve problems of taste and odors.

The process A can be an alternative treatment to clarification + disinfection.

Microfiltration

Microfiltration using the process B produces a water quality nearly identical to water produced by ultrafiltration as far as physico-chemical parameters are concerned: 20 % of total organic carbon is removed and permeate turbidity remains lower than 0.2 NTU, whatever the turbidity of the raw water.

Table 5a. Number of Cysts of Giardia per liter in Suresnes

Seine river	3,700	25,400	1,750	1,900
permeate	0,000	0,000	0,000	0,000
Sand filtered water	0,005	0,000	0,000	0,000
GAC filtered water	0,005	0,000	0,008	0,000
	1990/12/15 PAC + UF	1991/1/14 UF	1991/1/15 UF	1991/2/4 UF

Table 5b. Number of Phages of Shigella and *Escherichia Coli* per liter

Seine river	391,000 *169,000*	5975,000 *193,000*	2128,000 *131,000*
Permeate	0,000 *0,000*	0,000 *0,000*	0,000 *0,000*
	1990/12/10	1991/1/14	1991/1/28

Table 6. Efficiency of the PAC + UF treatment for the removal of micropollutants

	Raw	PAC + UF	EEC Standards
Total pesticides ng/l	800 1700 240	< 20 130 < 20	< 500
Total PAH ng/l	351 116 44	< 10 < 10 < 10	< 200

Concerning the disinfection ability, the results in Table 5 also show an identical removal of Giardia, Cryptosporidium, Escherichia Coli and Pseudomonas Aeruginosa. The major difference lies in the MS2 viruses removal which is less than 0.5 log for the MF process compared to more than 6 log for the UF process.

No process coupling has been suggested to improve the removal of TOC, THMFP nor pesticides. The process B can be an alternative treatment for clarification.

SWTR Requirements

Both processes A and B can meet the current SWTR filtration requirements and the disinfection requirements of 3 log removal of Giardia (Jacangelo and Adham, 1994; Coffey et al., 1994). These membranes should also meet the enhanced SWTR requirements that may regulate other micro-organisms such as Cryptosporidium. The UF process can also easily meet the current SWTR requirements of 4 log removal of virus. The MF process using higher cut-off membrane recover partially the viruses in the permeate and does not meet the SWTR disinfection requirement.

The Alternative Membrane Treatment Plants

Many comparative studies have been carried out on the choice and place of membrane processes for drinking water production (Anselme et al., 1993: Jacangelo and Adham, 1994). Due to its effects on water quality, the UF process can be an alternative to clarification + disinfection, the coagulation + mineral MF process an alternative to clarification and the previously described MF process is an alternative to clarification. Different ways of coupling chemical reaction and membrane separations have been proposed, like PAC adsorption and biological denitrification.

Figures 2 to 6 present the alternative membrane treatments for each water category which was previously specified.

We must point out the interest of membrane bioreactors (Fig. 5) which would encounter a significant development in the next years (5 plants will soon be under operation in France) and which can be coupled with PAC pretreatment for water containing pesticides and nitrates. This process has the advantage to include a biological treatment, while producing a concentrated eluate like ion exchange. On the other side, the micro-organisms are totally removed by the membranes, thus allowing reduced final chlorination.

Electrodialysis is another interesting and modular membrane process. A plant has yet been realized in Italy by Eurodia – France for drinking water denitrification.

Concerning the surface waters (Fig. 6), the alternatives could be either a replacement of clarification by processes including membranes operations (alternative 1 and 2) OR the use of membrane processes as refining treatment by PAC/UF OR nanofiltration after a nearly complete conventional treatment (alternative 3 and 4) to improve the quality of the distributed water.

The choice will depend on the resource composition, on the quality of the water desired and also on the nominal daily production and investment and treatment costs.

The Membrane Plants in Operation

The first membrane drinking production plant was operated in Keystone – Colorado in 1987. It uses MF polymeric membranes and its production capacity is about 105 m³/day. Then, in 1988, Lyonnaise des Eaux started the first UF plant on an underground water in Amoncourt – France. (240 m³/day).
    Presently we number a total of:

- 24 UF plants (process A), 21 being in France, representing a total production capacity of 31,400 m³/day
- 10 polymeric MF plants (2 in Australia, 7 in USA and 1 in France), representing a total production capacity of 21,900 m³/day (Tazi-Pain et al., 1994).

Many plants are under construction or consideration, mostly in France and in California. The starting up of 2 polymeric MF plants in France has yet been announced and 2 UF plants are under construction in France, also: a 28,000 m³/day plant for 1995 and a 55,000 m³/day for 1996.

Conclusion

Because of the more stringent water regulations and a decrease in resource quality, membranes processes are becoming an alternative to conventional treatments for drinking water production.
    The membrane alternative offers the following advantages:

- the treated water quality is independent of the raw water quality. Raw water quality variations and operating parameters influence only the plant productivity,
- it does not require reagent adjunction,
- it does not generate by-products,
- it is an absolutely separate step,
- it permits reduction of chlorination doses,
- it is compact,
- it is easily automated.

2.1.5
New Tendency II: Ozonation

In the field of water treatment, ozone is used for various purposes, including oxidation of compounds and disinfection. Several parameters must be considered for the design of ozone reactors:

- gas-liquid mass transfer efficiency,
- kinetic aspects,
- hydrodynamic behavior.

The main types of ozone contactors are described on Fig. 10.
    In taking into account the relationship between the kinetics of reactions and the hydrodynamic behavior of the liquid phase, it is possible to predict the performances of an existing reactor or, for a new reactor, to select the type of contactors as well as the

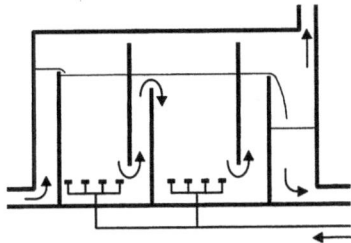

a Scheme of a conventional bubble contactor
   (according to Degrémont Handbook)

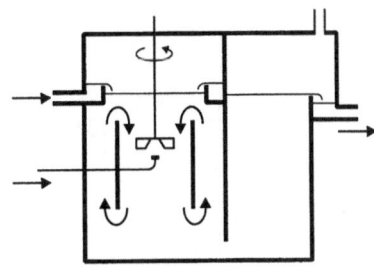

b Ozone contactor with turbine
   (according to Degrémont Handbook)

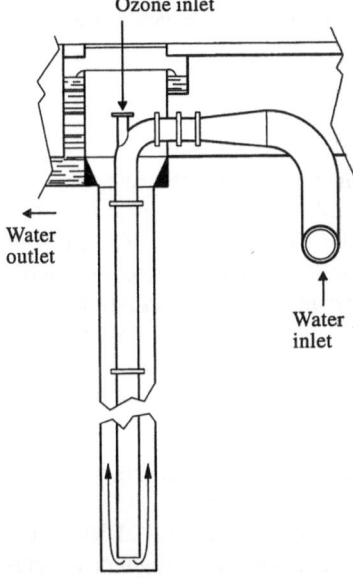

c Scheme of industrial Deep-U-tube

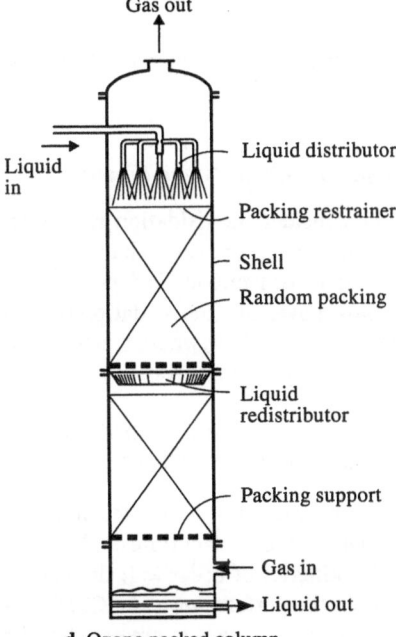

d Ozone packed column

Fig. 10. Main types of ozone contactors

design and operating parameters in order to achieve the selected water quality treatment objectives.

Residence Time Distribution Curves

The RTD's curves can be obtained from tracer tests performed either on pilot plants or on full scale plants (Roustan et al., 1993; Le Sauze et al.,1991; Martin et al., 1993). They can be modeled by different kinds of mathematical models: the tanks in series model characterized by number j (j = 1 for a CSTR, j → ∞ for a Plug Flow), or the axial dispersion model characterized by the Peclet number (Pe = 0 for a CSTR, Pe → ∞ for a Plug Flow). One example is given on Fig. 11.

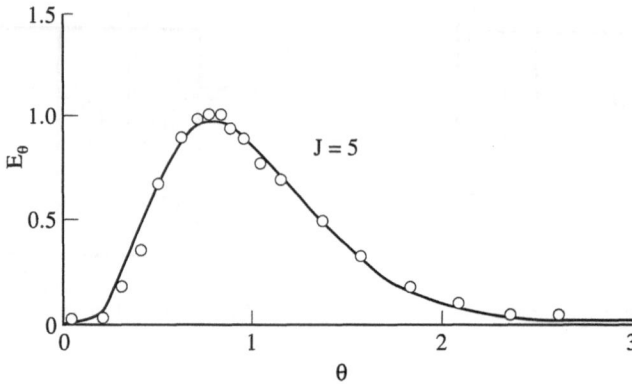

Fig. 11. RTD curve for Croissy post-ozonation (water flow rate: 1400 m³/h ; gas flow rate: 50 m³/h)J well mixed reactors in series

Coupling Hydraulics and Kinetics

If we consider the oxidation reactions with 0, 1st or 2nd order kinetic rates, the performance of the reactor, characterized by the rate of completion X of the reaction (fractional conversion), is dependent on the hydraulics of the reactor. Table 7 presents the laws governing the oxidation reaction completion rate, X in relation to the kinetics and the type of flow, where k is the apparent kinetic constant equal to k_0 [O₃] residual, C_0 is the initial concentration and τ is the hydraulic contact time (ratio of volume over flowrate).

In the water treatment field, models generally correspond to kinetics of first order. For these conditions, Table 7 shows that:

- for kinetics of order 0: the contactor hydraulics has no effect on the reaction completion; the contact time applied is the predominant parameter.
- for kinetics of order ≥ 1: the performance of a plug flow reactor is always better than the performance of a CSTR. The value of X is, in fact, always higher as in plug flow, since $1 - e^{-k\tau}$ is greater than $k\tau/(1 + k\tau)$. This is the case for most reactions involved in water treatment.

Table 7. Laws governing the reaction completion rate and hydraulics

	Type of flow	
Order of oxidation kinetics	CSTR	Plug flow
0	$X = k\tau / C_0$	$X = k\tau/C_0$
1	$X = k\tau / 1 + k\tau$	$X = 1 - e^{-k\tau}$
2	$X/(1 - X)^2 = k\tau\,C_0$	$X/1 - X = k\tau\,C_0$

Where k is the apparent kinetic constant equal to k_0 x [O₃] residual, C_0 is the initial concentration and τ is the hydraulic contact time (ratio of volume over flowrate).

If the tanks in series model is used the value of X is given by the equation:

$$1 - X = \frac{1}{\left[1 + \dfrac{k\tau}{j}\right]^{j}}$$ (2)

and the value of the hydraulic contact time τ by the equation:

$$\tau = \frac{j}{k}\left[\left(\frac{1}{1-X}\right)^{\frac{1}{j}} - 1\right]$$ (3)

Application I: Giardia Inactivation [Roustan et al., 1991]

When considering the case of Giardia inactivation by ozone at a temperature of 10 °C, the constant k_0 is equal to 4.8 l/mg·min. With an ozone residual of 0.4 mg/l, an apparent 1st order constant k, equal to 1.92 min^{-1} can be obtained. In these conditions, Table 8 gives the contact time (in minutes) for the water in the reactor in relation to the inactivation rate and the reactor hydraulics characterized by the number J of CSTRs in series.

If the goal defined is to achieve a disinfection resulting in a 0.5 log removal of Giardia, Table 8 shows that a plug flow reactor is not necessary and that disinfection can be implemented in a reactor where the value of J is between 2 to 4. These J values are typical for a conventional ozone contactor comprised with 2 to 3 contact chambers. For a higher J value, the gain in time or reactor volume is not significant. On the other hand, if the goal for Giardia inactivation is 3 log, Table 8 shows that a plug flow hydraulics becomes necessary in order to obtain reasonable contact times, in the region of 6 minutes. For these conditions, a reactor with hydraulics corresponding to a J value higher than 8 is required. It should be noted that this number cannot be achieved with a classical ozone contactor unless there are many compartments in series (about 8). This hydraulics can easily be achieved using a deep-U-Tube type ozone contactor. The plug flow effect is of greater interest when the treated water quality goals are more stringent (higher log number inactivation).

For a lower ozone residual concentration, the equation giving τ shows that the contact time τ required to achieve the same water treatment objectives will be increased by the

Table 8. Effect of reactor hydraulics on contact time (min) for a given quality goal

Removal J	0.5 log (68.4%)	1 log (90%)	2 log (99%)	3 log (99.9 %)
1 (CSTR)	1.13	4.70	51.6	520
2	0.81	2.25	9.4	31.9
3	0.73	1.80	5.7	14.1
4	0.69	1.62	4.5	9.6
6	0.66	1.46	3.6	6.7
8	0.64	1.39	3.2	5.7
10	0.63	1.35	3.0	5.2
20	0.62	1.27	2.7	4.3
infinite (PF)	0.60	1.20	2.4	3.6

same order as the decrease of ozone residual concentration. For example, the contact time is multiplied by 4 if the ozone residual can be changed from 0.4 to 0. 1 mg/l.

Application II: Micropollutant Oxidation

It is now well known that ozone in the aqueous phase reacts with micropollutants following two mechanisms: direct and selective reactions of the ozone molecules, and reactions of the hydroxyl radicals which are non-specific and extremely rapid. The radicals are produced by the transformation of the ozone under the influence of the natural organic matter and hydroxyl ions. Table 9 contains the reaction kinetic constants of the ozone and hydroxyl radicals with several potential contaminants of drinking water dealt with in WHO guidelines.

The Table clearly indicates that ozone is usually of low reactivity and that for the oxidation of these compounds, the formation of hydroxyl radicals needs to be promoted. Several types of oxidation process enable the production of radicals, for example UV photolysis of ozone hydrogen peroxide, or the combined use of ozone and hydrogen peroxide.

The required contact time depends on the reaction kinetics of the micropollutant to be eliminated as well as of all the other compounds of the water. It should be noted that the kinetics vary according to the quality of water. The presence of promoters of ozone decomposition (organic matter, hydrogen peroxide, etc.), allowing the production of hydroxyl radicals, favors the elimination of micropollutants. To illustrate the impact of reactor hydraulics from Table 9, three main reaction types can be selected, covering a large range of micropollutant kinetics.

Let us consider the case of the removal of organic micropollutants with various oxidation kinetics:

- "rather slow type" kinetic rates: Atrazine in buffered ultra-pure aqueous solution with a kinetic constant $k_0 = 10$ l/mole · s. Xylene or geosmine in the absence of promoters of ozone decomposition can also be included in the category.

Table 9. Rate constants for the reactions of ozone and hydroxyl radicals with selected organic contaminants according to YAO (1991)

	KO_3 l/mole · s	l/mole · s
Dichloromethane	< 0.1	2.4×10^{10}
Vinyl chloride	> 5000	1.2×10^{10}
Trichloroethene	15	4.2×10^{10}
Benzene	2	7.8×10^{10}
Orthoxylene	90	6.8×10^{10}
Monochlorobenzene	0.8	5.5×10^{10}
Metaldichlorobenzene	0.57	5.0×10^{10}
Acrylamide	1.0×10^{10}	5.9×10^{10}
Epichlorhydrin	< 0.003	3.0×10^{10}
Atrazine	10	2.6×10^{10}
Alachor	3.8	7.0×10^{10}
Lindane	< 0.04	5.1×10^{10}
Methoxychlor	300	NA
Phenol	1300	NA

- "medium type" kinetic rate: Methoxychlor (pesticide) in buffered ultra-pure aqueous solution with a kinetic constant $k_0 = 300$ l/mole · s. This can also be the case for geosmine or atrazine in the presence of a natural organic matrix which acts as a promoter of radical reactions.
- "rapid type" kinetic rate: Phenol in buffered ultra-pure aqueous solution with a kinetic constant $k_0 = 1,300$ l/mole · s.

Table 10 gives the required contact times to obtain a given removal rate for the various types of compounds. The plug flow effects is of even greater interest when the quality goals are more stringent (higher removal percentage). If the treatment goal is to ensure the oxidation of a micropollutant (phenol, atrazine, geosmine, trimethylbenzene, etc.), Table 10 shows that:

- For"rather slow" type kinetics, the completion rate of the reaction is little affected by the reactor hydraulics. In any case, with this type of kinetic, the contact times are very long whatever the type of flow. To reduce the contact times or reactor volumes, combined treatment with $H_2O_2 + O_3$ is recommended.
- For kinetics of the "fast type", the reactor hydraulics play a predominant role. Consequently, if 90% of the pollutant is to be eliminated, the reactor hydraulics must be such that the value of J is greater than 10 in order to obtain contact times in the vicinity of 4 minutes. This treatment objective cannot be achieved with a conventional contactor. Therefore, it is necessary to use other reactor types with a near Plug Flow hydraulics such a deep-U-Tube reactor or packing column, in order to maintain an acceptable contact time (4 minutes). If the requirements are lower (50% eliminated), the hydraulics have little effect on the contact time.

In these conditions, this goal can be achieved with a conventional ozone contactor.

In conclusion, depending on the goals to be attained, the hydraulics can have a decisive effect on the quality of the treated water. This implies that any reactor must be designed after establishing clear goals regarding the quality of water to be produced. In the case of natural water, the apparent kinetics are often higher than those specified above owing to the presence of promoters of radical reactions. It should be noted however that in the presence of hydrogen peroxide, it is the transfer of ozone in the

Table 10. Examples of required contact time (in minutes) for the elimination of various compounds in relation to the ozone contactor hydraulics

Removal J	"Slow" reaction $k_0 = 10$ l/mole·s Residual O_3 = 0.4 mg/l				"Medium" reaction $k_0 = 300$ l/mole·s Residual O_3 = 0.4 mg/l				"Rapid" reaction $k_0 = 1,300$ l/mole·s Residual O_3 = 0.4 mg/l			
	50%	80%	90%	99%	50%	80%	90%	99%	50%	80%	90%	99%
1 (CSTR)	200	800	1800	19800	6.67	26.7	60.0	600	1.54	6.15	13.9	152
2	166	494	865	3600	5.52	16.5	28.8	120	1.27	3.80	6.65	27.7
3	156	426	693	2185	5.20	14.2	23.1	72.8	1.2	3.28	5.33	16.8
4	151	396	623	1730	5.05	13.2	20.8	57.7	1.16	3.05	4.79	13.3
6	147	369	561	1385	4.90	12.3	18.7	46.2	1.13	2.84	4.32	10.7
8	145	357	534	1245	4.83	11.9	17.8	41.5	1.11	2.74	4.10	9.58
10	144	349	518	1170	4.78	11.6	17.3	39.0	1.10	2.69	3.98	9.00
20	141	335	488	1036	4.70	11.2	16.3	34.5	1.09	2.58	3.75	7.97
Infinite (PF)	139	322	461	921	4.62	10.7	15.4	30.7	1.07	2.48	3.54	7.09

water which becomes the limiting parameter in as much as the reaction of ozone with hydrogen peroxide is fast ($k_0 = 5.5 \times 10^6$ l/mole·s for $[O_3][HO_2^-]$). Owing to the fact that the ozone transfer is significantly greater in the case of a U-tube, the U-Tube will be more efficient than a classic contactor whatever the treatment objective.

2.1.6
Conclusion

The importance of the hydraulic behavior and the choice of the type of reactor are strongly dependant on the water treatment quality objectives. If the removal requirements are relatively low (50% removal) the reactor hydraulics do not play a predominant role. However, a plug flow reactor is of great interest when the quality goals are more stringent.

Conclusion

New regulations and guidelines on drinking water production in Europe and the US have clearly shown that the conventional treatment plants suffer from major drawbacks. In the last ten years, research has focused on new processes which could remediate these problems and serve as a good alternative to conventional treatments. Membranes and ozone disinfection are now widely integrated in drinking water plants. Some fundamental mechanisms have been identified and explained but the processes are still to be optimized. Studies are now focused on a better understanding of kinetics, hydraulics, fouling phenomena with the aim to optimize the existent processes and to develop new ones.

2.1.7
References

Anselme C, Bersillon JL, Mallevialle J (1991) The use of powdered activated carbon for the removal of specific pollutants in ultrafiltration. JAWWA, March 10–13, Orlando – Florida

Anselme C, Baudin I, Mazounie P, Mallevialle J (1992) Production of drinking water by combination treatments: ultrafiltration and adsorption on powdered activated carbon. JAWWA, Vancouver

Anselme C, Mandra V, Baudin I, Mallevialle J (1993) Optimum use of membrane processes in drinking water treatment. AIDE, Budapest

Baudin I, Anselme C, Mazounie P (1992) Optimum use of ultrafiltration in surface water treatment. The case of complex treatment lines. Euromembrane Oct 5–8, Paris

Bersillon JL, Anselme C, Mallevialle J (1991) Ultrafiltration in drinking water treatment – Long term estimation of operating conditions and water quality for three water production plants. AWWA, Membrane Processes Conf, Orlando-Florida

Bourdon F, Bourbigot MM, Faivre L (1988) Microfiltration tangentielle des eaux d'origine karstique. L'eau, l'industrie, les nuisances, 121:35

Coffey M, Steward H, Wattier K (1994) Evaluation of microfiltration for metropolitan's small domestic water systems. Microfiltration for Water Treatment Symposium, Irvine – California

Cabassud C, Anselme C, Bersillon JL, Aptel P (1991) Ultrafiltration as a nonpolluting alternative to traditional clarification in water treatment. Filtration and separation, May/June, pp 194–198

Cabassud C, Meireles M, Aptel P (1992) A new way to characterize the fouling ability of natural waters. Euromembrane Oct 5–8, Paris, pp 115–120

Degrémont, Mémento Technique de l'Eau, distributed by Lavoisier, Tec et Doc, 1988

Fane A (1994) An overview of the use of microfiltration for drinking water and waste water treatment. Microfiltration for Water Treatment Symposium, Irvine – California, pp 3–12

Jacangelo J, Lainé JM, Carns K, Cummings E, Mallevialle J (1991) Low-pressure membrane filtration for removing Giardia and microbial indicators. Jour AWWA, Sept., pp 97–106

Jacangelo J, Adham S (1994) Comparison of microfiltration and ultrafiltration for microbial removal, Microfiltration for Water Treatment Symposium. Irvine – California, pp 57–60

Le Sazze N, Laplanche A, Martin G, Langlais B, Martin N (1991) The residence time distribution of the liquid phas in a bubble column and it effect on the ozone transfer. Proceedings of the 10th Ozone World Congress IOA, Monaco, Vol. 2, pp 73–91

Martin N, Bellamy WD, Bourbigot MM, Brink D (1993) Full scale evaluation of ozone contact chambers for disinfection by ozone, Proceedings of the 11th Ozone World Congress IOA, San-Francisco, Vol 2, pp S14/39–57

Moulin C, Cote P, Mercier M (1993) Nanofiltration: l'avenir de l'eau. Biofutur, Déc., pp 37–43

Moulin C, Bourbigot MM, Tazi-Pain A, Bourdon F (1991) Environmental Tech., vol. 12, p 841

Moulin C, Tazi-Pain A, Barraud V, Faivre M (1993) Use of organic microfiltration membranes for ground water treatment. Conference/Membrane Technology for the water industry, Baltimore, August 1–4

Olivieri VP et al. (1991) Continuous microfiltration of surface water. Proc AWWA Conference Membrane Processes, Orlando – Florida

Roustan M, Beck C, Wabloe O, Duguet JP, Mallevialle J (1993) Modeling hydraulics of ozone contactors. Ozone Science and Engineering, vol 15, n°3, pp 212–226

Roustan M, Stambolieva Z, Duguet JP, Wable O, Mallevialle J (1991) Influence of hydrodynamics on Giardia Cysts inactivation by ozone. Study by kinetics and by CT approach, Ozone Science and Engineering, vol 13, n°4, pp 451–462, 1961

Tazi-Pain A, Faivre M, Barraud V, Morineau Y (1994) production d'eau potable par microfiltration sur la ressource karstique de Bernay-Ouest. L'eau, l'industrie, les nuisances, n°172, pp 37–40

Thebault P, Chevalier MR, Anselme C, Mazounie P (1992) Ultrafiltration in drinking water treatment – Long term estimation of operating cost and water quality. Euromembrane Oct 5–8, Paris

Vickers JC, Willinghan GA, Parker DY, Schrott D (1992) Treatment of surface water using continuous microfiltration. AWWA, Vancouver-Canada

Vickers JC, Cline G (1994) Drinking water treatment using microfiltration in conjunction with coagulation processes. Microfiltration for Water Treatment Symposium, Irvine – California, pp 51–54

Yao DC, Haag WR (1991) rate of constants for direct reaction of ozone with several drinking water contaminants. Water Research, Vol 25 n°7, pp 761–773

Yoo SR et al. (1994) Making the move to microfiltration experiences at a 5 MGD microfiltration drinking water treatment plant. Microfiltration for Water Treatment Symposium, Irvine – California, pp 25–27

2.1.8
Further Readings

Title: Membrane technology for water treatment.
Author: Comb Lee F
Corporate Source: Osmonics Inc, Minnetonka, MN, USA
Source: Filtration and Separation v 31 n 3 May 1994. p 223–225
Publication Year: 1994
CODEN: FSEPAA ISSN: 0015-1882
Abstract: The author explains how the use of reverse osmosis, nanofiltration and ultrafiltration membranes is to the advantage of water treatment plants. All three provide very fine filtration and are capable of producing a final water product of a purity unattainable through the more conventional means of water treatment.

Title: Effect of pretreatment strategies on the efficiency of microporous hollow fiber membranes to remove VOCs.
Author: Castro Kevin; Zander AK
Corporate Source: Clarkson Univ, Potsdam, NY, USA
Source: Journal of the New England Water Works Association v 107 n 3 Sep 1993. p 199–213
CODEN: JNEWA6 ISSN: 0028-4939

Abstract: The contamination of drinking water sources by volatile organic compounds (VOCs) has become a major water supply problem in this country. The use of micro-porous polypropylene hollow fiber membranes to remove VOCs by air stripping has been introduced as a potential treatment method. However, before this membrane process gains acceptance, the long-term durability of the membranes under typical operating conditions must be established. Various size and porosity polypropylene hollow fiber membranes were exposed to a range of pH levels and typical residual con-centrations of chemical oxidants. VOC removal rates were measured before exposure, compared to predicted values, and then compared to measured rates after periods of exposure. Preliminary results of this ongoing study indicate that siloxane-coated poly-propylene membranes were attacked by 0.19 mg/l ozone at pH 5, failing after 18 days of exposure. Uncoated polypropylene membranes under similar conditions were resistant to ozone attack. Siloxane-coated polypropylene membranes appeared to be unaffected by solutions in the range from pH 5 to pH 9 without ozone. (Author abstract) 16 Refs.

Committee Report: Membrane Processes in Potable Water Treatment
Am Water Works Assoc, Jan 92, v 84, n 1, p 59 (9) journal article
Abstract: Membrane technologies such as reverse osmosis, ultrafiltration, and micro-filtration can be used to good effect in drinking water treatment. Synthetic membranes can be made from organic polymers, porous carbon, ceramics, and other materials. Various aspects of membrane separation are discussed. Areas of interest include: colloidal fouling of membranes; organics, adsorption, and membrance fouling; in-organic precipitate fouling in reverse osmosis and nanofiltration; biological fouling; operational experiences for water treatment and particle removal; and membrane phase-contact processes. For each area of interest, research needs identified by the Membrane Technology Research Committee for the evaluation, improvement, and application of membrane processes in potable water treatment are highlighted (2 graphs, 1 photos, 41 references, 1 tables).

2.2
Adsorption of Organic Micropollutants onto Activated Carbon Fibers: Cloth and Felt

BERNARD DELANGHE · LIONEL MERCIER · PIERRE LE CLOIREC

Abstract

Adsorption of polluted solutions is performed by different kinds of activated carbon: grains, powder and fibers (cloth or felt). The adsorption is determined in batch reactor. Classic models are applied and kinetic constants are calculated. Results showed that performance of fibrous activated carbon (FAC) is significantly better than that of granular activated carbon (GAC), and is quite similar to that of powder activated carbon (PAC). Moreover, the adsorption capacities for phenol of FAC is markedly greater than GAC. Therefore the application of FAC adsorbers may lead to smaller adsorption reactors. The breakthrough curves obtained with FAC adsorbers are particularly steep, suggesting a smaller mass transfer resistance than for the GAC. The adsorption zone in the FAC bed is about 3.4 mm and is not dependent on the flow rate for the range 0.67–2.07 m · h^{-1}.

2.2.1
Introduction

The removal of organic matter and micropollutants with granular activated carbon (GAC) or powder activated carbon (PAC) is commonly performed for the treatment of natural waters, especially low molecular-weight compounds (Cheremisinoff and Ellerbush, 1978; Schulhof, 1979). GAC adsorbers have been proved effective in removing a large number of organic molecules (Clark and Lykins, 1989) and especially pesticides (Baldauf, 1993). However, one of the problems met in application of adsorption processes in water treatment is the slow kinetics of adsorption which are due to the slow intraparticle diffusion in granular adsorbents such as GAC (Bansal, 1988, Suzuki, 1991). PAC on the other hand provide faster adsorption kinetics, but the difficulties in industrial application of powder adsorbents (handling problems, difficuty of regenerative use, etc.) limit its applications.

Fibrous activated carbon (FAC) has received increasing attention in recent years as an adsorbent for purifying water. The raw materials of FAC are polyacrylonitrile (PAN) fibers, cellulose fibers, phenol resin fibers or pitch fibers and their cloths or felts. They are first pyrolysed and then activated at a temperature of 700–1000 °C in an atmosphere of steam or carbon dioxide (Seung-Kon Ryu,

1990). FAC has only micropore which are directly connected to the external surface of the fibers (Suzuki, 1991). The pore size distribution for PAN-based activated carbon fibers concentrate around 2.5 to 2.6 nm (Tse-Hao Ko and co-workers, 1992). Thus adsorbates reach adsorption sites through micropore without additional diffusion resistance of micropore which usually is the rate-controlling step in the case of granular adsorbents (Suzuki 1991). Moreover, the small diameter of the fibers results in the large external surface area exposed to the flowing fluid. Thus FAC adsorbents provide much faster adsorption kinetics compared with GAC (Baudu and co-workers, 1990, 1991). Most of the studies related to the adsorption onto FAC adsorbers and reviewed by Seung-Kon Ryu (1990) have been carried out with a gaseous phase. In this case, in situ regenerations by joule effect have been demonstrated (Le Cloirec et al., 1990a).

The main objective of the present paper is to assess the performance of the FAC adsorbents in water treatment. Phenol and its derivatives are the basic structural unit of a wide variety of synthetic organics including many pesticides. Phenol has been listed as a priority pollutant by the U.S. Environmental Protection Agency and by the E.E.C. and aromatics and pesticides were the model compounds used throughout the present adsorption studies.

2.2.2
Materials and Methods

Activated carbon materials

Activated carbon materials are commercial products. The main characteristics of the materials used in the present investigations are presented in Table 1. The thicknesses of the cloth and the felt are about 1.5 and 2 mm, respectively. Scanning electronic microscopy pictures of the different adsorbants have been shown elsewhere (Le Cloirec et al., 1990b)

Batch reactor studies

Kinetic studies. Activated carbon (about 0.6 g) in the form of GAC, FAC or PAC was continuously stirred with 1 liter of an aqueous solution at $20 \pm 1\,°C$ containing initially $100\ mg \cdot l^{-1}$ as micropollutants. Samples (1 ml) were withdrawn at regular times and then filtered (Whatman, 0.45 μm) (for analysis until a steady state was obtained, up to 300 minutes for GAC.

Table 1. Main Characteristics of Activated Carbons (PICY Company, Levallois France)

	GAC	PAC	FAC (cloth)	FAC (felt)
Commercial name	NC 60	NC 60	PICA 900	PICA 300
Origin	coconut	coconut	Polyacrylonitrile	
Size (mm)	1.5–2	0.1–0.2	–	–
Apparent density (kg m⁻³)	520	–	–	68.7
Porosity	microporous	microporous	micorporous	microporous
Specific surface area (m² · g⁻¹)	1,200	1,200	1,550	1,800

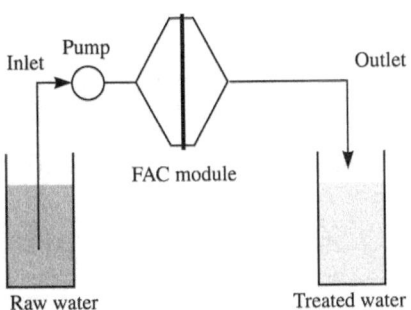

Fig. 1. Continous flow reactor

Equilibrium studies. Activated carbon was continuously stirred for 48 hours with 0.25 liter of an aqueous solution at $20 \pm 1\,°C$ containing initially $100\ mg \cdot l^{-1}$ as micropollutant. For each of the material studied, the activated carbon mass was varied from 0.05 to 0.5 g. The final solution was then filtered and analyzed.

Continuous Flow Reactor. A laboratory pilot unit was set up for the continuous flow study (Fig. 1). The raw water contained $50\ mg \cdot l^{-1}$ of micropollutant and was pumped to the adsorption line which was composed of four similar stages. Just one stage is shown on the Fig. 1. Each of the four modules contained 2 pieces of felt (i.e. a thickness of 4 mm) with a diameter of 2.5 cm. The dead volume inside and between the FAC modules was negligible. The water flowed through each FAC module with a given velocity of 0.62, 1.02 or $2.07\ m \cdot h^{-1}$. Samples were taken at regular times in the outflow of the FAC modules in order to determine the corresponding breakthrough times.

Solutes and analysis

Reagent-grade phenol or other aromatic compounds (Fluka) were used as the model pollutants in the experimental work. A solution of Humic substances (HS), coming from Aldrich, at $50\ mgCl^{-1}$ as raw water, were ultrafiltrated with a UF module of Pleiade Rayflow type (Tech-Sep Company) to get two fractions (< 40,000 or < 3,000 daltons). Aromatics were analyzed by U.V. radiation using a Shimadzu UV-160A spectrophotometer at maximum absorption wavelength. Humic substances were measured, as total organic carbon, on a Shimadzu TOC 5000 analyzer.

2.2.3
Results and Discussion

Adsorption kinetics in batch reactors

The adsorption kinetics could be described by the Adam, Bohart, Thomas' relation. The differential equation is the following:

$$dq/dt = K_1 C (q_m - q) - K_{2q} \tag{1}$$

with   q: adsorption capacity (mg · g⁻¹),
       C: solution concentrations (mg · l⁻¹),
       K_1: adsorption kinetic constant (1 · mg⁻¹ · s⁻¹),
       K_2: desorption kinetic constants (s⁻¹),
       q_m: maximal surface concentration (mg · g⁻¹),
       t: time (s).

At the initial stage of the adsorption reaction, i.e., when $t \to 0$, Then, $q \to 0$ and $C \to C_0$. The equation (1) could be then rewritten:

$$(dq/dt)_{t \to 0} = (d(C_0 - C/dt)_{t \to 0} V/m = K_1 C_0 q_m \tag{2}$$

where C_0: initial solution concentration (mg l⁻¹),
       V: volume of the solution (1),
       m: mass of the activated carbon in the batch reactor (g).

It is then possible to calculate the initial kinetic coefficient (Baudu et al., 1991):

$$\gamma = -K_1 q_m = -V/C_0 m (dc/dt)_{t \to 0} \tag{3}$$

The kinetic coefficient was calculated for the various activated carbon materials and different pollutants, taking into account the initial slopes of the curves presented in Fig. 2, 3, 4 and 5. the values of γ are reported in Table 2 and 3.

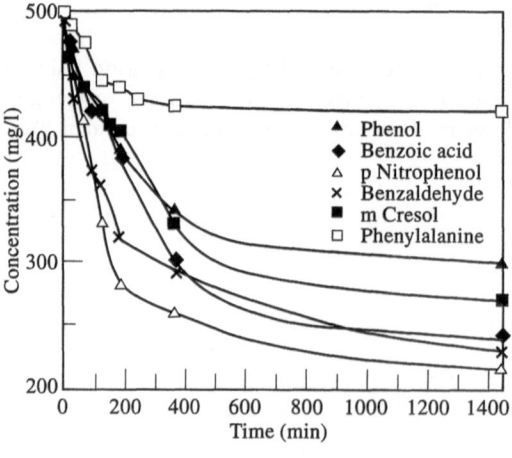

Fig. 2. Adsorption of pollutant onto activated carbon grains (Baudu, Le Cloirec and Martin, 1991)

Phenol
Benzoic acid
p Nitrophenol
Benzaldehyde
m Cresol
Phenylalanine

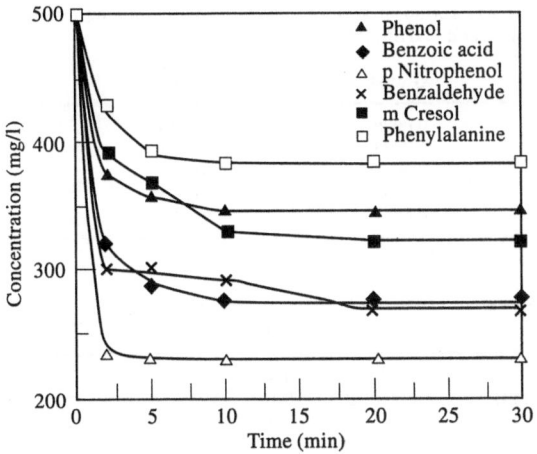

Fig. 3. Adsorption of pollutants onto activated carbon powder (Baudu, Le Cloirec and Martin, 1991)

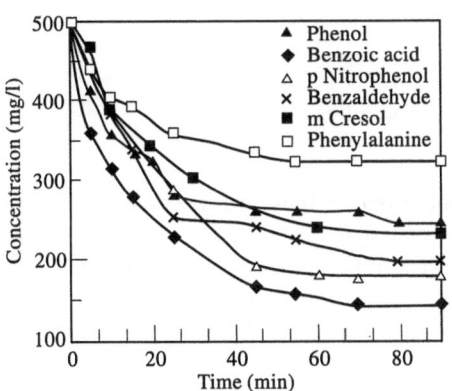

Fig. 4. Adsorption of pollutants onto activated carbon cloth (Baudu, Le Cloirec and Martin, 1991)

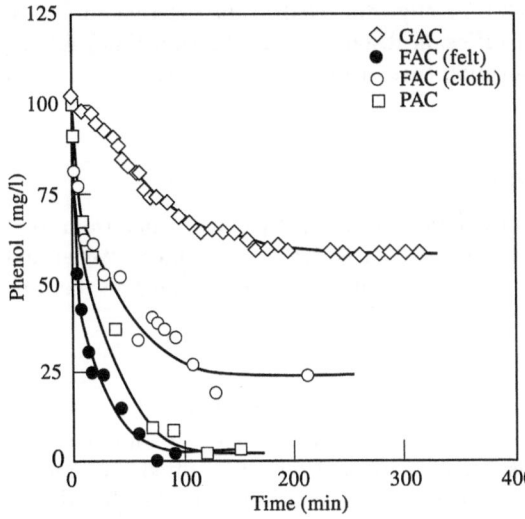

Fig. 5. Batch kinetic test data for the adsorption of phenol using different activated carbons (activated carbon dose: 0.6 g; liquid volume: 1 l)

Table 2. Initial kinetic coefficients for the adsorption of various aromatic compounds on different kinds of activated carbons (Baudu, Le Cloirec and Martin, 1991)

Molecules	Powder	Grains	Cloth
m-cresol	0.33	0.0028	0.030
p-nitrophenol	1.10	0.0033	0.029
Benzaldehyde	0.50	0.0035	0.045
Benzoic acid	0.45	0.0022	0.058
Phenylalanine	0.12	0.0012	0.0252

Table 3. Initial kinetic coefficients for the adsorption of phenol onto different activated carbons

	$\gamma \, (l \cdot g^{-1} \cdot min^{-1})$
GAC	0.006
FAC (felt)	0.157
FAC (cloth)	0.042
PAC	0.029

The values of the kinetic coefficient varied as follows:

Felt > (Cloth – Powder) > Grain

The kinetic coefficient for the felt was more than 5 times greater than that of the PAC, and more than 25 times greater than that of the GAC. However, Suzuki (1991) measured similar intraparticular diffusion coefficients with GAC and FAC. On another hand, Seung-Kon Ryu (1990) reviewed that the superposition of the adsorption forces generated by the opposite walls of the micropores causes an increase in the adsorption potential inside them. Therefore, one might hypothesize that the high kinetic coefficients obtained for the FAC are due to the homogeneous pore distribution with a large volume of micropores (Le Cloirec et al., 1990; Abe and co-workers, 1992) directly connected to the external surface of fibers. There is a relatively small diffusion distance (and diffusion time) for the adsorbates to be fixed on the adsorption sites of the FAC micropores, which could explain the faster kinetics obtained with FAC.

Adsorption isotherms in batch reactors

Adsorption isotherms were conducted for the adsorption of phenol from synthetic solutions. All of the isotherms reflected a favorable adsorption (Weber and Smith, 1987). The equilibrium surface concentration varied according to the type of activated carbon used with the following order:

(Felt, Powder) > Cloth > Grain.

For instance an equilibrium concentration of 30 mg · l^{-1} corresponded to a surface concentration of 230 mg · g^{-1} for the felt and the PAC, 150 mg · g^{-1} for the cloth and 120 mg · g^{-1} for the GAC. Adsorption of phenol onto FAC fitted the Freundlich model better than the Langmuir model, which is not the case for the adsorption of phenol onto other forms of activated carbon (Xing and co-workers, 1994). Therefore the Freundlich isotherm equation was used to describe the equilibrium data:

$$q_e = k C_e^{1/n} \tag{4}$$

in which q_e: adsorption capacity at the equilibrium (mg g^{-1}),
C_e: equilibrium solution concentration (mg l^{-1}),
k: equilibrium constant (mg$^{(1 - 1/n)}$ l$^{1/n}$ g^{-1}),
1/n: constant.

The Freundlich equilibrium constants of all isotherms are provided in Table 4 and 5.

The Freundlich equilibrium constants were higher for FAC than for GAC probably because of the higher specific surface area and the microporous structure of the activated carbon fibers. Therefore the FAC materials are more interesting than GAC from the standpoint of adsorption capacities. The felt was the material that exhibited the best adsorption kinetics and capacities. Therefore it was used for the breakthrough experiments.

Breakthrough of a FAC adsorber

Breakthrough experiments were carried out for the adsorption of phenol onto felt material with different flow rates through the FAC modules. Typical breakthrough curves for a given flow rate of 2.07 m · h^{-1} are presented in Fig. 6. Very steep breakthrough curves were obtained for all of the flow rates used in the present investigation. This characteristic shape has already been mentioned in the review of Seung-Kon Ryu (1990). Again, Suzuki (1991) showed drastic differences between the breakthrough curves obtained in the same experimental conditions with either GAC or packed FAC. The sharper breakthrough curves observed for

Table 4. Freundlich equilibrium constants of phenol adsorbed onto the different activated carbons

	k*	l/n	r
GAC	51.5	0.213	0.933
FAC (felt)	69.7	0.345	0.980
FAC (cloth)	53.2	0.311	0.958
PAC	44.8	0.463	0.971

Table 5. Freundlich equation parameters for different solutes (Baudu, Le Cloirec and Martin, 1991)

Molecules	Powder			Grains			Cloth		
	K	l/n	r	K	l/n	r	K	l/n	r
m-cresol	11.4	0.451	0.957	46.5	0.289	0.981	34.1	0.387	0.988
p-nitrophenol	16.6	0.629	0.977	43.8	0.346	0.961	77.5	0.283	0.974
benzaldehyde	7.4	0.605	0.980	48.0	0.350	0.992	43.4	0.363	0.983
benzoic acid	–	–	–	57.9	0.269	0.982	50.4	0.360	0.981
phenylalanine	0.72	0.944	0.944	11.4	0.325	0.975	22.1	0.365	0.990

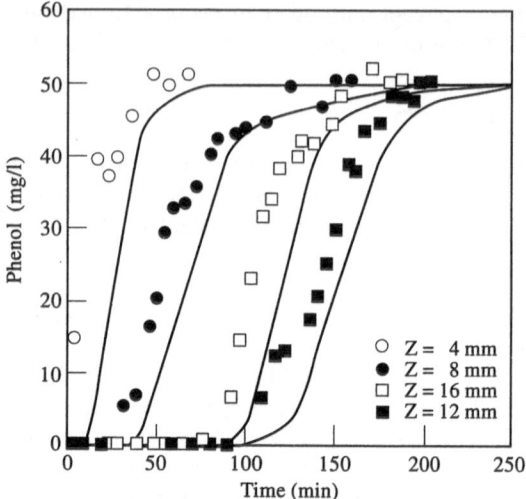

Fig. 6. Breakthrough curves at
different felt thicknesses (flow rate:
2.07 m · h⁻¹; raw water
concentration: 50 mg phenol · 1⁻¹)

Fig. 6. Breakthrough curves at different felt thicknesses (flow rate: $2.07 \text{ m} \cdot \text{h}^{-1}$; raw water concentration: 50 mg phenol $\cdot \text{ l}^{-1}$)

the FAC suggested smaller mass transfer resistance than for the GAC. In the same way, Seung-Kon Ryu (1990) concluded that the adsorption rates of FAC are much higher than those of GAC. Therefore the mass transfer zones are much smaller in the case of adsorption onto FAC.

The breakthrough times were measured when the phenol concentration (C) reached 0.05 C_0 (initial concentration). The breakthrough time values obtained for the various thicknesses and flow rates used in the study were introduced in the Bed Depth Service Time relation developed by Hutchins (1973) and currently used now:

$$t_b = N_0/v C_0 (Z - Z_0) \qquad (5)$$

in which t_b: breakthrough time (min),
        v: flow rate $(\text{m} \cdot \text{h}^{-1})$,
        C_0: raw water concentration $(\text{mg} \cdot \text{l}^{-1})$,
        N_0: maximal surface concentration $(\text{mg} \cdot \text{l}^{-1})$ at given v and C_0 values
        Z: FAC bed thickness (m)
        Z_0: mass transfer zone thickness (m) at given v and C_0 values.

The plotting of t_b against Z values gave straight lines with a slope of $N_0/v\,C_0$ and an origin of $N_0 Z_0/v\,C_0$. Therefore the slope and the origin of the lines $t_b = f(Z)$ for given v and C_0 values allowed the calculation of the parameter $N_0 Z_0$. These parameters are reported in Table 6. The N_0 and Z_0 values were not really strongly dependent on the flow rate within the range $0.62 - 2.07 \text{ m} \cdot \text{h}^{-1}$. One might hypothesize that the adsorption reaction was not significantly influenced by the external mass transfer of the solutes through the hydrodynamic boundary layer and into the macroporosity. The main resistance to the mass transfer might be due to the diffusion through micropores inside the activated carbon fibers. The adsorption capacities (N_0) were recalculated as a function of the activated carbon weight

Table 6. Adsorption zone and maximal surface concentration at different flow velocities

	Z_0 (mm)	N_0 (mg · l^{-1})	N_0 (mg · g^{-1})
0.62	3.5	9210	134
1.02	3.3	8925	130
2.07	3.4	8625	126

Fig. 7. Breakthrough curves ($v = 0.6 \cdot m \cdot h^{-1}$) process with an aqueous solution of phenol (100 mg · l^{-1}) and solution of humic acid (50 mg · l^{-1} in raw water) ultrafiltrated at a cut-off diameter of 40,000 or 3,000 daltons

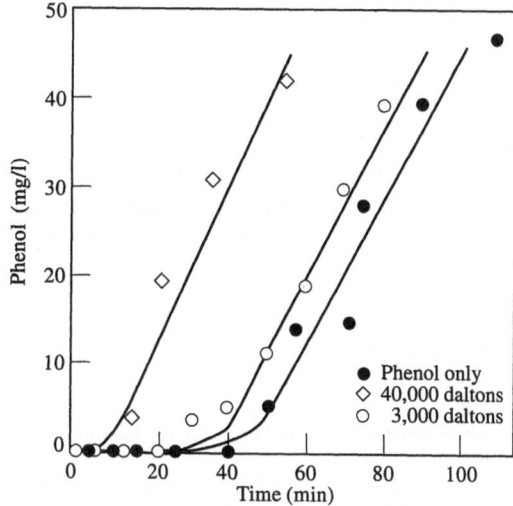

(Table 6). A nice adsorption capacity (about 130 mg/g), close to the maximum surface concentrations determined in batch reactor, is found with this dynamic system.

A second experiment was carried out with a mixture of humic substances and phenol at concentrations of 50 mg · l^{-1} and 100 mg · l^{-1} respectively. The commercial humic substances were ultrafiltrated with a UF membrane with a cut-off diameter of 400,000 and 3,000 daltons (Le Cloirec and Delanghe, 1994). The breakthrough curve obtained under these experimental conditions is presented in Fig. 7 (black points) and the breakthrough time (t_b when $C = 0.05\ C_0$) was found to be about 50 minutes.

When an aqueous solution of phenol and humic substances (HS) was used as raw water, the FAC service time was strongly dependent on molecular weight of the humic substances (MWHS) used in the experiments. With MWHS < 40,000, phenol breakthrough time was reduced at 25% of its value found with phenol only. This drastic decrease in service time could be explained by a competition for the adsorption sites on the fibers between phenol and other organic compounds. When MWHS < 3000 the HS concentration is very low and then there is, of course, no real adsorption competition between phenol and low molecular weight of HS. The breakthrough curve could be compared with the curve obtained with the pure phenol solution.

2.24
Conclusion

Results showed that performance of FAC is significantly better than that of GAC, and is quite similar to that of PAC. Because of the small diffusion distance in the fibers, FAC adsorbers provide much faster adsorption kinetics compared with GAC or PAC. Moreover, the adsorption capacities for phenol or other aromatic molecules of FAC is markedly greater than GAC. Therefore the application of FAC adsorbers may lead to smaller adsorption reactors. The breakthrough curves obtained with FAC adsorbers are particularly steep, suggesting a smaller mass transfer resistance than for the GAC. The adsorption zone in the FAC bed is about 3.4 mm and is not dependent on the flow rate for the range $0.67 - 2.07 \text{ m}^3 \cdot \text{m}^{-2} \cdot \text{h}^{-1}$. As a conclusion, FAC have an advantage of fast adsorption kinetics, high adsorption capacities, and easiness of handling when compared with granular or powdered adsorbents. However, studies have to be performed for complex mixtures of pollutants. New reactors have to be proposed in order to use these new adsorbants in the water or wastewater treatments.

2.2.5
References

Abe M, Kaneko Y, Agui W and K Ogino (1992) removal of humic substances dissolved in water with carbonaceous adsorbents. Sci Tot Environ, *117/118*, 551–559

Adham SS, Snoeyink VL, Clark MM and Bersillon J-L (1991) Predicting and verifying organics removal by PAC in an ultrafiltration system. J Am Wat Works Ass *83*, 12, 81–91

Adham SS, Snoeyink VL, Clark MM and Anselme C (1993) Predicting and verifying TOC removal by PAC in pilot-scale UF systems. J Am Wat Works ASS *85*, 12, 58–68

Anselme C, Chevalier MR, Mazounie P and Mallevialle J (1992) Applications industrielles de l'ultrafiltration pour la production d'eau potable – Bilan de fonctionnement des installations – Perspectives d'évolution. Tech Sci Municip-L'Eau, *87*, 9, 403–408

Baldauf G. (1993) Removal of pesticides in drinking water treatment. Acta Hydrochim Hydrobiol *21*, 4, 203–208

Bansal RC, Donnet JB, Stoeckli N (1998) Active Carbon, Marcel Dekker Inc, NY, USA

Baudu M, Le Cloirec P, and Martin G (1990) Caractéristiques et performances des fibres de charbon actif – Applications au traitement d'eau. Tech Sci Municip-L'Eau *85*, 12, 621–625

Baudu M, Le Cloirec P, and Martin G (1991) Pollutant adsorption onto activated carbon membranes. Wat Sci Technol *23*, 1659–1666

Cheremisinoff P.M., Ellerbush F (1978) Carbon adsorption Handbook, Ann Arbor Sci, Ann Arbor Mi, USA

Clark MK, and Lykins BW (1989) Control of trihalomethane and synthetic organics. In RM Clark (Ed), Granular activated carbon – Design, operation and cost. Lewis Publisher, Michigan, pp 257–293

Hutchins R (1973) New method simplifies design of activated carbon systems – Water bed-depth service time analysis. Chem Eng *20*, 133–138

Le Cloirec P, Baudu M and Martin G (1990a) European Patent

Le Cloirec P, Baudu M, Martin G, Dagois G (1990b), Membrane, toiles, fibres ou feutres: des charbons actifs d'utilisations prometteuses, Rev Sci Tech Défense *2*, 111–123

Le Cloirec P, and Delanghe B (1994) Ultrafiltration-Fibres de charbon actif – Dispositif de mise en œuvre, et composites. French patent n° 9411625

Seung-Kon Ryu (1990) Porosity of activated carbon fibre. High Temp- High Pressures *22*, 345–354

Schulhof P. (1979) An evaluating approach to activated carbon treatment, J Am Water Works Ass *71*, 648–661

Suzuki M (1991) Application of fiber adsorbents in water treatment. Wat Sci Technol *23*, 1649–1658

Tse-Hao Ko, Chiranairadul P, Chung-Hua Lin (1992) The study of polyacrylonitrile-based activated carbon fibres for water purification: Part I. J Materials Sci Letters *11*, 6–8

Weber WJ, and Smith EH (1987) Simulation and design models for adsorption processes. Environ Sci Technol *21*, 11, 1040–1050

Xing B, McGill WB, Dudas MJ, Maham Y, and Hepler L (1994) Sorption of phenol by selected biopolymers: isotherms, energetics, and polarity. Environ Sci Technol *28*, 466–473

2.2.6
Further Readings

Title: Influence of heat or chemical treatment of activated carbon onto the adsorption of organic compounds.
Author: Mazet, M., Farkhani, B.; Baudu, M.
Corporate Source: Lab de Genie chimique Traitement des Eaux, Limoges, Fr
Source: Water Research v 28 n 7 Jul 1994. p 1609–1617
CODEN: WATRAG ISSN: 0042-1354

Abstract: Adsorption on activated carbon is an attractive method of removal of organic compounds as micropollutants from natural waters and wastewaters. The physico-chemical nature of the surface of carbon is an important factor in the adsorption process and should be considered in selection or preparation of activated carbon for specific adsorption. The purpose of the present study is to quantify the effect of PAC, heat treated PAC(T), acid treated PAC (OC) of powder activated carbon (PAC less than 50 μm; Chemviron F400) and to compare the removal of organic compounds (humic acids, phenol, phthalic acid, salicylic acid, benzoic acid, atrazine, sodium dodecyl sulphate). Mechanism of the adsorption will be discussed on the basis of the results: adsorption capacities zeta potential of particles and physical characteristics of PAC. (Author abstract) Refs.

Title: Predicting and verifying TOC removal by PAC in pilot-scale UF systems.
Author: Adham, Samer S.; Snoeyink, Vernon L.; Clark, Mark M.; Anselme, Christophe
Source: Journal of the American Water Works Association v 85n 12 Dec 1993. p 58–68
CODEN: JAWWA5 ISSN: 0003-150X
Language: English

Abstract: A modeling technique was developed to predict organics removal by powdered activated carbon (PAC) added as a pretreatment to hollow-fiber ultrafiltration (UF). Model predictions were successfully verified for large- and small-scale PAC-UF pilot plants. Experiments were conducted using different water sources, PAC dosages, and backwashing frequencies. Natural organic matter in the water sources, represented by total organic carbon, was the target compound studied. The modeling procedure can be used in the design and evaluation of PAC-UF systems. (Author abstract) 17 Refs.

2.3
Removal of Micropollutants in Some Ozone Contactors: Efficiency and Simulation

ALAIN LAPLANCHE · CHRYSTELLE LANGLAIS · DOMINIQUE WOLBERT · VINCENT BOISDON

Introduction

A number of recent studies have shown that micropollutant contamination of raw waters exists on a wide scale. Among these pollutants, the most frequently identified are pesticides (particularly atrazine and simazine in Europe), but also common are aryloxyacids, carbamates, amide and urea derivatives. European standards stipulate that the maximum contaminant level (M.C.L.) for each individual pesticides is 100 ng/l and 500 ng/l for the total concentration of the substances. Alternatively, the World Health Organization directives set maximum permissible concentrations for each pesticide (2 µg/l for atrazine, simazine, lindane and MCPA; 5 µg/l for carbofuran; 9 µg for isoproturon).

Both the European community and the United States have ground and surface waters with pesticide contamination higher than the M.C.L. Suitable methods are therefore required to treat drinking water. Among the Processes for micropullutant removal from water, ozone oxidation and advanced oxidation processes (H_2O_2/O_3), along with activated carbon adsorption, are the most commonly used.

Today, ozonation reactors are mainly bubble towers, but sometimes it is possible to use static mixers. Industrial parameters are quite different and it is very interesting to compare the removal efficiency of the micropollutants on the two systems. First we here present the experimental results obtained on the two reactors, micropollutant model being the atrazine pesticide. Secondly, we describe the parameters and equations necessary to compute simulated data. Finally we show the difference in the concept of the model to simulate the two types of reactors.

2.3.1
Experimental results on atrazine removal

Among the pesticides, atrazine was chosen by researchers as the analogue for micropollutant. Especially in France and in some European countries, atrazine is responsible for exceeding the drinking water standards because it is very persistent in the environment. Studies that consider atrazine degradation by ozone and by hydrogen peroxyde – ozone systems are presented here. The simplified diagram obtained is illustrated in Fig. 1 (Legube et al., 1987; Kearney et al., 1988; Hapeman-Somich et al., 1992; Adams and Randtke, 1992; Husley et al., 1993; Chramosta, 1993).

Moreover, some kinetic studies allow for some kinetic constants, either for ozone or for hydroxy radicals, that react on atrazine (Paillard et al., 1991; Yao and Haag, 1992; Xiong and Graham, 1992; Beltrane et al., 1993; Tace, 1993; Chramosta, 1993). Recently,

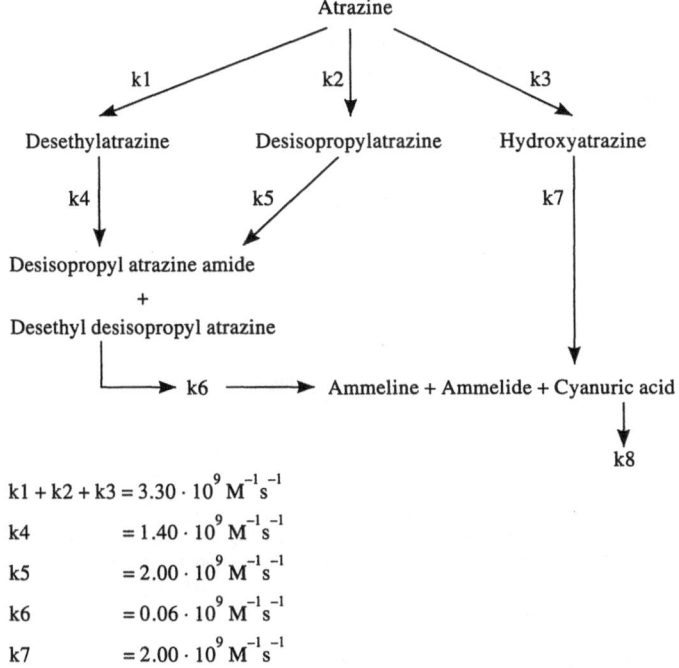

$$k1 + k2 + k3 = 3.30 \cdot 10^9 \, M^{-1} s^{-1}$$
$$k4 \qquad = 1.40 \cdot 10^9 \, M^{-1} s^{-1}$$
$$k5 \qquad = 2.00 \cdot 10^9 \, M^{-1} s^{-1}$$
$$k6 \qquad = 0.06 \cdot 10^9 \, M^{-1} s^{-1}$$
$$k7 \qquad = 2.00 \cdot 10^9 \, M^{-1} s^{-1}$$

Fig. 1

Chramosta (1993) determined the kinetic constants for the reactivity of OH° upon some metabolites.

2.3.1.1
Efficiency of bubble towers for atrazine removal

Bubble towers are the most used reactors in drinking water treatment plants. The contact times are several minutes and in the case of pesticides removal, the ozone amount can reach 5 g/m³. Free radical reactions are enhanced by the addition of hydrogen peroxide (Ratio H_2O_2/O_3 from 0.2 g/g to 0.6 g/g) (ozonation conditions are reported in Table 1).

Table 1

Authors	T (°C)	pH	Alcalinity (mg/l)	COT (mg/l)	Treatment rate (g/m³)	Contact time (s)	H_2O_2/O_3 ratio (g/g)
Paillard and al (1990)	22	7.9	205–230	1.8–2	1.5 to 4.5	240 to 900	0–1
Duguet and al (1991	13	7.7	220	3.4	1 to 4	900	0.3
Chramosta and al. (1991)	20	8.3	46	1.8	1.32	288	0–2

For instance, we can report some results gathered from data in Figs. 2a and 2b. Paillard and al. (1991) worked on a pilot plant treating filtrated Seine river water spiked with atrazine. Duguet and al. (1991) reported on Mont Valerien plant results where a treatment strategy on pesticides removal is tested. Chramosta et al. (1991) studied pesticide degradation in a continuous flow laboratory reactor by O_3/H_2O_2 system with Poitiers tap water spiked with pesticides.

With classical treatments rates of 2 mg/m³, the percentage of degradation ranges from 15 to 30% for ozone treatment and raises by 20% with addition of hydrogen peroxyde. To obtain significant removals, it is necessary to use O_3/H_2O_2 system with higher applied ozone dose.

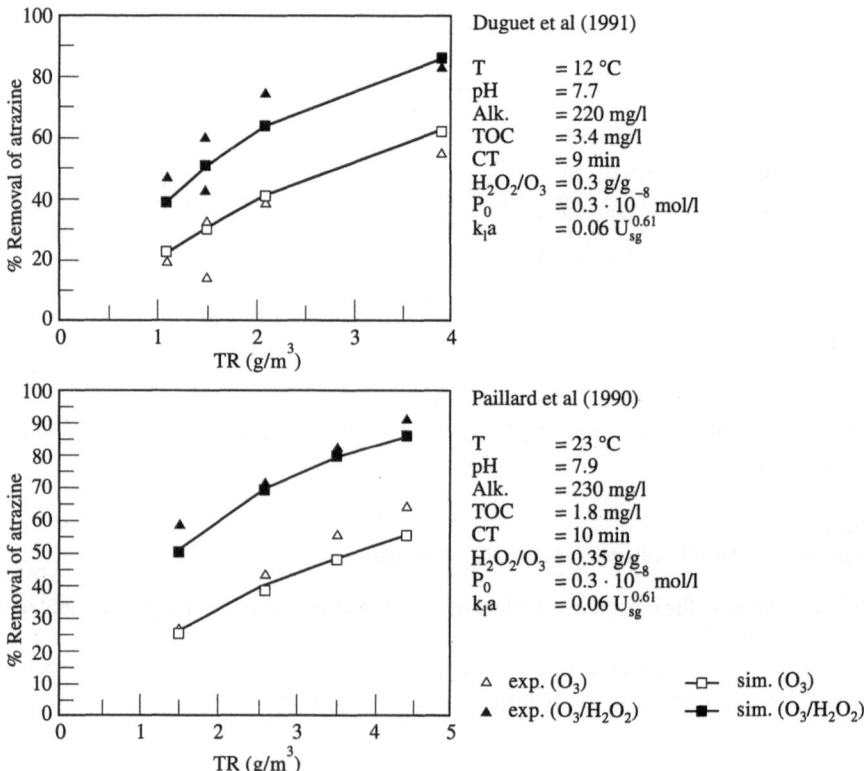

Fig. 2a. Atrazine pesticide removal in bubble column, influence of treatment rates. Experimental data and comparison with simulation

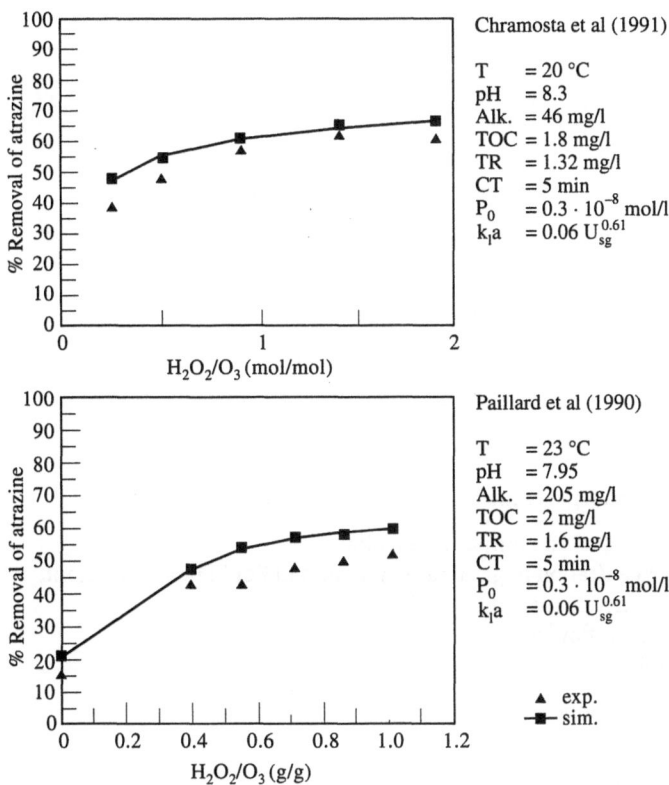

Fig. 2 b. Atrazine pesticide removal in bubble column, influence of H_2O_2/O_3 ratio. Experimental data and comparison with simulation

2.3.1.2
Efficiency of static mixers for atrazine removal

Conceived, at first, to transfer oxidant from gaseous phase to liquid phase, static mixers are characterized by very short residence times (of about one second) but with in return high pressure drops of liquid phase (up to 0,2–0,4 bars). Data exists that allows the calculation of pressure drops, k_La, interfacial area, dissipated power. Nevertheless, the literature doesn't report any illustration of static mixers used as chemical reactors to remove pesticides; thus the laboratory scale experiments about these possibilities that follow.

a) The pilot unit

The pilot plant is shown by the Fig. 3. Reactor is constituted by six static mixers (Sülzer – SMV 4). They are fitted together two by two, separated by leer parts (with the same volume).

The reactor is in vertical downward position. The used water is Rennes tap water spiked with atrazine.

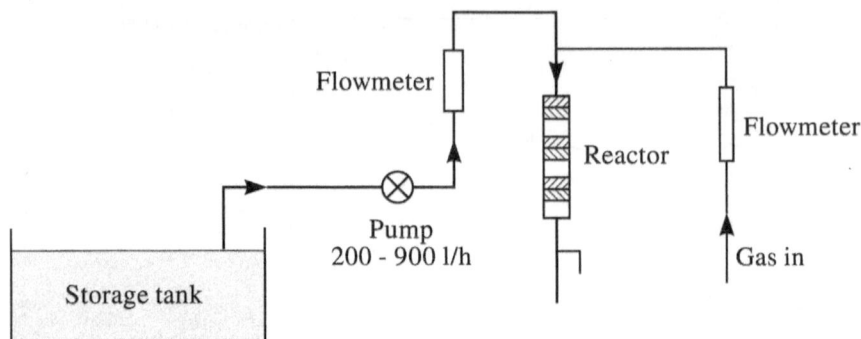

Fig. 3

Different parameters are measured:

- atrazine in input and output of the pilot,
- ozone concentration in gas phase input and in liquid phase in output,
- water parameters (pH, TOC, TAC, T),
- liquid and gas flowrates,
- headlosses.

Experimental parameters are:

- treatment rate (g/m^3): 3 – 6 – 9
- ratio H$_2$O$_2$/O$_3$ (g/g): 0 – 0.2 – 0.4 – 0.6 – 1
- liquid flowrates (m^3/h): 0.3 and 0.5
- gas flowrates (m^3/h): 26.4 · 10^{-3} – 40.5 · 10^{-3} – 54.1 · 10^{-3} – 68.4 · 10^{-3}

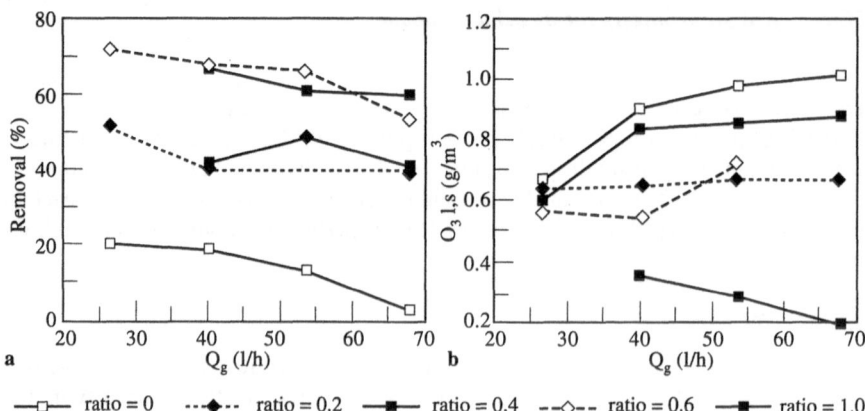

Fig. 4 a, b. a Variation of atrazine removal versus gas flowrate for different ratios H$_2$O$_2$/O$_3$ – Treatment rate = 3 g · m^{-3}, Ql = 300 l · h^{-1}. **b** Variation of O$_3$ l,s versus gas flowrate for different ratios H$_2$O$_2$/O$_3$ – Treatment rate = 3 g · m^{-3}, Ql = 300 l · h^{-1}

b) Experimental results

Results are shown in Figs. 4, 5, 6, 7. They affect atrazine removal (%) and the ozone residual concentration (g/m³) in the liquid phase in output.

c) Discussion

Results indicate the best efficiency lies with O_3/H_2O_2 system. It can reach 50 to 65% with a treatment rate of 3 g/m³. With higher treatment rates, atrazine removal is at least

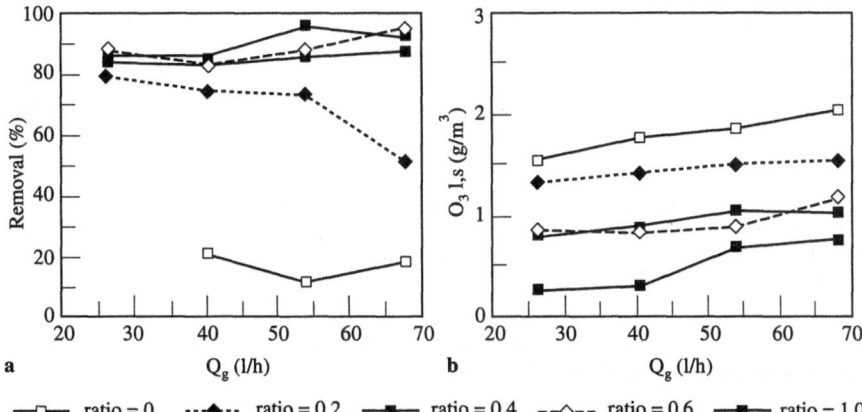

—□— ratio = 0   ···◆·· ratio = 0.2   —■— ratio = 0.4   --◇-- ratio = 0.6   —◆— ratio = 1.0

Fig. 5 a, b. a Variation of atrazine removal versus gas flowrate for different ratios H_2O_2/O_3 – Treatment rate = 6 g · m⁻³, Ql = 300 l · h⁻¹. **b** Variation of O_3 l,s versus gas flowrate for different ratios H_2O_2/O_2 – Treatment rate = 6 g · m⁻³, Ql = 300 l · h⁻¹

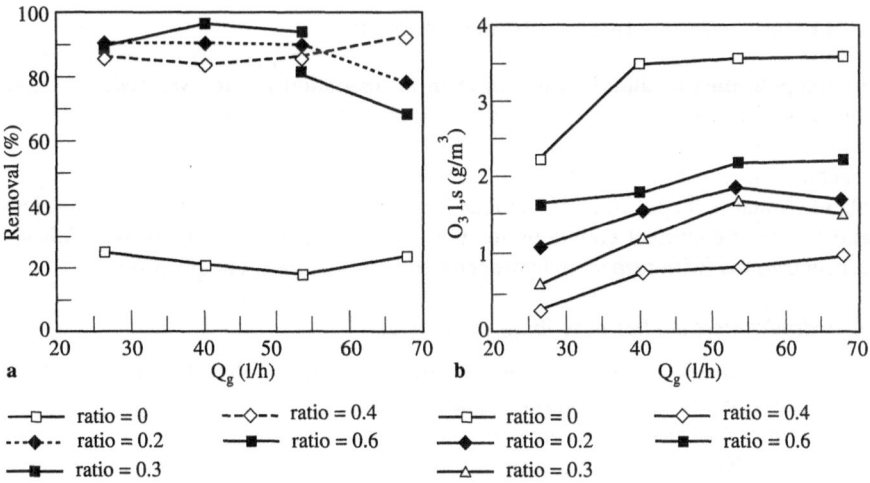

—□— ratio = 0   --◇-- ratio = 0.4   —□— ratio = 0   —◇— ratio = 0.4
···◆·· ratio = 0.2   —■— ratio = 0.6   —◆— ratio = 0.2   —■— ratio = 0.6
—■— ratio = 0.3   —△— ratio = 0.3

Fig. 6 a, b. Variation of atrazine removal versus gas flowrate for different ratios H_2O_2/O_3 – Treatment rate = 9 g · m⁻³, Ql = 300 l · h⁻¹. **b** Variation of O_3 l,s versus gas flowrate for different ratios H_2O_2/O_3 – Treatment rate = 9 g · m⁻³, Ql = 300 l · h⁻¹

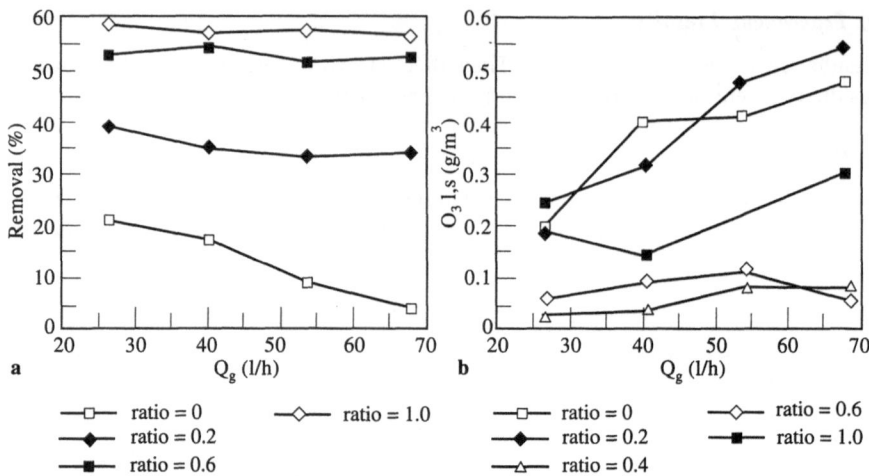

Fig. 7 a, b. Variation of atrazine removal versus gas flow rate for different ratios H_2O_2/O_3 – Treatment rate = 3 g · m⁻³, Ql = 500 l · h⁻¹. b Variation of O_2 l,s versus gas flow rate for different ratios H_2O_2/O_3 – Treatment rate = 3 g · m⁻³, Ql = 500 l · h⁻¹

equal to 80 % whereas residence time are less than two seconds. Static mixers behave particularly efficiently as chemical reactors.

The main parameters that have an effect upon the treatment are the contact time and the ratio H_2O_2/O_3 (Fig. 8), the treatment rate (Fig. 9) and the ozone concentration in gaseous phase at the input of the reactor (an increase of this concentration promotes micropollutant degradation).

2.3.2
Parameters necessary to compute simulated data

To compute micropollutant elimination in an oxidation reactor, we need to know the chemical reactivity between oxidant species and micropollutants, the residence time distribution in the reactor with the hydrodynamic model describing the real reactor, the equation for the mass transfer of ozone from gaseous to liquid phase (which is the location of the reaction between pesticide and oxidant species) and the semi-empirical equations for both consumed ozone and hydroxy radicals concentrations (with respect to hydrogen peroxide and water composition).

a) Hydrodynamic models describing the reactors

Ozonation reactors used in drinking water treatment are mainly bubble towers, where ozonated air is distributed by porous glass. The hydrodynamics of these reactors are very complex. Working on two full scale plants (Neuilly/Marne and Mery/Oise), Martin (1992) and Martin et al. (1992) observed that the ozonated air and water were heterogeneous in flow and that back mixing occured inside the reactor.

For the total reactor, the Peclet number is defined by uL/D (D being a dispersion coefficient, u, the fluid velocity and L, the depth of the reactor) and obtained by

Fig. 8. Influence of contact time and ratio H_2O_2/O_3 on removal of atrazine $Qg = 40,5$ l/h – TR = 3 g/m³

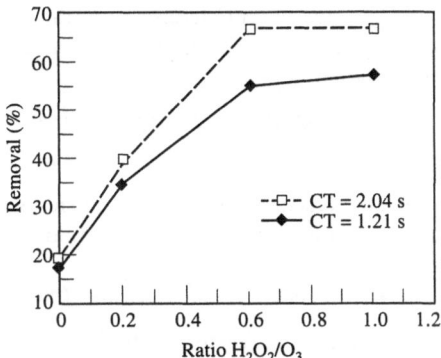

Fig. 9 Influence of treatment rate $Qg = 40,5$ l/h – Ql = 300 l/h

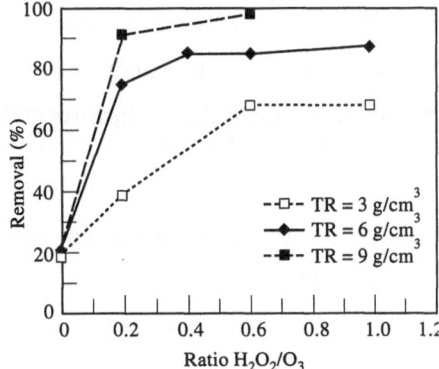

calculations from residence time distribution. It varies between 6 and 12. For a pilot plant (without heterogeneous phenomena), Le Sauze (1993) showed that the reactor can be considered in three parts, each zone being characterized by a number J of completely stirred tank reactors in series.

For the static mixers, hydrodynamic is assumed to be plug flow for liquid and gas. It is also possible to modelise them by the CSTR in series model, J being a very large number (J → ∞ for the ideal plug flow reactor).

In any one CSTR, when the oxidation reactions have occured in the bulk, the mass balance on ozone is:

$$Q_L(O_3)_E + k_L a \, \Delta v \, ([O_3]^* - [O_3]_R) = Q_L[O_3]_R + [O_3]_R \, \Delta v \, \{k_D[P]_R$$
$$+ k_2[H_2O_2]_R \cdot 10^{pH-11.6} + k_8[OH°]_R\}$$

$k_L a$ is computed from semi-empirical relations (Akita, 1974; Roustan et al., 1987; Le Sauze et al., 1992) and $(O_3)^*$ from the Henry law.

In the same manner, the mass balance on hydrogen peroxide is:

$$Q_L[H_2O_2]_E = Q_L[H_2O_2]_R + [H_2O_2]_R \, \Delta v \, \{k_2[H_2O_2]_R \cdot 10^{pH-11.6}[O_3]_R$$
$$+ (k_6 \, 10^{pH-11.6} + k_7) \, [OH°]_R\}$$

Finally, to compute hydroxy radical concentration in each CSTR from stationary state theory and 75 experiments carried out on natural water (Orta de Velasquez, 1992), we can say that:

$$[OH^\circ]_R = k' [O_3]_R = [O_3]_R \{2 k_1 10^{pH-14} + 3 \cdot 16 \cdot 10^{-7} \cdot 10^{0.42 pH} COT$$
$$+ 2 k_2 10^{pH-11.6} [H_2O_2]_R\}/k_{ID}(P)/[HCO_3^-] (k_9 + k_{10} 10^{pH-10.25})$$

The equations include the mass transfer of ozone from the gaseous phase to the liquid phase and the same empirical equations to compute ozone consumption.

b) Chemical reactivity of pesticides towards oxidants

In an ozonation reactor (with or without hydrogen peroxide addition), hydroxy radicals are produced. Publications show that, for the reaction between ozone and hydroxy radicals with respect to organic compounds, the total kinetic order is two (Doré, 1989; Langlais et al., 1991; Yao and Haag, 1990, 1992). Consequently, the reaction rate giving the degradation of a micropollutant P is:

$$R_P = - (k_D [O_3] [P] + k_{ID} [OH^\circ] [P])$$

Our experience shows that the hydroxy radical concentration is proportional to ozone concentration ($[OH^\circ] = k' [O_3]$). The reaction rate can then be written:

$$R_P = - (k_D + k_{ID} k') [O_3] [P] = - k_G [O_3] [P]$$

If the reaction has occured in the bulk, ozone concentration is constant in a CSTR and

$$R_P = - k_T [P] \quad \text{with} \quad k_T = k_G [O_3]_R$$

Considering the Δv volume for the CSTR, the mean residence time of water is $\tau = \Delta v/QL$ and the micropollutant removal is:

$$[P]_R/[P]_E = 1/(1 + k_G [O_3] \tau)$$

2.3.3
Modelization of micropollutant removal

a) In a bubble column, by using previous concepts, a calculation program allows us to estimate the reactive oxidant species concentrations in each CSTR. Then, it is possible to use this for inferring the evolution of the compound and sometimes of the oxidation byproducts.

For example, by referring to the reported works and to the work of Husley et al. (1993) with two bubble columns in series (contact time: 12 min), Fig. 2 and Fig. 10 show a comparison between the measured and calculated values for different treatment rates.

In the same way, Fig. 11 shows the comparison of these values in the case of a treatment by H_2O_2/O_3 system on a semi-continuous laboratory pilot.

b) If we apply the previous concept to static mixers (calculation of the concentrations of oxidants and reactions in bulk liquid), as contact times are very short, we obtain very low removal rate, that do not connect to the results of the experiments.

To explain the great reactivity in static mixers, we submit the hypothesis that the reaction doesn't take place in bulk liquid, like in bubble columns, but in liquid film

Fig. 10 Comparison Simulation/Experiments
Hulsey et al. (1993)
T° = 17°C
pH = 7.5
COD = 3.0 mg C/l
Alk. = 140 mg CaCO₃/l
DO = 0.05
Contact Time = 12 min

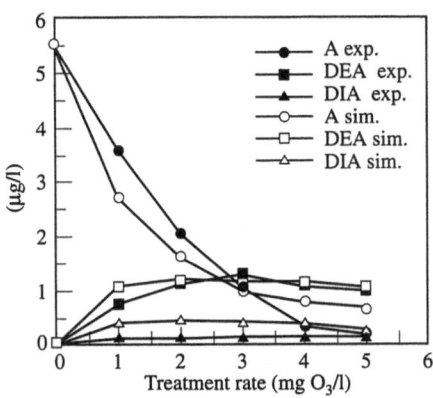

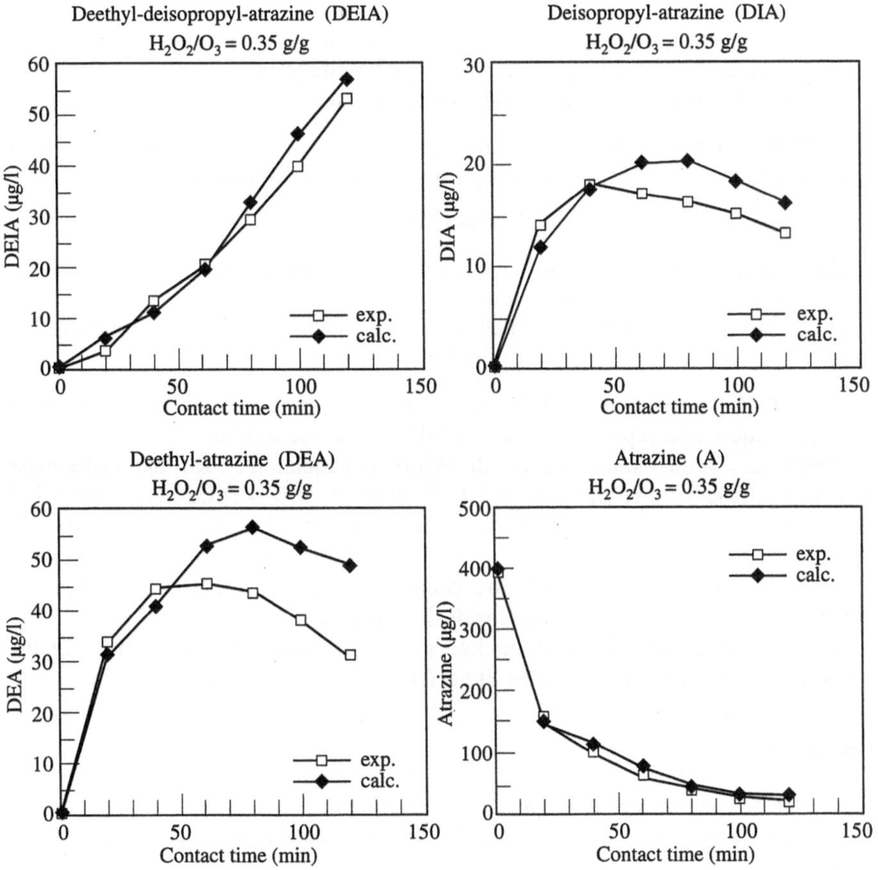

Deethyl-deisopropyl-atrazine (DEIA)
$H_2O_2/O_3 = 0.35$ g/g

Deisopropyl-atrazine (DIA)
$H_2O_2/O_3 = 0.35$ g/g

Deethyl-atrazine (DEA)
$H_2O_2/O_3 = 0.35$ g/g

Atrazine (A)
$H_2O_2/O_3 = 0.35$ g/g

Fig. 11. Comparison Simulation/Experiments Chramosta (1993)
T° = 20 °C; pH = 8.1; COT = 1.8 mg C/l; Alk. = 560 mg CaCO3/l; DO = 0.05; O₃ = 0.34 mg/l · min;
$H_2O_2/O_3 = 0.35$ g/g; {Atrazine} initiale = 1.06 µmol/l

Fig. 12

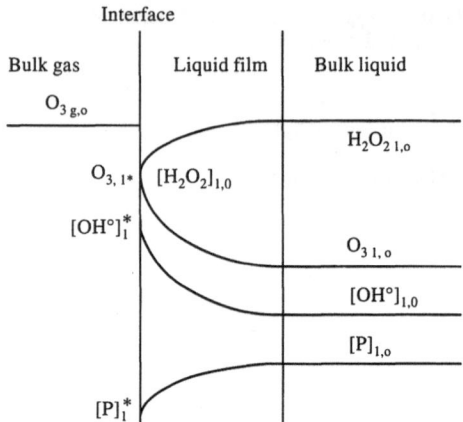

near the interface between gas and liquid. As static mixers are high energy transfer reactors, liquid film should represent a great part of the phases.

This hypothesis comes from several establishments.

- pesticide removal is in the same region of importance in static mixers than in bubble columns. It means that $k\,[O_3]$. τ should be nearly identical in the two cases.

$\tau_{SM} < \tau_{BC}$ implies necessary that $[O_3]_{SM} > [O_3]_{BC}$

If we consider that $[O_3]_{BC}$ is the concentration in bulk liquid, higher concentrations are possible only near the interface.
- In the same way, the fact that the efficiency varies with the ozone concentration in gaseous phase is in harmony with this hypothesis.

In a first step, by making simplified calculation by using logarithmic means, the result is that the liquid film represents 20% of total liquid phase volume.

We obtain a similar value by using the Whitman's equation (1923) that relates mass transfer coefficient k_L to film thickness L and diffusivity $D:k_L = D/L$ and semi empirical equation giving $k_L\,a$ (Zhu et al., 1992; Roes et al., 1984) and a. A more accurate simulation (work in progress) should allow the calculation of different oxidant species and pesticides contours in the film by taking into account the transfer to the interface, the different chemical reactions and the diffusion inside the film (Fig. 12).

The parameters are the film thickness and the adjustment constants are the measured or calculated values to the limits ($[O_3]^*$, $[H_2O_2]_{bulk}$, $[O_3]_{bulk}$, $[P]_{bulk}$).

Conclusion

This work shows how to model pesticide removal efficiency in ozonation processes. First, it is necessary to take into account the type of reactor. Two points are particularly important:

- what is the hydrodynamic model describing the real reactor?
- where is the reaction located: film or bulk?

Secondly, if it is possible to answer these questions, the knowledge of chemical reactivity and reaction rate becomes the limit of the calculations.

If today, several studies give the kinetic constants of ozone and hydroxyl radicals towards numerous pesticides; except for atrazine, few data exist on oxidation schemes and the behavior of oxidation metabolites.

The comparison between bubble towers and static mixers is an example where, chemical reactions being known, we have demonstrated how to obtain simulated data in agreement with experimental results, however two different approaches are necessary. In bubble columns, chemical reaction takes place in bulk liquid whereas in static mixers, it is mainly located in the liquid film.

Annexe

* Atrazine:
2-chloro-4 éthylamino-6-isopropylamino 1-3-5 triazine

* DEA ou deséthylatrazine:
2-amino-4-chloro-6-isopropylamino-s-triazine

* DEIA ou deséthyldesisopropyl atrazine:
2-4-diamino-6-chloro-s-triazine

* DIA ou desisopropylatrazine:
2-amino-4-chloro 6-éthylamino-s-triazine

$$\begin{array}{c} Cl \\ | \\ C \\ N \diagup \diagdown N \\ \| \qquad \| \\ C \qquad C \\ H_2N \diagup \diagdown N \diagdown NHC_2H_5 \end{array}$$

* DIAA ou deisopropylatrazine amide:
2-acétamido-4-amino-6-chloro-s-triazine

$$\begin{array}{c} Cl \\ | \\ C \\ N \diagup \diagdown N \\ \| \qquad \| \\ C \qquad C \\ H_2N \diagup \diagdown N \diagdown NHCCH_3 \\ \| \\ O \end{array}$$

* HA ou Hydroxy atrazine:
2 éthylamino-4-hydroxy-6-isopropylamino-s-triazine

$$\begin{array}{c} OH \\ | \\ C \\ N \diagup \diagdown N \\ \| \qquad \| \\ C \qquad C \\ C_3H_7NH \diagup \diagdown N \diagdown NHC_2H_5 \end{array}$$

* HDEIA ou Ammeline:
2-4-diamino-6-hydroxy-s-triazine

$$\begin{array}{c} OH \\ | \\ C \\ N \diagup \diagdown N \\ \| \qquad \| \\ C \qquad C \\ \diagup \diagdown N \diagdown \\ NH_2 \qquad NH_2 \end{array}$$

* Ammelide:
2-amino-4,6-dihydroxy-s-triazine

$$\begin{array}{c} OH \\ | \\ C \\ N \diagup \diagdown N \\ \| \qquad \| \\ C \qquad C \\ HO \diagup \diagdown N \diagdown NH_2 \end{array}$$

References

Adams CD, Randtke SJ (1992) Ozonation by products of atrazine in synthetic natural waters. Env Sci Technol, 26(11), 2218-2227

Akita K, Yoshida F (1974) Ind Eng Chem Proc Des Dev 13(1), 84

Beltrane FJ, Ovejero G, Acedo B (1993) Oxidation of atrazine in water by ultraviolet radiation combined with hydrogen peroxide. Wat Res 27(6), 1013-1021

Chramosta N, (1993) Etude de l'oxydation de s-triazines en milieu aqueux par ozonation en absence et en présence de peroxyde d'hydrogène. sous-produits de dégradation de l'atrazine et réactivité de quelques s-triazines vis-à-vis des radicaux hydroxyles. Thèse Poitiers n°25.

Chramosta N, DeLaat J, Dore M, Suty H, (1991) Effects of alcalinity and dissolved organic matter on the degradation of atrazine by O_3 and H_2O_2/O_3. 10th Ozone World Congress Monaco, 1, 291- 300

Dore M, (1989) Chimie des oxydants et traitement des eaux, Lavoisier, Paris, p 505

Duguet JP, Wable O, Bernazeau F, Malevialle J (1991) Removal of atrazine from the Seine river by the ozone - hydrogen peroxide combination in a full scale plant. 10th Ozone World Congress, Monaco, 1, 301

Hapeman-Somich CJ, Gui-Ming Z, Lusby WR, Muldoon MT, Waters R (1992) Aqueous ozonation of atrazine. Products identification and description of the degradation pathway. J Agr Food Chem 40(11), 2294-2298

Husley RA, Randtke SJ, Adams CD, Long BW (1993) Atrazine removal using ozone and GAC. A pilot plant study. Ozone Sci Eng 15(3), 227-244

Kearney PC, Muldoon MT, Somich CJ, Ruth JM, Voaden DJ (1988) Biodegradation of ozonated atrazine as a wastewater disposal system. J Agri Food Chem 36, 1301-1306

Langlais B, Reckow D, Brink D (1991) Ozone in water treatment, Application and Engineering. Lewis Publishers

LeSauze N, Laplanche A, Martin N, Martin G (1993) Modelling of ozone transfer in a bubble column, Wat Res 27(6), 1071-1083

LeSauze N, Laplanche A, Orta de Velasquez MT, Martin G, Langlais B, Martin N (1992) The residence time distribution of the liquid phase in a bubble column and its effect on ozone transfer. Ozone Sci Eng 14(3), 245-262

Legube B, Guyon S, Dore M (1987) Ozonation of aqueous solutions of nitrogen heterocyclic compounds: benzotriazoles, atrazine and amitrole. Ozone Sci Eng 9, 233-246

Levenspiel O (1972) Chemical reaction engineering. John Wiley and Sons, New York

Martin N (1992) Etude de l'application de l'ozone en traitement des eaux. Stratégie de qualification des cuves industrielles; emploi des mélangeurs statiques. Thèse Sciences Chimiques de l'Université de Rennes, N° 929

Martin N, Benezet-Toulze, Laplace L, Faivre M, Langlais B (1992) Design and efficiency of ozone contactors for disinfection. Ozone Sci Eng 14(15), 391-405

Orta de Velasquez MT (1992) Elimination des micropolluants dans les filières d'ozonation du traitement de l'eau potable. Thèse Sciences Chimiques de l'Université Rennes I, n° 827

Paillard H, Legube B, Gilbert M, Dore M (1990) Removal of nitrogenous by direct and radical type ozonation, dans „Organic micropollutants in the aquatic environment". Commission of the European Communities, Kluwer Academic Publishers VII, 234

Paillard H, Renoux L, Carbonnier F (1991) Elimination des pesticides azotés par oxydation combinée O_3/H_2O_2, comparaison avec l'ozonation. Journal Francais d'Hydrologie 22(2), 147-162

Roes AWM, Zeeman AJ, Bukkens FHJ (1984) High intensity gas/liquid mass transfer in the bubbly flow region during co-current upflow through static mixers. I. Chem Symp Ser, N° 87, pp 231-238

Roustan M, Duguet JP, Mallevialle J (1987) Mass balance analysis of ozone in conventional bubble contactors. Ozone Sci Eng 9(3), 289-298

Tace M (1993) Etude cinétique de la dégradation de chloroéthanes et de s-triazines en milieu aqueux par irradiation U.V. en absence et en présence de peroxyde d'hydrogène. Thèse Poitiers N° 24

Whitman WG (1923) The two film theory of absorption. Chem and Met Eng 29, 147

Xiong F, Graham NJD (1992) Rate constants for herbicide degradation by ozone. Ozone Sci Eng 14(3), 283-301

Yao CCD, Haag WR (1990) Rate constants for ozonation of pesticides, Solvents, PCB's and other organics in water. Am Chem Soc Preprints, Div of Environ Chem 15

Yao CCD, Haag WR (1992) Rate constants for reaction of hydroxyl radicals with several drinking water contaminants. Envir Sci Technol 26(5), 1005

Zhu ZM, Hannon J, Green A (1992) Use of high intensity gas-liquid mixers as reactors. 47(9-11), pp 2847-2852

Further Readings

Title: Advanced oxidation of atrazine in water – II. Ozonation combined with ultraviolet radiation.
Author: Beltran, Fernando J.; Garcia-Araya, Juan F.; Acedo, Benito.
Corporate Source: Universidad de Extremadura, Badajoz, Spain
Source: Water Research v 28 n 10 Oct 1994. p 2165–1274
CODEN: WATRAG ISSN: 0043-1354
Language: English

Abstract: The combination of ozone with ultraviolet radiation (254 nm) to oxidise atrazine is investigated. The importance of the three routes of atrazine elimination: direct photolysis, direct ozonation and hydroxyl radical oxidation is established in percentages. At neutral pH and 20 degree C 87 % of the oxidation rate is due to the radical way while direct ozonation only represents 1 %. Mass transfer and kinetic data obtained in previous works [Beltran F.J., Ovejero G. and Acedo B. (1993) Wat. Res. 27, 1013–1021; Beltran F.J., Garcia-Araya J.F. and Acedo B. (1994) Wat. Res. 28, 2153–2164] have been applied to mol. balance equations of atrazine, ozone (both in the gas and water) and hydrogen peroxide to obtain their corresponding concentrations in different conditions. It was necessary to include a hydroxyl radical initiating rate factor to account for radical ways not included in the basic mechanism of ozone decomposition. Comparison between different ways of oxidation and photolysis of atrazine (ozone, ozone/u.v. radiation, hydrogen peroxide/u.v. radiation) is also presented. (Author abstract) 20 Refs.

Title: Advanced oxidation of atrazine in water – I. Ozonation.
Author: Beltran, Fernando J.; Garcia-Araya, Juan F.; Acedo, Benito.
Corporate Source: Universidad de Extremadura, Badajoz, Spain
Source: Water Research v 28 n 10 Oct 1994. p 2153–2164
CODEN: WATRAG ISSN: 0043-1354
Language: English

Abstract: The ozonation of atrazine in water has been studied under different conditions of ozone partial pressure, pH, temperature and presence of hydroxyl radical scavengers. The process mainly develops through radical reactions, even at pH 2, probably due to the presence of traces of impurities in the starting material. The rate constant of the direct reaction between ozone and atrazine at pH 2 has been determined and expressed as a function of temperature. Thus, at 20 degree C the rate constant was found to be 4.5 M^{-1} s^{-1} the energy of activation being 18.8 kJ center dot mol^{-1}. A model formed by mol balance equations of atrazine and ozone, the latter in the liquid and gas phases, with the ozone decomposition rate based on the Staehelin and Hoigne mechanism allowed the determination of the concentrations 'of these species at different experimental conditions. The aqueous solution of atrazine is characterized by a hydroxyl radical initiating rate factor which accounts for other radical reactions not included in the basic ozone decomposition mechanism. With this factor deviations between experimental and calculated concentrations of atrazine are less than plus or minus 15 % in most of the cases. (Author abstract) 27 Refs.

Title: Prediction and verification of atrazine adsorption by PAC.
Author: Qi, Shaoying; Adham, Samer S.; Snoeyink, Vernon L.; Lykins, Ben W..
Corporate Source: Univ of Illinois, Urbana, IL, USA
Source: Journal of Environmental Engineering v 120 n 1 Jan-Feb 1994. p 202–218
CODEN: JOEEDU ISSN: 0733-9372
Abstract: A procedure was developed to predict the removal of trace organic compounds from natural water by powdered activated carbon (PAC) adsorption systems, which function as a batch reactor or continuous stirred tank reactor (CSTR). The procedure uses the equivalent background compound method coupled with the ideal adsorbed solution to qualify the competition between trace organics and background organic matter in water, and uses the pseudo single-solute homogeneous surface diffusion model to describe the adsorption kinetics of the target compound under the influence of background organic matter. The parameters required as input data can be independently determined from adsorption isotherms and a batch of kinetic test data. Good agreement between predicted and actual performance was found for adsorption of atrazine from Central Illinois ground water at different initial concentrations and different carbon doses and using a batch reactor as two CSTRs, one of which was a PAC/ultrafiltration system. (Author abstract) Refs.

2.4
Pervaporation and Membrane Stripping: Potentialities on Micropollutants Removal from Water

Edmond Julien · Yves Aurelle

Abstract

Adsorption on granular activated carbon or resin and air stripping are conventional processes used to reduce the amount of organic compound in drinking water. Pervaporation has been recently suggested as an alternative technique. This paper relates experimental studies on pervaporation through dense membranes and stripping on microporous membranes in order to remove trace amounts of halogenated compounds from drinking water. We observe that concentration polarization may develop at the liquid-membrane interface. The experimental data fit the film theory model very well. Further investigations on mass transfer were made in order to test the potential of these techniques. Pervaporation and stripping on microporous membranes is comparable in operating cost to air-stripping. However, additional research is required to find ways to lower the membrane cost in the pervaporation system.

2.4.1
Introduction

Today's environmental problems cannot be ignored or left unsolved because of economic considerations, especially those regarding potentially toxic or carcinogenic organic contaminants. Current technology for the removal of these compounds includes packed-tower air stripping and granular activated carbon adsorption (GAC). Volatile organic compounds (VOC's) are efficiently removed by air stripping and semi-volatile organic compounds (SOC's), such as some herbicides and pesticides that are readily adsorbed onto GAC.

There is a need for a single treatment option for removal of both volatile and less volatile organic aqueous contaminants. This can be performed by pervaporation or stripping on a microporous membrane. Pervaporation consists of splitting up homogeneous liquid solutions by vaporizing the solutes through a dense membrane. Stripping on a membrane is actually an air stripping technology with a fixed exchange surface area that does not vary with either the liquid or the gas flow rates. Mass transfer depends on the chemical potential variation across the membrane for both processes.

Pervaporation was first applied to the separation of azeotropic mixtures by Binning (Binning, 1960). His pioneer work clarified the possibility of water/ethanol separation by the use of pervaporation membranes. Pervaporation gained importance with research and development on composite membrane (Neel, 1982), but the process has been left to more fundamental or laboratory scale work. A breakthrough came in 1982 when GFT first established the commercial base pervaporator in Brazil using a cross-linked PVA composite membrane. The biggest installation producing refined ethanol at present is situated in France (the sugar refinery of Bethéniville, Marne, near Reims). It produces 155 m³ per day of refined ethanol and allows 2620 tons oil energy equivalent to be saved in comparison with the azeotropic distillation process. Later, Zhu (Zhu, 1983) clearly pointed out the potential of pervaporation for removing chlorinated solvents from aqueous solutions. As a result of this, much research has been done on this problem. The aim of this paper is to describe the contribution of our laboratory to the subject in testing the ability of pervaporation and the stripping on a microporous membrane to remove trace amount of pollutants from drinking water. Other authors' contributions will also be related.

2.4.2
VOC Removal Survey

The contamination of groundwater supplies by VOC's, particularly halogenated compounds, is now recognized as a major problem. About half of the 129 USEPA priority pollutants are VOC's and each is known to be toxic and/or carcinogenic. VOC's are emitted in large quantities (1,600,000 to 5,000,000 metric tons per year) from waste treatment, storage and disposal facilities (Shen, 1988), a clear indication that conventional waste treatment technologies are inadequate to destroy or remove and contain VOC's. More effective techniques like adsorption or air stripping have to be used.

Adsorption on to activated carbon or resin is based on the physico-chemical inter-action of the component to be and the surface of the adsorbent particles. Because of the extremely high specific surface area of these adsorbents, a high loading capacity is obtained. Concerning activated carbons, only the granular form gives good perfor-mances. With this material, organic compounds compete for adsorption sites and therefore the removal efficiency decreases as these sites become saturated. Carbon adsorption is effective for low concentration applications but becomes expensive at higher concentrations because spent carbon must be disposed of or regenerated. The cost of the regeneration procedure can be very high, depending on the kind of the pol-lutant, the adsorption system and the regeneration technique.

Air stripping can also be applied to remove VOC's from aqueous solutions. The effi-ciency ranges from 60% up to 90% when using aeration or waterfall processes. En-hanced removal can be obtained using packed towers with plastic packing. Three chlorinated solvents have been studied in our institute: chloroform, trichloroethylene and tetrachloroethylene (Roustan, 1986). A column, 0.2 m in internal diameter and 2 m in height, was used. The packing material was 0.0159 m diameter polypropylene Pall rings. The water was sprayed in at the top of the column, while air was introduced at the bottom. Two liquid flow rates were used, 1.2 and 1.8 Nm³ h⁻¹. The gas flow rate ranged from 11 up to 60 Nm³ h⁻¹. Although nearly independent of the liquid flow, the efficiency of the process greatly depends on the gas flow rate and temperature, thus bringing out the importance of the Henry's law coefficient of the compound to be removed in this process. The stripping factor is also an important parameter. There is an optimal value of this factor at which packing volume and total dissipated power are both minimal. Air stripping is a simple and cheap process. However it is limited to the removal of compounds that significantly partition to air over water (high Henry's law coefficient). Furthermore, groundwater that contains iron promotes fouling of the stripping columns due to iron oxidation and or carbonate precipitation, reducing process efficiency and resulting in increased maintenance cost. In addition to the above limitations, unless the off-gas is treated, air stripping merely turns a water pollution problem into an air pollution problem. The most popular method for off-gas treatment is carbon adsorption.

Pervaporation is a low-pressure membrane process in which a liquid mixture is brought into contact with a dense membrane at the upstream side and the permeate is removed as a vapor at the downstream side. Transport through the membrane is induced by maintaining the vapor presssure on the permeate side of the membrane lower than the vapor pressure of the feed liquid side using either an inert carrier gas or a vacuum pump. The permeate vapor is then condensed.

In the pervaporation process three steps can be distinguished in the membrane:

- selective uptake of one of the components from the feed solution,
- selective transport (diffusion) through the membrane,
- desorption (evaporation) at the permeate side.

The first two steps are the most important and determine the transport rate observed in the membrane.

Pervaporation was rarely employed to remove VOC's in drinking water until the eighties. Then pervaporability was demonstrated (Zhu, 1983) for a somewhat arbitrary selection of chlorinated hydrocarbons. The recurring species in all studies is chloroform, a typical VOC's with bp 61.2 °C and water solubility of about 8000 ppm. Other chlorinated species which have been enriched by pervaporation to various degrees, include methylene chloride, 1,2-dichloroethane, trichloroethylene, tetrachloroethylene, 1,1,2-trichloroethane and chlorobenzene. The standard material for the separation is polydimethylsiloxane (PDMS), which is very selective with halogenated compounds even though they are present as traces in water. According to the published studies, it can also be pointed out that with this type of membrane and within the used experimental conditions:

- the pervaporate flux of a particulate solute does not depend on the presence of the other compounds, at least when concentration levels are low,
- mass fluxes are very low, less than $0.2 \text{ kg h}^{-1} \text{ m}^{-2}$.

All these results concern solute concentration levels between 10 and 5000 ppm. This is a far cry from the real concentrations existing in groundwater and does not take in account a possible liquid mass transfer limitation. Therefore our laboratory, with Aptel and Aurelle, initiated research to fill this gap. That was the starting point of numerous studies on pervaporation and VOC removal from drinking water.

2.4.3
Experimental Set-up

A classical pervaporation device was used (Fig. 1). The treated water was pumped from a tank to a hollow fiber membrane module, through the lumen of the fibers, and

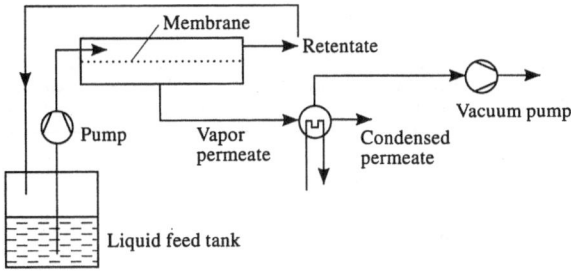

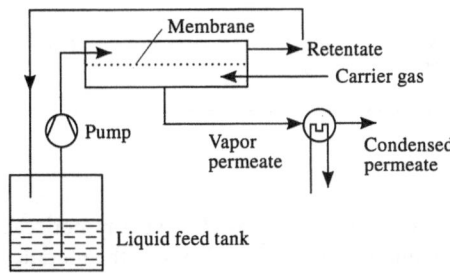

Fig. 1. Schematical representation of the pervaporation processes

back to the tank whose contents were continuously mixed. The permeate was evacuated using either vacuum pump (residual vapor pressure less than 400 Pa) or dry nitrogen gas countercurrent flow. The volumic gas flow to volumic liquid flow ratio lay between 12 and 500. The inlet solute concentration in the liquid phase ranged from 10 up to 300 µg l⁻¹. Table 1 presents characteristics of the hollow fiber bundles we used in the experiments. When using a microporous membrane it is better to talk about membrane stripping or membrane contactor rather than pervaporation although there is, in both cases, a partial vaporization of the solutes and the solvent.

2.4.4
Mass Transfer Model

Initial pervaporation studies emphasize membrane selectivity as the key to the separation process. In that case, authors correlate experimental data using the "solution-diffusion" model. However, if the faster permeant is present as a trace in the feed mixture, the solute concentration-gradient in the vicinity of the upstream interface cannot be very high, and the membrane is not fed rapidly enough with the solute. In this case, concentration polarization may develop at the liquid-membrane interface. Polarization effects were observed by Psaume (Psaume, 1986, 1988) during his attempt to extract, by pervaporation through silicone-rubber hollow fibers, small amounts of trichloroethylene or other chlorinated solvents, from polluted water.

Table 1. Hollow Fiber Bundle Characteristics

Membrane Type	Number of Fibers	Internal Diameter (μm)	Thickness (μm)	Porosity (%)	Active Length (m)	Exchange Area (m²)
Silicone (Dow Corning)	15	305	165	Dense film	0.35	0.005
Silicone on P.S.F.	20	625	2	Internal skin composite	0.25	0.010
Polypropylene (Celanese)	60	240	15	0.3–0.4	0.38	0.017
P.S.F. (Lyonnaise des Eaux)	25	560	100		0.36	0.016
	25	1000	250		1.06	0.083

Whatever his experiments, Psaume did not observe any noticeable temperature variations between inlet and outlet feed. The efficiency of the module is:

$$Y = 100 \cdot \frac{(C_e - C_s)}{C_e} \tag{1}$$

This is the same, using either sweeping gas or vacuum, and is independent of the gas flow if the latter is more than $8 \cdot 10^{-7}$ Nm³ s⁻¹. When axial Reynolds number value is a constant, a linear relationship exists between solute flux density and solute feed concentration (Fig. 2), which is in agreement with previous results (Brun, 1985), (Nguyen, 1987). It is clear (Fig. 3) that solute permeation is enhanced by higher channel velocities (axial Reynolds number values range from 10 up to 60). Furthermore the efficiency does not appear to depend on the inlet solute concentration. Similar results have been reported with polypropylene membrane contactors (Julien, 1987), (Semmens, 1989). The theoretical efficiency of a module, based on a concentration polarization model, is given by the expression:

$$Y_{Th} = 1 - \exp\left[-5.516\, D_i^{2/3} Q_1^{-2/3} (nL)^{2/3}\right] \tag{2}$$

This yield is independent of the nature of the fibers. The difference between experimental and calculated efficiency does not exceed 10 % with silicone fibers. For polypropylene fibers the gap rises up to 20 % when axial Reynolds number value is high.

According to these results, a better model for the organic compound removal is obtained using the classical resistance-in-series model which takes gas-side and liquid-side hydrodynamic parameters into account. This model assumes ideal behavior of the membrane (low swelling), which is true in our case. It is now widely described in published works, so we shall give only its major features when it is applied to hollow fiber bundles. We consider only the gas-liquid countercurrent flow case, the liquid flow being inside the lumen of the fibers.

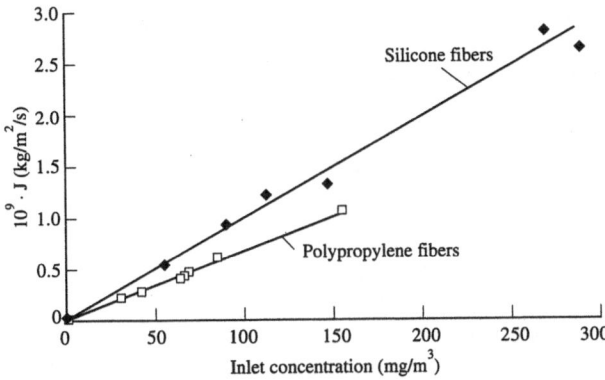

Fig. 2. Trichloroethylene flux density as a function ot the inlet feed concentration (Reynold number Re = 22)

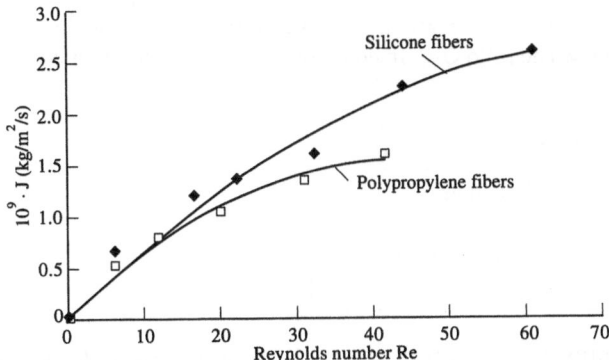

Fig. 3. Trichloroethylene flux density as a function ot the channel Reynolds number. (Inlet feed concentration Ce = 150 mg/m³)

Solute flux density changes all along the fibers. In a particular point between the inlet and the outlet of a fiber, we have:

$$J = K \left[C_1 - (C_g/H) \right] \tag{3}$$

The flux density of the solute is low, so we can assume that the overall volumetric gas and liquid flow rates have a nearly constant value all along the fibers. At steady state, the local mass transfer balance is:

$$(Q_1/n) \cdot dC_1 = - J \cdot \Pi \cdot d_i \cdot dz \tag{4}$$

When pervaporation under vacuum is used, $C_g = 0$, so we can arrange equations (3) and (4). Since the feed is a very dilute solution, we can assume that the overall mass transfer coefficient K is a constant. Integrating between $z = 0$ and $z = L$, we obtain:

$$K = \frac{Q_1}{n \Pi d_i L} \cdot Ln \left(\frac{C_e}{C_s} \right) = \frac{Q_1}{A} \cdot Ln \left(\frac{C_e}{C_s} \right) \tag{5}$$

When a sweeping gas is used, a mass balance constructed over a module length from the clean gas input to any cross section, provides, assuming the incoming gas contains a negligible amount of the investigated contaminants, the general relation:

$$Q_l C_l = Q_l C_s + Q_g C_g \tag{6}$$

Taking equations (3) and (6) into account, the integration of the equation (4) leads to the expression of the overall mass transfer coefficient:

$$K = \frac{Q_l R}{A(R-1)} \cdot Ln \left[\frac{(R-1) C_e + C_s}{RC_s} \right] \tag{7}$$

R is the stripping factor, that is:

$$R = \frac{Q_g H}{Q_l} \tag{8}$$

Equation (5) or (7) indicate that VOC removal will be enhanced by increasing the mass transfer coefficient K, the length of the module L, or the membrane exchange area A. A decrease in velocity or increase in the water residence time will also improve the removal efficiency. The ratio Q_l/A in equation (7) can be made explicit:

$$Q_l = \frac{n \cdot \Pi \cdot (d_i^2/4) \cdot v_l}{n \cdot \Pi \cdot d_i \cdot L} = \frac{d_i \cdot v_l}{4L} = \frac{v_l}{aL} \tag{9}$$

Hence, expression [7] becomes:

$$L = \left[\frac{v_l}{Ka} \right] \cdot \left\{ \frac{R}{R-1} \cdot Ln \left(\frac{(R-1) \cdot C_e + C_s}{RC_s} \right) \right\} \tag{10}$$

where the quantity in square brackets is the HTU, an estimate of the efficiency of the equipment, while that in the braces is the NTU, representative of the efficiency of the separation.

Analytical gas-phase chromatography allows us to determine the solute concentrations C_e and C_s. So we have access to the experimental value of the overall mass transfer coefficient. This coefficient includes the effect of three separate resistances: the liquid film resistance, the membrane resistance and the gas phase resistance. The overall resistance to mass transport is computed by adding these individual resistances. Table 2 resumes the various expressions of the individual resistances when the treated water circulates inside the lumen of the fibers. Individual mass transfer resistances may be estimated using classical correlations which are presented in Table 3. In Table 4 we have summarized some characteristics of the halogenated compounds we have investigated. One can compare the experimental overall mass transfer coefficient values with the theoretical predicted values and so obtain interesting information on the limiting step in the process.

Table 2. Individual Mass Transfer Resistances

Resistance (s m^{-1})	Dense Membrane	Microporous Membrane
Liquid	$\dfrac{1}{k_1}$	$\dfrac{1}{k_1}$
Membrane	$\dfrac{d_i \cdot Ln\,(d_o/d_i)}{2 \cdot H \cdot P_m}$	$\dfrac{e \cdot \tau}{\varepsilon \cdot D_g \cdot H}$
Gas	$\dfrac{d_i}{d_o} \cdot \dfrac{1}{H \cdot k_g}$	$\dfrac{1}{H \cdot k_g}$

Table 3. Empirical Correlations Employed to Estimate the Individual Mass Transfer Resistances in Hollow Fibers

Liquid	Re < 2000	$0.617 \cdot \left[\dfrac{L \cdot d_i}{v_1 \cdot D_1^2}\right]^{1/3}$
Resistance	Re > 2000	$38.46 \cdot \dfrac{d_i^{0.2} \cdot v_1^{0.47}}{v_1^{0.8} \cdot D_1^{0.67}}$
Gas Resistance		$45.45 \cdot \dfrac{d_h^{0.4} \cdot v_g^{0.27}}{v_g^{0.6} \cdot D_g^{0.67} \cdot H}$

Table 4. Some Characteristics of Investigated Compounds

	Trichloroethane	Trichloroethylene	Tetrachloroethylene
H (Dimensionless)	0.032	0.327	0.594
$D_1 (m^2 \ s^{-1})$	8.13 10^{-10}	8.86 10^{-10}	8.20 10^{-10}
$D_g (m^2 \ s^{-1})$	7.97 10^{-6}	8.12 10^{-6}	7.41 10^{-6}
$P_m (m^2 \ s^{-1})$		5.7 10^{-8}	

2.4.5
Results and Discussion

When the treated water circulates inside the lumen of the fibers, which is the most usual case, the relative importance of the individual mass transfer resistances changes with the nature of the membrane, the hydrodynamics conditions and particularly with the solute volatility. Because of the interdependence of the parameters, it is not easy to formulate simple rules about pervaporation or stripping on a microporous membrane.

Most authors do not take into account a possible gas-phase resistance or have found it to be minor over most of the practical operating conditions. We agree with these findings. As an example, Fig. 4 and 5 illustrate the change of the efficiency and mass transfer resistances with volumetric gas flow rate in the case of microporous poly-

70   Edmond Julien · Yves Aurelle

sulfone fibers. One can see that the contribution of the gas phase resistance to the overall resistance is negligible as soon as the stripping factor value is more than 1. This is
not to say that process performance is not influenced by gas flow rate for all VOC's. It
is important to recognize the inverse dependence of the gas phase resistance on the
Henry's law constant value of the solute. A compound with a Henry's law constant value
of about 0.003, which is an order of magnitude smaller than that of 1,1,2-trichloro-
methane, would be gas phase controlled for an air to water ratio of 100:1 (R = 0.3)
(Semmens, 1989). Vacuum stripping may prove beneficial in accelerating the transfer
of less volatile organic compounds, the separation of which is gas phase controlled.

A survey of the literature clearly emphasizes the importance of the solute volatility
as a pervaporation parameter (Côté, 1988). So we have studied the removal of three
volatile compounds whose Henry's law constant value is given in the Table 4 (Atlan,
1993). In these experiments, the Reynolds number has a constant value of 20. When
microporous fibers are used, the overall mass transfer resistance does not change

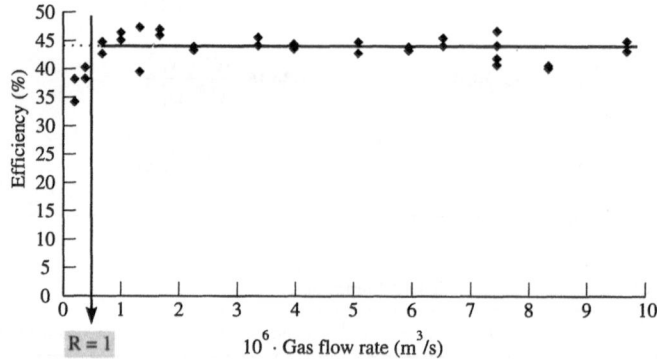

Fig. 4. Efficiency of a polysulfone hollow fiber module as a function of the gas flow rate. (TCE
inlet concentration Ce = 250 mg/cm³)

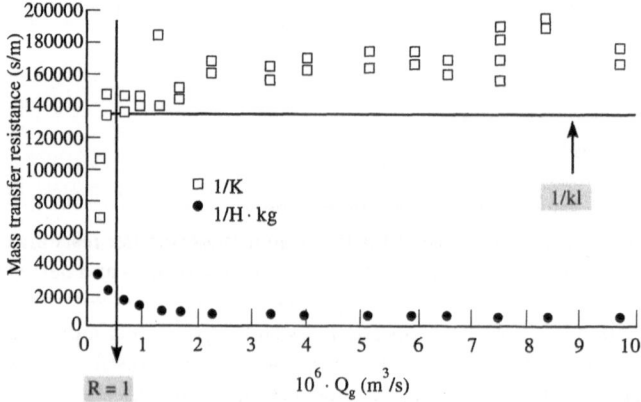

Fig. 5. Mass transfer resistances as a function of the flow rate in a polysulfone hollow fiber
module. (TCE inlet concentration Ce = 250 mg/m³)

(Fig. 6) although the Henry's law constant varies. This result means that the membrane and the gas phase resistances are insignificant, owing to the higher diffusion coefficient values of the solute in the gas phase. The case of the dense silicone membrane is different; the overall mass transfer resistance decreases as the Henry's law constant of the solute increases (Fig. 7). This suggests that the controlling factor in determining the relative importance of the liquid and the membrane resistances is the volatility of the solute. Assuming the gas phase resistance is negligible, one can easily compute the Henry's law constant value which corresponds to an equality for the theoretical liquid and membrane resistances. The results we obtain are:

$$H_{th} = 0.0072\ (Re)^{1/3} \qquad \text{for silicone rubber fibers}$$

$$H_{th} = 0.0010\ (Re)^{1/3} \qquad \text{for microporous fibers}$$

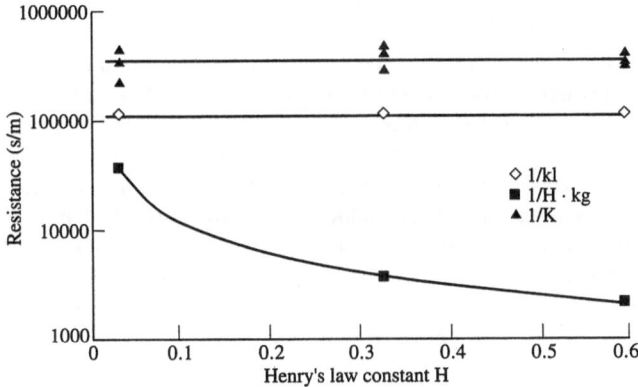

Fig. 6. Mass transfer resistances as a function of the volatility for microporous polysulfone hollow fibers

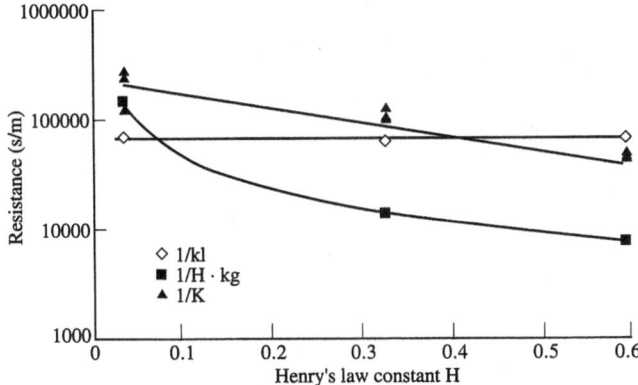

Fig. 7. Mass transfer resistance as a function of the volatility for dense silicone hollow fibers

The computation concerns a module which contains 25 fibers whose internal diameter is 500 μm, external diameter is 830 mm and length is 0.3 m. A mean value of the gas permeation coefficient of the solutes, that is $5.7 \cdot 10^{-8}$ m^2 s^{-1}, has been used in the case of the silicone membrane. With more volatile compounds ($H > H_{th}$), the organic flux density can only be increased by improving the hydraulic conditions because the diffusion in the liquid phase is the limiting step. With less volatile compounds ($H < H_{th}$), thinner membranes could be used to improve the flux density but to the detriment of the selectivity. Microporous fibers appear less sensitive to a membrane limitation of the flux density than dense silicone fibers. However, we must remember that the gas phase resistance may be, in some cases, equal or higher than the liquid or the membrane resistance. This may be encountered with low volatile compounds, but also when using a module with too little packing.

The nature of the membrane is also an important parameter, at least when dense polymeric membranes are used. This effect is included in the gas permeation coefficient P_m. When the treated water contains a small quantity of pollutant, the dense membrane does not swell and behaves as an ideal membrane whose physical-chemical characteristics have constant values. In these conditions, the efficiency of the membrane depends on the affinity between the solutes and the membrane (thermodynamical aspect) and on the importance of the values of the solute diffusion coefficients within the membrane (kinetical aspect). In order to remove a known solute, the choice of the membrane may be made using mutual solubility parameters values. Nevertheless, this procedure only gives an indication and experimental tests are a necessity. Fig. 8 illustrates this question. It concerns the kinetic of the diffusion of solutes across two dense plane polymeric membranes: a polyethersulfone membrane and a silicone rubber membrane. The slope of the straight lines provides an estimate of the diffusion coefficient of the solute within the membrane. Clearly, the silicone membrane behaves better than the polyethersulfone one in the removal of trichloroethylene or tetrachloroethylene.

The results we obtained (Psaume, 1986), (Julien, 1987), (Atlan, 1993) and those reported in the literature (Côté, 1988), (Semmens, 1989), (Nijhuis, 1990) suggest that mass transfer is often controlled by the liquid film resistance. This situation is gener-

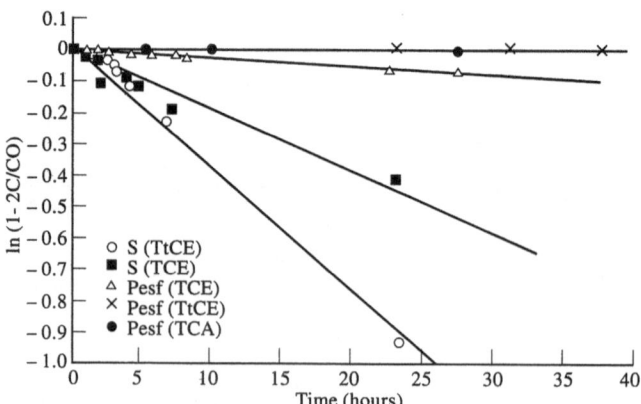

Fig. 8. Halogenated compound diffusion across a dense polyethersulfone and silicone plane membrane

ally met with volatile contaminants and with normal operating conditions. One can think it is due to the low solute concentrations we used as suggested by the fact that the flux density of the water is constant and is controlled by the membrane resistance in all the experiments, water whose concentration is evidently high. So we have studied the pervaporation under vacuum of trichloroethylene and dichloromethane depending on the solute concentration, using dense silicone hollow fibers. As an example, Fig. 9 and 10 give the results we obtained with the trichloroethylene. The experimental overall resistance values are always of the same order of the value of the computed liquid resistance and so with solute concentration values high as much as 1.1 g l^{-1} which is the limit of the solubility of TCE in the water. Similar results were observed with the dichloromethane for which the limit of the solubility in water is 8 g l^{-1}. In both cases, the solute flux density increased linearly with the solute concentration. This is an indication that the characteristics of the membrane do not vary even with high level of solutes in the treated water.

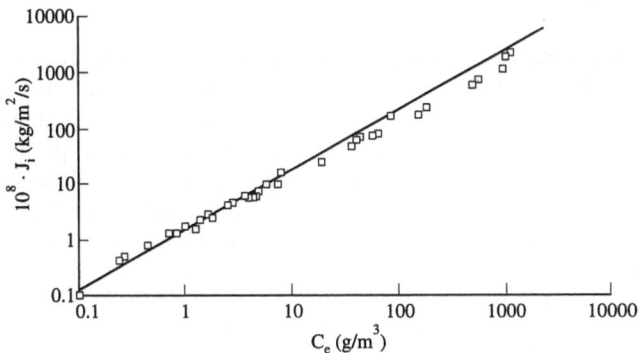

Fig. 9. Mean flux density as a function of the TCE inlet feed concentration in a dense silicone hollow fiber module

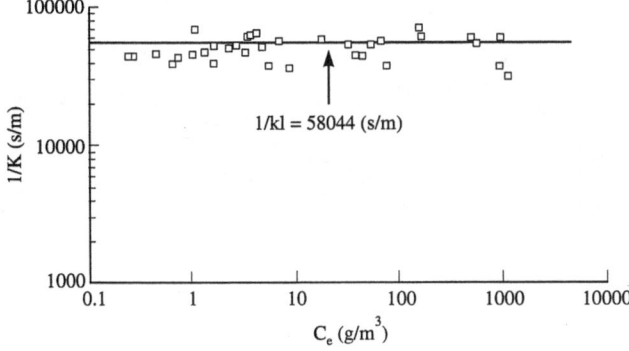

Fig. 10. Mass transfer resistance as a function of the TCE inlet feed concentration in a dense silicone hollow fiber module

Consequently, when we try to remove trace amounts of solutes in water, the nature of the membrane is a minor parameter, all the mass transfer resistance being in the liquid film for normal hydrodynamic conditions. To improve this situation we have presented the dependence of the overall mass transfer coefficient on the liquid flow rate in a dimensionless relation between the Sherwood number Sh and the Peclet Number Pe. The data collected for various cases are plotted in such a manner in the Fig. 11. Experimental data compare favorably with the correlation of Lévêque, improving the liquid mass transfer limitation of the process. So the choice of the membrane will depend on what is desired. If the objective is the recovery of the organic compound, a thick dense polymeric membrane is preferable because of its better selectivity. On the opposite side, if the aim of the process is only the removal of solutes from water, a microporous membrane is much more suitable because it is not selective and so allows the elimination of a larger range of pollutants (halogenated compounds, alcohols, ketones, etc.).

2.4.6
Module Design

As with packed-tower air stripping, several module designs can provide the required level of removal of a specified compound at a given water flow rate. Because countercurrent contacting gives the most complete separation, it is the geometry most frequently used. Occasionally crossflow is used, in which the gas flows perpendicularly to the feed. In this case, the concentration differences are not as efficient but the mass transfer coefficient K is much larger and hence the height of a transfer unit may be smaller. However, it should be noted that high performance levels are achieved with outside fiber water flow and with a module having careful, exact fibers spacing made (Yang, 1986). So we will discuss here only parallel countercurrent contacting where the treated water circulates inside the lumen of the fibers.

The design variables include the air-to-water ratio, R, the number, n, of fibers and their length, L. Examination of equations (7) or (10) indicates that once an air-to-water

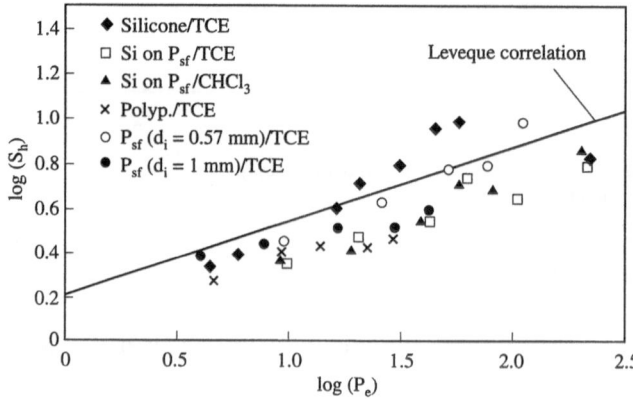

Fig. 11. Mass transfer correlation for halogenated compound removal on various membranes. (Sherwood number versus Peclet number)

ratio is specified, R is fixed and the efficiency of a hollow fiber module is controlled by the dimensionless product $Ka\,(L/v_l)$. Previous results suggest that, in the particular case of organic solvent trace removal, the mass transfer is limited by film diffusion. Combination of the correlation of Lévêque (Table 3), the expression for the interfacial area per unit of volume, a, and the expression for the superficial velocity, produces, for a laminar flow:

$$Ka \cdot \left(\frac{L}{v_l}\right) = 6.48 \cdot \left(\frac{\Pi D_1}{4\,Q_1}\right)^{2/3} \cdot (nL)^{2/3} \tag{11}$$

This interesting result indicates that removal performance is independent of the internal diameter of the fibers and simply depends on the number and length of the fibers. Thus, the module should be designed to minimize operating cost. Pumping water through the lumen of the fibers is likely to be a major expense, and the cost is strongly dependent on the internal diameter of the fibers, that is in the case of dilute aqueous solutions:

$$E = 0.512 \cdot \left[\frac{LQ_i^2}{n\,\Pi^2 d_i^4}\right] \tag{12}$$

It is therefore desirable to use short length and large diameter of the fibers to reduce this cost.

In order to situate pervaporation or stripping on microporous membranes in comparison with air-stripping, we have estimated the energy consumption in the following practical case of TCE removal (Aptel, 1988):

- inlet concentration feed: 100 µg l^{-1}
- outlet concentration feed: 1 µg l^{-1}
- treated water flow: 2400 m^3 per day.

Table 5 sums up the estimations we obtain for air stripping and various hollow fibers. The computation is based on an internal diameter of the column value of 1.2 m and a void fraction value of 0.55 when using hollow fibers. Obviously, only polypropylene and composite silicone-PSF fibers compare favorably with air stripping. We must remember that, when the efficiency process and the liquid flow rate are fixed, then the product nL is also fixed. Increasing the internal diameter of the fibers with a view to reducing energy consumption will in fact lead, for a given column diameter, to a decrease in the total number, n, of the fibers, which implies an increase of the length of the fibers and therefore an increase in the energy consumption, to keep the efficiency at the initially fixed value.

An interesting study has been made on the economical evaluation of the pervaporation process for removing organic compounds from water (Côté, 1990). The major objective of this work was the determination of the optimum design conditions for classical pervaporation under vacuum, taking into account capital and operating cost. Simulation was used to evaluate the removal, with 99% efficiency, of TCE from water. The inlet solute concentration was 10 mg l^{-1} and the feed flow rate was 10 m^3 h^{-1}. The material was dense silicone membrane and three kinds of modules were used:

- Inside flow module (IFM), with narrow bore inside fiber diameter of 500 µm when using laminar hydraulic flow, or with wide bore inside fiber diameter of 1000 µm when using turbulent hydraulic flow; the length of the fibers was 1 m.

Table 5. Energy Consumption Estimation for Trichloroethylene Removal (Efficiency is 99 % and liquid flow rate is 100 m^3 h^{-1})

	Air Stripping	Hollow Fiber Columns (Diameter is 1.2 m and void fraction is 0.55)				
	(Column) Diameteris 1.2 m	Dense silicone	Silicone on polysulfone	Microporous polypropylene	Microporous polysulfone	
d_o (µm)		605	629	270	760	1500
d_i (µm)		305	625	240	560	1000
n	–	$1.77\ 10^6$	$1.64\ 10^6$	$8.89\ 10^6$	$1.12\ 10^6$	$0.29\ 10^6$
L (m)	4.9	13.5	14.6	2.69	21.3	83.0
V (m^3)	5.6	15.3	16.5	3.04	24.1	93.9
A (m^2)	–	22916	46960	18032	42075	75135
E (kW)	1.5	35.3	2.34	3.65	7.73	11.54

- Transversal flow module (TFM), with 500 µm inside diameter hollow fibers; the fibers were packed into a 10 cm shell using a staggered geometry with transversal spacing of 1.5 mm, center to center.
- A spiral wound module (SVM) 4″ · 40″ (Filmtec FT 30 module).

The authors found that the pervaporation unit treatment cost was $ 0.56/m^3 for the TFM, $ 1.10/m^3 for the SVM, $ 1.41/m^3 for the wide bore fibers IFM and $ 3.80/m^3 for the narrow bore fibers IFM. As a comparison, the cost of conventional technologies ranged from $ 0.10/m^3 for air stripping alone (an environmentally unacceptable solution), to $ 0.80/m^3 for treatment trains including stripping and granular activated carbon on the aqueous or off-gas streams.

2.4.7
Conclusion

Pervaporation is a technically feasible and economically attractive method for treating VOC contaminated water. The process is compact, continuous and it does not use exhaustible materials. The selective membrane layer can be chosen from a wide variety of polymers, giving the flexibility needed to remove specific group of organic contaminants. When used to remove trace amounts of halogenated contaminants from water, the material effect of the nature of the membrane polymer is usually insignificant and microporous membranes can advantageously take the place of the dense membranes. However, progresses are still needed to develop hollow fiber bundles at low cost. Presently, the best use for pervaporation appears to be solvent recuperation or aroma extraction, the valorization of the product making up for cost treatment.

2.4.8
Notations

Symbols

a	Ratio of the exchange area to the liquid phase volume, $[m^{-1}]$
A	Membrane exchange surface area, $[m^2]$
C	Concentration, $[mol\ m^{-3}]$
d	Diameter, $[m]$
D	Diffusion coefficient, $[m^2\ s^{-1}]$
E	Pumping power, $[J\ s^{-1}]$
H	Henry's law constant, [dimensionless]
J	Flux density, $[mol\ m^{-2}\ s^{-1}]$
k	Mass transfer coefficient, $[m\ s^{-1}]$
K	Overall liquid side mass transfer coefficient, $[m\ s^{-1}]$
L	Active length of the fibers, $[m]$
n	Number of the fibers in a module, [dimensionless]
P	Gas permeation coefficient, $[m^2\ s^{-1}]$
Q	Volumetric flow rate, $[m^3\ s^{-1}]$
R	Stripping factor, [dimensionless]
v	Superficial velocity, $[m\ s^{-1}]$
Y	Efficiency, [dimensionless]

Subscript

e	module entrance,
g	gas phase,
i	inside,
l	liquid phase,
m	membrane phase,
o	outside,
s	module exit,
th	theoretical.

References

Aptel P, Julien E, Ganne N, Psaume R, Aurelle Y, Roustan M (1988) Pervaporation Situation among other Competitive Techniques in Halogenated Solvents Removal from Drinking Water. Proc. 3rd Int. Conference on Pervaporation Processes, Nancy, Bakish R. Ed., Englewood, 463–475

Atlan I (1993) Pervaporation et Stripping dans des Contacteurs à Fibres Creuses. Elimination des Micropolluants des Eaux. Thesis, INSA Toulouse (France)

Binning RC, Lee RJ (1960) US Patent 2,953,502 Sep 20

Brun JP, Larchet C, Bulvestre G, Auclair B (1985) Sorption and Pervaporation of Dilute Aqueous Solutions of Organic Compounds through Polymer Membranes. J of Membrane Science, 25, 55–100

Côté P, Lipski C (1988) Mass transfer Limitation in Pervaporation for Water and Wastewater Treatment. The Third International Conference on Pervaporation Process in the Chemical Industry, Nancy, France, September 19–22

Julien E, Psaume R, Aurelle Y, Aptel P (1987) Removal of Chlorine-containing Hydrocarbons from Wastewater by Pervaporation. Fourth Mediterranean Congress on Chemical Engineering, Barcelona, Spain, November, Vol. II, 712–713

Lipski C, Côté P (1990) The Use of Pervaporation for the Removal of Organic Contaminants from Water. Environmental Progress, 9(4), 254–261

Neel J, Aptel P (1982) La Pervaporation. 1ere partie: Principe de la Technique. Entropie, n° 104, 15–26
Nguyen TQ, Nobe K (1987)) Extraction of Organics Contaminants in Aqueous Solutions by Pervaporation. J of Membrane Science, 30, 11–22
Nihjuis HH (1990) Removal of Trace Organics from Water by Pervaporation. A Technical and Economic Analysis. Thesis, Twente University, Holland
Psaume R (1986) Application de la Pervaporation au Traitement de l'Eau Potable. Elimination de Dérivés Halogénés à l'Etat de Traces. Thesis, n° 34, INSA Toulouse (France)
Psaume R, Aptel P, Aurelle Y, Mora JC, Bersillon JL (1986) Pervaporation: Importance of Concentration Polarization in the Extraction of Trace Organics from Water. J of Membrane Science, 36, 373–384
Roustan M, Ganne N, Faucher G, Brodard E (1986) Elimination des Solvants Chlorés de l'Eau: Methodologie de Dimensionnement des Colonnes à Garnissages à Contre-courant. Environmental Technology Letters, n° 7, 273–282
Semmens MJ, Qin R, Zander A (1989) Using a Microporous Hollow Fiber Membrane to Separate VOC's from Water. Journal AWWA, 81(4), 162–167
Shen TT, Sewel GH (1988) Control of VOC's Emissions from Waste Managements Facilities. J Env Engng, 114, 1392
Yang MC, Cussler EL (1986) Designing Hollow Fibers Contactor. Aiche J, 32, 1910
Zhu CL, Yuang CW, Fried JR, Greenberg DB (1983) Pervaporation Membranes. A Novel Separation Technique for Trace Organics. Environmental Progress, Mai, N° 2, 132–143

Further Readings

Title: **Modeling of multicomponent pervaporation for removal of volatile organic compounds from water.**
Author: Ji Wenchang; Sikdar Subhas K; Hwang Sun-Tak.
Corporate Source: CeraMem Corp, Waltham, MA, USA
Source: Journal of Membrane Science v 93 n 1 Aug 8 1994. p 1–19
CODEN: JMESDO ISSN: 0376–7388

Abstract: A resistance-in-series model was used to study the pervaporation of multiple volatile organic compounds (VOCs)-water mixtures. Permeation experiments were carried out for four membranes: poly(dimethylsiloxane) (PDMS), polyether-block-polyamides (PEBA), polyurethane (PUR) and silicone-polycarbonate copolymer (SPC) membranes. Three VOCs, i.e., toluene, 1,1,1-trichloroethane and methylene chloride were studied. Both organic and water permeabilities of the PEBA membrane for 1 VOC-water, 2 VOCs-water and 3 VOCs-water mixtures were found to be comparable with each other. Coupling effects for trace organic transfer through the membrane were not observed when the downstream pressure was close to zero. However, at high downstream pressure, if the downstream side mass transfer resistance dominated the overall mass transport, coupling effects might occur within the vapor phase. The downstream pressure effect for the PDMS membrane was determined. The experimental results were correlated very well by a simple mass transfer equation. The downstream pressure may have positive or negative effects on the separation factor, depending on the ratio of overall organic permeability over water permeability, beta **p**e**r**m. The value of beta **p**e**r**m is a function of the intrinsic organic and water permeabilities, liquid boundary layer mass transfer coefficient as well as membrane thickness. The vapor phase mass transfer resistance was found to be negligible at low downstream pressure (less than 15 mmHg). It was clearly shown in this work that the resistance-in-series model could be used effectively to describe the pervaporation of dilute multiple VOCs-water mixtures through polymeric membranes. (Author abstract) 35 Refs.

Title: Pervaporation.
Author: Baker RW. Auth. Address: Membrane Technology and Research, Inc, Menlo Park, CA.
Source: Membrane Separation System: Recent Developments and Future Directions, Noyes Data Corporation, Park Ridge; NJ. 1991. p151–188, 18 fig, 7 tab, 18 ref.

Abstract: Pervaporation is a membrane process used to separate mixtures of dissolved solvents. A liquid mixture contacts one side of a membrane; the permeate is removed as a vapor from the other side. Transport through the membrane is induced by the difference in partial pressure between the liquid feed solution and the permeate vapor. This partial pressure difference can be maintained in several ways. In the laboratory a vacuum pump is usually used to draw a vacuum on the permeate side of the system. Industrially, the permeate vacuum is most economically generated by cooling the permeate vapor, causing it to condense. The components of the feed solution permeate the membrane at rates determined by their feed solution vapor pressures, that is, their relative volatilities and their intrinsic permeabilities through the membrane. Pervaporation has elements in common with air and steam stripping, in that the more volatile contaminants are usually, although not necessarily, preferentially concentrated in the permeate. However, during pervaporation no air is entrained with the permeating organic, and the permeate solution is many times more concentrated than the feed solution, so that its subsequent treatment is straightforward. (See also W92–07341) (Lantz-PTT)

Title: Use of Pervaporation for the Removal of Organic Contaminants from Water.
Author: Lipski C; Cote P Auth. Address: Zenon Environmental, Inc, Burlington (Ontario).
Source: Environmental Progress ENVPDI; Vol. 9, No. 4, p 254–261, November 1990. 3 Fig. 7 tab, 14 ref.

Abstract: Pervaporation is a new membrane process for the removal and concentration of volatile organic compounds from contaminated water. It can be used to treat groundwater, leachate, and wastewater. A resistance-in-series model was formulated and validated using silicone rubber hollow fibers for the treatment of water containing trichloroethylene. This process model was coupled with a costing model to assist in the design of pervaporation systems. High fluxes and separation factors were obtained for the removal of trichloroethylene from water using a thick silicone rubber membrane. Two different membrane configurations were tested using the same hollow fibers. Pervaporation was described by a resistance-in-series model: a liquid film resistance and a membrane resistance. Estimation of the two resistances showed that, for the trichloroethylene, mass transfer was limited by the liquid film resistance from an inside flow module. Membrane resistance only became apparent for a transversal flow module. A pervaporation case study of trichloroethylene demonstrated that the flux of trichloroethylene was a function of both the membrane thickness and module hydrodynamic conditions. A case study found that the pervaporation unit treatment cost was $0.56/m^3 for the transversal flow module, $1.10/m^3 for the spiral-wound module, $1.41/m^3 for the inside flow module (wide bore fibers) and $3.80/m^3 for the inside flow module (narrow bore fibers). For comparison, the cost of conventional technologies ranged from $0.10/m^3, for air stripping alone, to $0.80/m^3, for treatment trains including stripping and/or granular activated carbon on the aqueous or off-gas streams. In addition to the fact that pervaporation has definite technical advantages over conventional processes, it appear pervaporation has a economic advantage as well. (Mertz-PTT)

Title: Treatment of Organic-Contaminated Wastewater Streams by Pervaporation.
Author: Wijmans JG; Kaschemekat J; Davidson, JE; Baker RW. Auth. Address: Membrane Technology and Research, Inc , Menlo Park, CA.
Source: Environmental Progress ENVPDI; Vol. 9, No. 4, p 262–268, November 1990. 10 Fig. 3 tab, 6 ref.
Abstract: Pervaporation is a membrane process in which a permselective membrane is used to separate mixtures of dissolved solvents. The removal and recovery of organic contaminants from aqueous streams by pervaporation membrane systems is a viable and economical treatment for many waste streams. Three broad areas of industrial application have been identified as good opportunities for the technology: pollution control of dilute solutions of hydrophobic solvents; solvent recovery from process wastewaters; and volume reduction of mixed-solvent hazardous waste streams. The process is particularly suited to the treatment of water containing relatively hydrophobic volatile organics such as chlorinated solvents, naphthas, toluene, benzene and the like. Industrial demonstration of the technology is underway. (Mertz-PTT)

3 Air Pollution

3.1
Industrial Air Pollution: Removal of Dilute Gaseous Vapors

Marie-Hélène Manero

Abstract

This article deals with the removal of volatile organic compounds (VOC) and odors from contaminated air generated by industrial operations. First, general information is given about air pollution: the main gaseous air pollutants are classified (NO_x, SO_2, CO, CO_2, etc.) according to the targeted receptive area and the nuisance created. A point is made on the principal manufacturing sources involved in this pollution.

Then, a state of the art on main removal techniques is given. Five cleaning processes are discussed: adsorption, scrubbing, condensation, biological removal and oxidizing systems. The principal fields of use, advantages and disavantages are studied for each of them.

Activated carbon adsorption is the most common technique for solvent removal thanks to an easy use and an economical way to recover a wide range of gas streams. But improvements must be made to reduce the slowness of operations, the treatment of mixtures, etc.

Scrubbing is seldom used for low VOC content, while it is very efficient for odor removals, with chemical oxidizing absorption. Major problems are linked with the oxidants use.

Condensation can be energy intensive and is generally used for gas streams with high levels of VOC contamination.

Biological processes can achieve very good efficiencies for very low VOC and inorganic compounds contents. Caking problems and slow residence time are the main disadvantages of this technique.

Thermal combustion and catalytic oxidation are total destruction methods. Low valuable VOC and high concentrations are required to be treated by these expensive processes.

3.1.1
Introduction

Fighting air pollution is a relatively recent political preoccupation. In the 60's, the principal aim was the reduction of dust emissions and it is only in the 70's that some pollutants such as SO_2 or NO_x began to be treated. During the 80's, photochemical pollution was widely studied and other compounds were inplicated, such as Volative Organic compounds (VOC). The removal of VOC has now become a concern in many industrial processes. Usually, these compounds are found in dilute concentrations in the air.

Besides the nuisance for health, forest degradation and materials corrosion, air pollution also generates other unpleasant phenomena such as odors. These phenomena are less and less permitted, encouraging the development in recent years of treatments against pollutants such as NH_3 or H_2S. This article describes the main processes used to remove these low level of contaminations, with the benefits and drawbacks in each case.

What are the principal pollutants?

A lot of atmospheric compounds can be dangerous for the humans and their environment. This pollution can come from natural emissions or be created by human activities such as transport, and industry, which are always on the increase. Several kinds of pollution can be distinguished, according to their localization in the Atmosphere and the problems they generate (Elichegaray, 1993).

- Near from emission sources, in urban or industrial areas, the principal nuisances are material corrosion and risk of serious health problems. The main compounds responsible for this are NO_x, VOC., SO_2 and the polyaromatic hydrocarbons.
  For the problem of bad odors, which also take place in these areas, mercaptans, H_2S, aldehydes and cetones are incriminated.
- Far from urban and industrial areas (from 10 to 100 km), we find two phenomena. First the oxydation of SO_2 and NO_x in the atmosphere which generates acid rains. These are compounds responsible for the destruction of forests, the death of lakes and underground water contamination.

The second type of pollution is generated in the tropospheric area. The reaction of sun light on compounds like VOC, NO_x or CO creates photochemical pollutants such as ozone. The latter is very harmful for health (breathing problems, eyes irritations, etc.), plants and the climate. From a planetary point of view, air pollution is mainly the expression of two phenomena. The first of these is the depletion of the stratospheric ozone. This is essentially due to the action of CFC and NO_x against O_3. The second phenomena is the risk of global climate change, due to the worrisome increase of compounds like CO_2, CH_4, N_2O and O_3 in the low levels of the atmosphere.

We will deal here with the principal treatment that can be found to reduce dilute pollution, with a particular interest in the vapor emissions of volatile organic compounds (VOC) and odorous ones. We will not talk about the reduction of CFC, SO_2 or NO_x pollution.

What are the industrial activities involved in VOC emissions?

Transports activities are the most important sources of pollution, with 51% of VOC emissions (Bouscaren, 1989). As far as industry is concerned the greatest responsibility comes from the users of solvents. In France 19% of the emissions are solvents, as shown in Table 1.

Among industrial processes using solvents, painting activities, metal coating, printing, adhesive and paste industry are responsible of 54% of emissions, in

Table 1. Sources of VOC
non natural emissions,
in France, in 1989

Transport	51%
Industrial use of solvent	19%
Combustion	8%
Petrochemical chemistry	3%
Raffineries	1%
Varied (coal mines, domestic use, gas distribution…)	18%
Total	2.3 Mt/year

Europe (Bouscaren, 1989). In many cases, these solvents are on a per-unit basis. For example:

- Exit of car painting cabin: 100 to 500 mg/Nm³ (ketones, toluene, xylene…),
- Plastic material production: 500 to 10000 mg/Nm³ (alcohols),
- Drying shop: 120 mg/Nm³ (isopropyl alcohols).

Regulations

All over the world, requirements are increasingly stringent and more and more industries are concerned by new regulations. In Europe, for instance, next regulatory requirements will involve new industrial activities such as car painting, plane painting or metal coating. Two examples are given: VOC streams will be controlled next year with a limit of 50 mg/Nm³ for solvent emissions in drying rooms of car painting; VOC emissions from metal coating operations will be limited to 20 to 150 mg/Nm³.

Odors

Unpleasant odors are also sources of annoyance. They generally arise from macromolecules decomposition (proteins, sugars, etc.) in food processing, wastewater treatment, and the solvent industry.

Odorous compounds are divided in four categories:

- chemical containing nitrogen: ammonia (NH_3), amines
- sulfur compounds: hydrogen sulfide (H_2S), mercaptans (CH_3SH), methyl sulfide (($CH_3)_2S$)
- volatile fatty acids
- aldehydes, ketones and esters.

The specificity of odorous emissions is their very low load in contaminated air: from 10 to 30 mg H_2S/Nm³ for gaseous emissions in a sewage treatment plant for instance (Martin, 1991). These levels of concentrations are generally below the limit of toxicity. Some examples are shown, in Table 2, of both human detection limits and theoretical toxical limits (Paillard, 1988).

3.1.2
Removal of VOC and odors by adsorption techniques

The gaseous stream is passed through a granular bed where solvent vapors are captured. The most common system is comprised of two reactors that work alternatively in

Table 2. Limits of toxicity
and of detection for some
contaminants

	olfactive detection (mg/Nm³)	toxicity limit (mg/Nm³)
H₂S	0.0001 to 2	14
ethyl mercaptan	0.001 to 0.03	1.25
NH₃	0.03 to 37	18
formaldehyde	0.065 to 12	3
Acetic acid	0.025 to 6.5	25

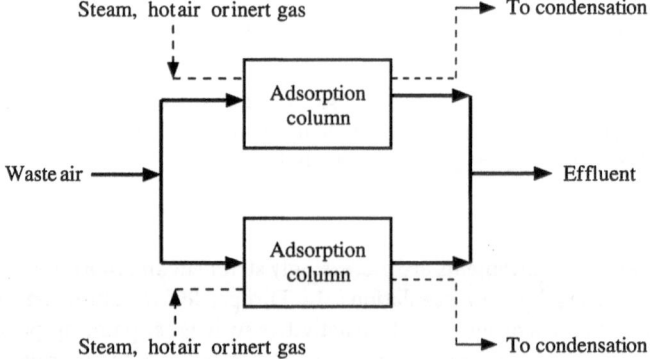

Fig. 1. Cyclic adsorption process

adsorption or desorption (Figure 1). They are generally filled with activated carbon in fixed beds. The collected vapors are periodically desorbed from the carbon through regeneration by steam, hot air or inert gas.

Principal data on commercial carbons (Martin, 1991)

specific surface: 1000 to 2000 m²/g
apparent specific gravity: about 0.5
particule size: 1 to 2 mm for 1 or 2 meter bed height
gaseous velocity: 500 to 3000 m/h

The adsorption technique is almost always used when the air is polluted with a valuable contaminant, that is recoverable. Indeed, all the interest of adsorption is this possibility of recovering the products thanks to the regeneration of the adsorbent material. If mixtures are adsorbed, it is necessary to add a separation technique behind. This is generally not made because of the supplementary cost.

Adsorption is the most used method to remove air pollution in solvent and food industry, thanks to its simplicity of use and to its good results. For compounds like styrene or chlorinated solvents, better efficiencies than for other technologies are noted.

Adsorption is seldom used for the removal of unpleasant odors, except in some cases such as kitchen hoods. When this method is used, carbon is permeated with a solution which favours the chemical absorption phenomena.

Regeneration of activated carbon

- Thermal regeneration
  The necessary heat for the desorption is brought by steam, inert gas or hot air. Solvent vapors are desorbed, highly concentrated, and then recovered by condensation. This operation is generally considered as efficient for contaminations higher than 500 mg/Nm^3, because of its important energy-saving costs (Woronoff, 1991)
- Other regeneration methods
  New techniques have been developed using the "Joule effect". That type of regeneration works on granular beds, but seems to be more developed with new types of structure like fibers of activated carbon.

Other recent investigations (Baudu, 1991) involve electromagnetic induction too regenerate and recover solvents.

Fields of use of activated carbon adsorption

- removal of solvents except for polymerizable or highly volative ones or with high molecular weights,
- industries concerned: printing, painting, metal coating, adhesive industry, chemical processes,
- low flow rates: less than 20,000 m^3/h,
- concentrations: less than 50 g/m^3 to avoid a quick saturation and the problems due to the exothermicity of the adsorption.

Main benefits of this method

- effective for solvent recovery: around 95%,
- low capital investment,
- applicable to a wide range of flow, concentration and contaminant,
- flexible to flow variations,
- simple to operate,
- efficient for chlorinated solvents.

Principal disadvantages

- temperature and moisture constraints (20 to 40 °C, humidity lower than 50%),
- pretreatments required to remove particulates,
- slowness and high cost of the regeneration operation,
- flammability of the carbon: risks of fire with some solvents such as ketones or aldehydes, or at high concentrations,
- bad profitability for mixtures or not valuable solvents,
- bad efficiency with aromatic compounds, styrene and solvents with low molecular weight,
- low velocity through the bed: around 0.8 m/s,
- steam regeneration: risk of pollutant products with chlorinated solvents and esters,
- hot air regeneration: oxidation and self-ignition risk.

Other adsorbent materials

Natural zeolites have been used since the 18th century in adsorption. Fixed bed technology has been replaced by a system with a concentrative wheel (Wedford, 1993). The main advantage of these molecular sieves is their stability on a large field of temperature (40 to 900 °C) and their hydrophobic properties. But a very high investment cost prevents their large development for the moment. Scientists have been working for a long time on synthetical materials, but efficiencies are not good enough.

3.1.3
Removal of VOC and odors by absorption techniques/Scrubbing

This removal technique involves the transfer of pollutants from the gas stream into a liquid. Physical absorption is less commonly used than adsorption for organic solvents, but it can be used for valuable products, with a separation technique below. For inorganic contaminants such as odorous ones, absorption process is more often found. In this case, the mass transfer is accompanied by a chemical reaction. This chemical absorption is very often used to deodorize in sewage treatment plants and in food industries.

Absorption system

Several versions of contactors are available. Packed towers are used in most cases for odorous and VOCs removals. The gaseous stream flows upward from the bottom of the tower and contacts the liquid absorbent, which flows downward. Sieve-plate towers are used in specific cases, like when the air flow rate is too low to adequately wet the packing. Spray towers or venturi systems can be found for dusty streams.

Odorous reduction in wastewater treatment

Reducing odorous nuisances means essentially removing nitrogen and sulfur compounds. As the malodorous air is a quite complexe mixture, two or three series-mounted columns are generally used to deodorize (Fig. 2). In the first one, the gas is scrubbed with an acid solution (pH $\approx$ 2.5) in order to remove the nitrous compounds (NH_3 and amines). In the second, the washing liquid is a basic oxidant solution (pH $\approx$ 9) designed to capture hydrogen sulfide (H_2S). In the last one, the other sulfurous components (mercaptans) are absorbed thanks to an oxidant solution that is more basic (pH $\approx$ 10.5). Oxidizers are used to regenerate scrubbing solutions.

Oxidizers

Chlorine and ozone are the most powerful oxidizers and the most commonly used. However, their use presents some problems. The principal drawback of the use of ozone is that an on site production unity is necessary, which increases the cost of the equipment. In the same way, stability of chlorine storage and the high cost of an on site electrochemical production are important disadvantages of chlorine use. More, some toxic chlorinated chemicals can be produced by this oxidation. Other oxidizers have been tried but none offer good efficiency.

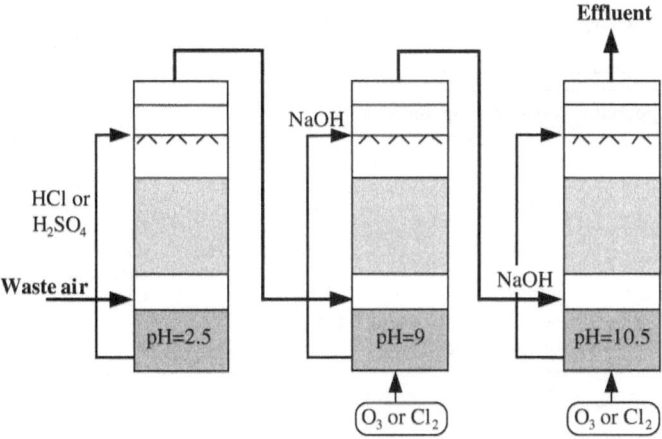

Fig. 2. Chemical oxidative scrubbing

Principal fields of use of absorption techniques

- organic or inorganic compounds that can be dissolved: H_2S, NH_3, HCl, aldehydes, ketones, mercaptans,
- industries concerned: sewage treatment plants, food processing, canning, slaughtering and rendering animals, tanning animal hides,
- concentrations in VOC above 200 ppm.

Main benefits of scrubbing

- the best technique for odor removals,
- easy use,
- system can be regulated,
- calculation methods available.

Disadvantages

- ineffective for hydrocarbon and organic insoluble vapors,
- waste liquid solution must be treated,
- existing calculation methods are based upon "savoir-faire" more than scientific laws. So some problems are found for the automatization (oxidizing chemical absorption).
- chemical oxidizers are corrosive and toxic,
- investment and operating expenses (for oxidizing chemical absorption).

3.1.4
Removal of VOC and odors by biological processes

This recently developed method, which is very efficient for very low contaminant levels, has been used for a few years in deodorization and, more recently, to remove

Table 3. Velocity of degradation by biological processes

Fast biological degradation		slow degradation	very slow degradation
inorganic compounds	organic compounds		
H_2S	alcohols	aromatic hydrocarbons	1, 1, 1 trichloroethane
amines	aldehydes	phenols	halogenated hydrocarbons
NH3	ketones	dichloromethane	polyaromatical compounds
methyl mercaptan	ethers	methane	CS_2
PH_3	esters	pentane	
SiH_4	organic acids	cyclohexane	

VOC from waste gas. The degradation efficiency is about 90–95% (Ottengraf, 1986) for amines, ammonia, alcohols, ethers, aldehydes, organic sulfides and hydrogen sulfide. But many other compounds can be degraded by this method (Table 3). Numerous researchers are working to find the best microbes in each case. This technique can be used alone or behind a conventional treatment, adsorption or absorption, to handle more concentrated streams.

Principles of the biofiltration systems

Biological materials, fixed on an inert support, degrades organic or inorganic compounds in CO_2, H_2O and salts. Several steps are found in the biological process: first, contaminants are adsorbed in the material's pores; as these materials are moisturized, contaminants are transferred in the liquid phase, by absorption. Then, the degradation is made by the microbes.

The biofilters are very often beds of compost and tree bark, peat or soil. These natural materials are cheap and have an interesting microbial flora. The bed is generally about 1 meter deep and the air is blown from the bottom of the reactor (Figure 3). The inlet air must be dust-free, saturated with water and not too hot. A constant feed of the biological bed is necessary to achieve good efficiencies: it needs to be humi-

Fig. 3. Biofilter process

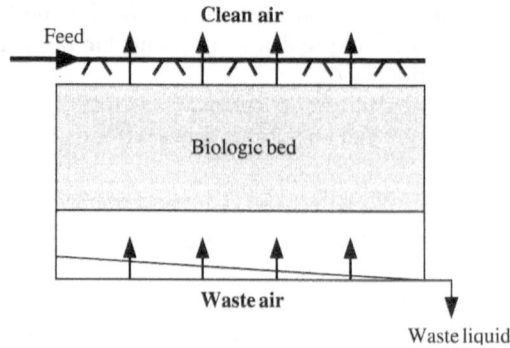

difyed continuously and the nutritional equilibrium must be kept (ration $DBO_5/N/P$ about 100/5/1). Early designs were open to the atmosphere and experienced problems with the rainfall, the hotness and dust. Now modern units are sealed in containers.

Fields of use of biological filtration

- Food processings, sewage treatment plants, rendering, plastic processing, aroma extraction, slaughtering,
- low contaminants level: less than 1 g/Nm^3,
- flows less than 30000 m^3/h,
- temperature less than 40 °C.

Main benefits

- the cheapest technique: capital and operating costs,
- very efficient for very low concentrations, until less than 1 mg/m^3 (Fouhy, 1992),
- little residue,
- able to treat a wide range of compounds,
- very adaptable to flows variation.

Drawbacks

- long reaction time which implies very big units and large areas occupation,
- pretreatments may be required to remove dust and/or to cool the air,
- gas flow distribution often becomes irregular,
- bed's feed must be seriously controlled,
- high pressure drop: beds must be stirred periodically to prevent caking.

Biological scrubbers

This technique hasn't yet been very developed in industry but numerous laboratory designs have been testing. Absorption and biological degradation are associated in one or two steps in order to be more effective. This process is destinated to treatments where biofilter is not efficient, like for compounds insoluble in water, or when the products would harm the biofilter bed, like the acids created by hydrogen sulfide and ammonia degradation. The contaminants are absorbed in a packed column by a solvent made of water and sludge, which flows in a countercurrent way (Figure 4). The

Fig. 4. Biological scrubbing

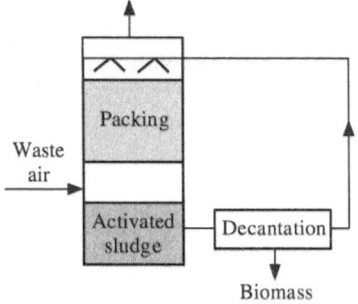

biodegradation goes on in a sedimentation tank and the solution is continuously recycled back to the column. This looks hopeful for styrene, chlorinated and aromatic compounds (Martin, 1991).

3.1.5
Removal of VOC and odors by condensation systems

This process allows the recovery of VOC when the concentration is not too low. It is often used upstream of absorbers, carbon beds or incinerators to reduce the solvent load and recover some cooled VOC. This system involves cooling the waste stream to a temperature below the dew-point of the mixture and collecting the condensed VOC as a liquid. That mixture requires a separation treatment to be recovered or a waste treatment prior to discharge.

This process has been used by small industries because of its simple operation, usually for low flow rates and high concentrations, with cooling water in a simple heat exchanger. More recent processes have been used, in pharmaceutic industries for instance, to recover low-boiling and low molecular weight solvents. This cryogenic process require two condensers and a gas such as nitrogen, with very low boiling point, which passes through the condensers in a countercurrent flow (Fig. 5). VOC removal efficiency by condensation can be fixed at any level desired. The only constraint is the cost. Settling cost and energy costs can be prohibitive for low concentrations. Therefore this process is not commonly used for VOC removal.

Fields of use

- printing, spray painting, polymer processing, pharmaceutical industries, petrochemical,
- low flow rates: less than 2000 m³/h,
- high levels of concentrations: more than 40 g/Nm³ (economical reasons).

Main benefits

- easy and well known use,
- recovery of solvents,
- heat exchanger cost is cheap,
- any chosen efficiency can be achieved with an appropriate cooling liquid,
- can be used as a pretreatment.

Fig. 5. Condensation process

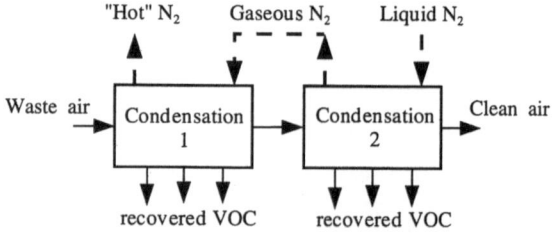

Main disadvantages

- refrigeration unit cost or inert gas cost,
- occupation area can be enormous if there are two or more heat exchangers, refrigeration unit.

3.1.6
Removal of VOC and odors by combustion systems

The gaseous stream is oxidized at high temperature to destroy organic and inorganic compounds. Excellent efficiencies are achieved with a complete oxidation, which means a good control of temperature, residence time and turbulence. Combustion-based methods might be, in many cases, the ideal technique because of its total removal of the contaminants.

No waste to treat and an easy means of use were very often reasons that explain the success of this process. Indeed, combustion is the most commonly used technology to treat gaseous emissions for high levels of contamination. In the case of low pollution, the costs are raised by a necessary addition of fuel to maintain the oxidation. So, for VOC or odor removal, the combustion systems are often considered as the last solution.

Thermal oxidation systems

In these technologies the waste air is burned on a flame. Three types of systems are available. The simplest one is the "direct flame incinerator". It is comprised of a combustion chamber and a fuel reserve. It is quite efficient but needs a high energetical contribution: oxidation temperature must be maintained to 800 to 1000 °C. So more recent systems use heat exchangers to recover heat and to preheat the inlet gas (Fig. 6).

The third type of thermal oxidation system is called "regenerative oxidation system" and is used for low levels of contamination. The gas flow passes through a hot bed and then is oxidized in a combustion chamber. Specific materials such as ceramic are used to keep the heat of the combustion and thus no need of supplementary fuel is required. The system is able to maintain the oxidation thanks to a cyclic change of direction of the gas stream (Kumar, 1993). In all cases certain compounds must not be present in the inlet stream. Sulfur, chlorine and fluorine combustion would producte dangerous compounds such as sulfur dioxide, hydrogen chloride or fluoric acids.

Fig. 6. Thermal oxidation

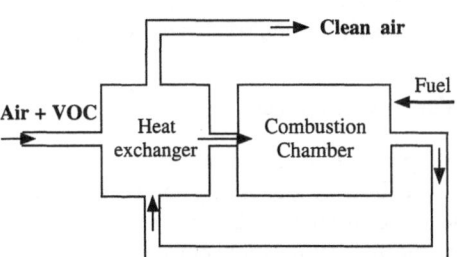

Fig. 7. Catalytic oxidation

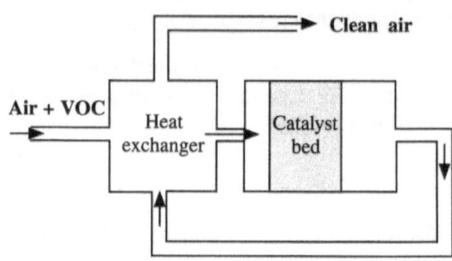

Catalytic oxidation systems

Instead of having a flame, the combustion chamber is comprised of a catalyst bed (Fig. 7). The waste gas is first preheated in a mixing chamber and then oxidized on the granular bed. Catalysts such as oxide of platinum, copper or chromium are used to enhance the rate of VOC oxidation. Thus, reactions can take place at lower temperatures than for thermal combustions: from 200 to 500 °C.

The oxidation takes place in the catalyst pores, where VOC and O_2 are adsorbed. Reaction products are immediately desorbed from these active sites. Good efficiencies can be reached with this system but problems caused the failure of the process in several cases. The catalysts are very sensitive to substances such as sulfurs, halogens, mercury, arsenic or lead and can be poisoned by them. Furthermore dust, condensed VOC or polymerized hydrocarbons are responsible for bad efficiencies.

Fields of use of oxidation systems

- solvent, chemical industries,
- foundry coke ovens,
- low valuable solvents (toluene, heptane, etc.),
- high levels of contamination.

Main benefits

- no residue,
- very efficient process,
- recovery of heat,
- easy use,
- available for non soluble or non adsorbable substances.

Main disadvantages

- operating costs: fuel, catalyst,
- catalysts: risk of caking and poisoning,
- toxic products with sulfur and halogens oxidation,
- thermal combustion: emissions of NO_x.

Fig. 8. Advanced oxidation process

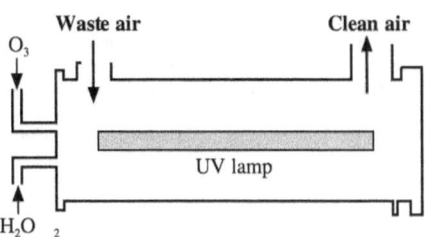

3.1.7
Removal of VOC and odors by advanced oxidation process

This process has been developed in the United States for the treatment of VOC present in wastewater, at very low levels of contamination (a few µg/l). The removal is made by air stripping. Thereby, VOC are transferred to the gaseous phase which must be treated. This is made in a reactor (Fig. 8) where the VOC are oxidized by ozone or hydrogen peroxide in conjunction with ultraviolet light (Johnson, 1991, Wekhof, 1991). Excellent efficiencies are found for laboratory tests (Nielsen, 1989), even for halogenated compounds: odors and VOC can be totally removed, which is the main benefit of this process. But on another hand, the use of oxidizers such as ozone and UV light implies prohibitive operation costs. This can explain the low development of this process, except in the states. Furthermore, its use is reserved for special cases of very low flow rates with very low concentration levels.

3.1.8
Conclusion

There is not a single process able to remove all kinds of contamination. Each case needs a special study, taking in account scientific and technique factors as well as economical ones.

Adsorption, scrubbing and condensation systems remove contamination with a recovery of solvents, whereas biological, thermal or chemical oxidation processes remove contaminations by destroying them. If the scientific feasability is proven, the technique chosen will depend on several factors: the kind of compounds present in the gaseous mixture, its level of contamination, its flow rate and the desired efficiency.

In economical terms, the cheapest process is the biological one; then activated carbon adsorption is generally less expensive than chemical scrubbing. All incinerations, thermal or catalystic, are considered expensive processes for low levels of contamination.

3.1.9
References

Baudu M, Le Cloirec P, Martin G (1992) Environ Technology, vol 13, pp 423–435
Bouscaren R, Allemand N (1989) Inventaire des emissions de COV dans l'atmosphere, CITEPA, Paris, France

Elichegaray C (1993) Les Pollutions de l'air, du local au global, Les entretiens de la Technologie, Ecole Centrale de Nantes, France, May 1993

Fouhy K (1992) Cleaning Waste Gas, Naturally. Chem Engineering, pp 42–46

Johnson GJ, Saarenvirta G, Brown PM (1991) Industrial odour control, the deodair process, 10th Ozone World Congress, Proceed vol 1, pp 387–401

Kumar KS, Pennington RL, Zmuda JT (1993) Chemical Engineering, pp 12–17

Le Cloirec P, Fanlo JL (1991) Traitement des odeurs, Désodorisation industrielle. Tech Tendances Innovation 128, Paris, France

Martin G, Laffort P (1991) Odeurs et désodorisation dans l'environnement, Tec & Doc, Lavoisier Paris, France

Manero M-H (1993) Elimination de constituants gazeux contenus à l'état de traces, FIRTECH report, GFGP, France

Nielsen SB, Underbrink TK, Wong E (1989) VOC removal by air phase advanced oxidation processes. 9th Ozone World Congress, Proceed vol 1, p 766

Ottengraf SPP, Meesters JJP et al. (1986) Biological elimination of volatile xenobiotic compounds in biofilters. Bioprocess Engineering, 1, p 61–69

Paillard H, Blondeau F, L'Eau TSM (1988) 83° année, n° 2, pp 79–88

Wedford J, Ford H (1993) Chemical Engineering. p 89–90

Wekhof A (1991) Environmental Progress. p 241–247

Woronoff G (1991) Technical and economical comparison of VOC control techniques. STEP SA, Belgium

3.1.10
Further Readings

Title: **Optimal design of multicomponent VOC condensation systems**
Author: Dunn, Russell F.; El-Halwagi, Mahmoud M.
Corporate Source: Auburn Univ, Auburn, AL, USA
Source: Journal of Hazardous Materials v 38 n 1 Jul 1994. p 187–206
Publication Year: 1994
CODEN: JHMAD9   ISSN: 0304-3894

Abstract: The objective of this work is to address the problem of optimally selecting and designing condensation systems for the recovery of volatile organic compounds (VOCs) from gaseous emissions. In a typical VOC condensation system, numerous refrigerants may be technically feasible for achieving the desired VOC recovery. The task of selecting the most economical refrigeration system is a challenging one. All potentially applicable refrigerants ought to be simultaneously screened so as to select the best separation system. Such an optimal system may be a hybrid process which includes more than one refrigeration unit. This paper presents a systematic design technique for optimizing the selection of refrigeration systems for gaseous emissions with multiple VOCs. The technique developed involves a three-stage targeting approach. The first stage identifies a VOC condensation network which possesses the minimum annual operating cost. The second stage involves the minimization of the fixed cost of the refrigeration units so as to achieve the minimum operation cost determined in stage one. Stage three entails the determination of the final refrigeration network, whereby tradeoffs between operating and fixed costs are considered. A case study covering the recovery of multiple VOCs from the gaseous emission of a magnetic tape manufacturing plant will be solved using the design methodology. (Author abstract) 21 Refs.

Title: Consider biofiltration for decontaminating gases
Author: Bohn, Hinrich
Corporate Source: Univ of Arizona, Tucson, AZ, USA
Source: Chemical Engineering Progress v 88 n 4 Apr 1992 p 34–40
CODEN: CEPRA8 ISSN: 0360-7275

Abstract: This article focuses on one type of biological treatment that chemical engineers may be less familiar with – the use of soil beds, or biofiltration, to treat contaminated gases. Biofiltration is the removal and oxidation of organic gases (volatile organic compounds, or VOCs) from contaminated air by beds of compost or soil. Many chemical process industries (CPI) operations could utilize biofiltration to treat waste gases, just as they use biological treatment for liquid and solid wastes. The removal and oxidation rates depend on the biodegradability and reactivity of the gases. The half-lives of contaminants range from minutes for the most reactive gases to months for the least biodegradable. A general classification system is shown. This list is much the same for all of the bioreclamation processes because they all rely on the same mechanism. 9 Refs.

Catalytic Oxidation for VOCs
Keith J. Herbert,
Allied-Signal Inc, Tulsa, OK;
Natl Environ J, Mar–Apr 93, v3, n2 p 46 (3) journal article
Catalytic oxidation is a proven technology for the destruction of hazardous air pollutants. Catalysts used in the systems for VOC destruction are usually noble metals, such as platinum or palladium. The VOC destruction efficiency is a function of the inlet temperature to the catalyst bed, volume of catalyst, and the quantity and type of noble metal. The advantages of catalytic oxidation include lower fuel consumption and lower nitrogen oxides, carbon monoxide, and carbon dioxide emissions. Technological improvements that address catalyst deactivation are described. (2 diagrams, 6 graphs, 1 photo)

3.2
Development of Trickle-Bed Air Biofilter

MAKRAM T. SUIDAN · FRANCIS L. SMITH · GEORGE A. SORIAL · RICHARD C. BRENNER

Abstract

The 1990 Amendments to the Clean Air Act have stimulated strong interest in the use of air biofilters for the control of volatile organic compounds (VOCs) in effluent air streams. Biofilters are specially suited for the treatment of gas streams contaminated with low to moderate concentrations of VOCs. The effectiveness of three biofilter media was compared; a peat mixture, a channelized medium, and a pelletized medium. Toluene was used as the model VOC. The performance of the peat biofilter was found to be very sensitive to air temperature, and feed toluene concentration, while the channelized medium biofilter suffered from short circuiting induced by uneven biomass accumulation. Furthermore, biomass removal from this medium was not practical. The pelletized medium appeared to be the medium of choice. It provided for a resilient and effective trickle bed air biofilter (TBAB) that combined efficient treatment at high organic loads and ease of biomass control.

3.2.1
Introduction

Since enactment of the 1990 amendments to the Clean Air Act, the control and removal of volatile organic compounds (VOCs) from contaminated air streams has become a major public concern (Lee, 1991). Consequently, considerable interest has evolved in developing more economical technologies for cleaning contaminated air streams, especially dilute air streams. Biofiltration has emerged as a practical air pollution control (APC) technology for VOC removal. In fact, biofiltration can be a cost-effective alternative to the more traditional technologies, such as carbon adsorption and incineration, for removal of low levels of VOCs in large air streams (Ottengraf, 1986). Such cost effectiveness is the consequence of a combination of low energy requirements and microbial oxidation of the VOCs at ambient conditions.

Preliminary investigations (Sorial et al., 1993) were performed on three media: a proprietary peat mixture; a synthetic, monolithic, straight-channeled (channelized) medium; and a synthetic, randomly packed, pelletized medium. These media were selected to offer a wide range of microbial environments and attachment surfaces and different air/water contacting geometries. The results of this preliminary work demonstrated that 95+% VOC removal efficiency could be sustained by all three media at a toluene loading of 0.725 kg COD/m³-d, but at different empty bed contact times (EBCTs). For the pelletized medium, this performance could be achieved at an EBCT of 1 min., for the channelized medium at 4 min., and for the compost medium at 8 min. Both synthetic media developed headloss over time, with the pelletized medium showing a pressure drop in excess of several feet of water after sustained, continuous operation. These results left open the question of which medium could provide the optimum combination of high VOC elimination efficiency at high loading with minimum pressure drop.

3.2.2
Experimental Apparatus

The biofilter apparatus used in this study consists of three independent parallel biofilter trains, each containing 4 ft. of attachment medium: Biofilters A, B, and C. A detailed schematic of the experimental apparatus is given in Fig. 1, while a detailed description of the system and the operating procedure is given elsewhere (Sorial et al., 1993). Biofilter A was filled with a proprietary compost mixture, B with a Corning Celcor® channelized medium, and C with a Manville Celite® pelletized medium. Biofilters A and B were square and have an inner side length of 5.75 in., and biofilter C was round, with an inside diameter of 5.75 in. The air supplied to each biofilter was highly purified for complete removal of oil, water, CO_2, VOCs, and particulates. After purification, the air flow for each biofilter was split off, the VOCs were injected into it, and then it was humidified and fed to the biofilters. The air feed was mass flow controlled, and the

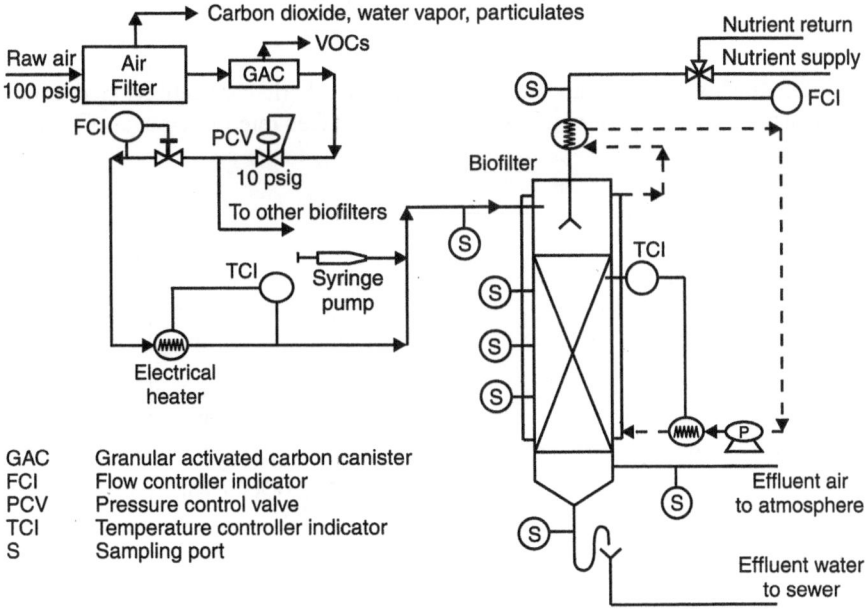

GAC	Granular activated carbon canister
FCI	Flow controller indicator
PCV	Pressure control valve
TCI	Temperature controller indicator
S	Sampling port

Fig. 1. Schematic of the experimental setup

VOCs were metered by syringe pumps. The flow direction of the air and nutrient inside each biofilter was downward. Each biofilter was insulated and independently temperature controlled.

Buffered nutrient solutions were fed to biofilters B and C. A detailed description of the nutrient composition is given elsewhere (Sorial et al., 1993). Each of these biofilters independently received a nutrient solution containing all the necessary macro- and micro-nutrients, with a sodium bicarbonate buffer. The nutrients required in biofilter A were included as part of the original compost mixture.

3.2.3
Results

Biofilter A: This biofilter run on the peat medium was made to evaluate the effect of temperature, and then loading, on toluene removal efficiency. Figures 2 and 3 summarize the biofilter performance. The biofilter was started up and, after some operational difficulties, stabilized by day 10 at 52° F, 50 ppmv toluene, 2 min EBCT, and a removal efficiency of about 58%. On day 17, the temperature was raised to 60° F, resulting in a rise in efficiency to about 75%, which decreased after day 24 into the 60's, and after day 32 into the 50's. On day 41, the temperature was increased to 70° F, resulting in a gradual increase in efficiency to about 75% by day 47. On day 53, the temperature was increased to 80° F, resulting in an increase in efficiency into the low 80's. On day 61, the temperature was increased to 90° F, resulting in a further increase in efficiency to the mid 90's (Fig. 2). After day 77, the feed was increased slowly to about 95 ppmv toluene, resulting in a drop in efficiency to about 88%. Further increases in the feed concentration to a maximum of 180 ppmv toluene on day 139 resulted in a further decline in efficiency to about 58% (Fig. 3). The run was terminated on day 215.

Biofilter B: This biofilter run was made on the synthetic channelized medium to evaluate the effect of temperature, and then nutrient feed rate, on removal efficiency.

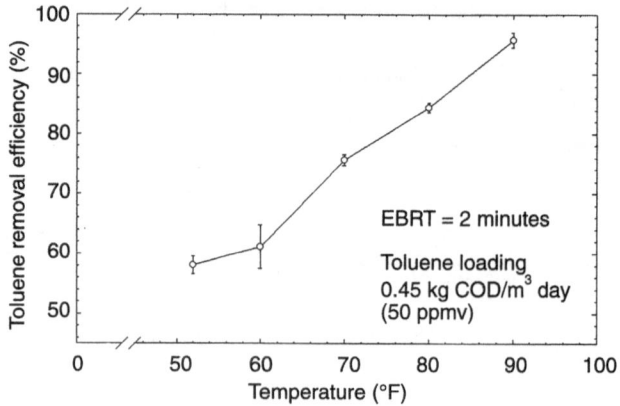

Fig. 2. Effect of temperature on the performance of peat biofilter

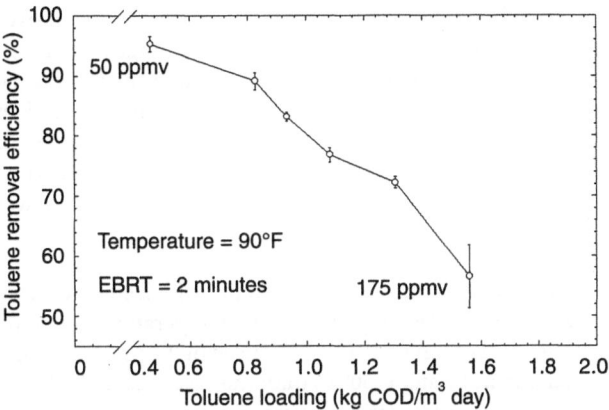

Fig. 3. Effect of toluene loading on the performance of peat biofilter

The biomass in the channels of the medium remaining from the previous run was removed by hydroblasting the eight 6 in. high medium blocks from top and bottom. The corners of these square blocks were filled with grout to provide a round active block. This last step was taken to match a round block cross section with the round pattern of the nutrient delivery spray nozzle.

Figure 4 shows the biofilter performance as a function of time. The biofilter was started up at 52° F, 50 ppmv toluene, and 2 min EBCT. By day 36, the removal efficiency had drifted over a range from about 62% to 80%. On day 36, the nutrient feed rate was increased from 30 to 60 L/day while keeping the mass loading of the nutrients constant. The increased nutrient flow rate effectively doubled the wetting cycle from 20 sec/min to 40 sec/min. An immediate increase in efficiency to 99% was observed, which then quickly dropped, and ranged by day 50 between about 30% and 70%. On

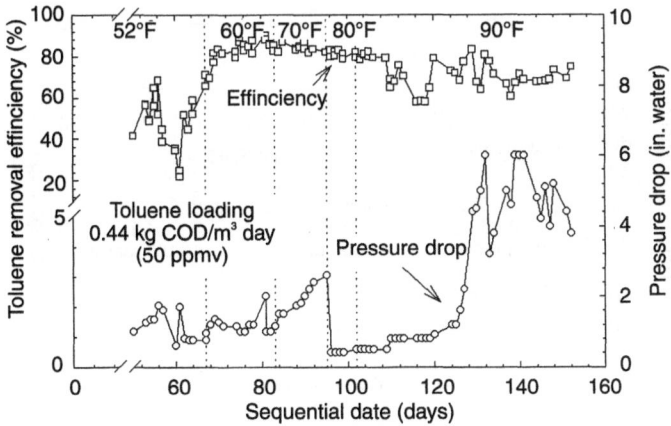

Fig. 4. Performance of channelized biofilter at an EBRT of 2 minutes

day 50, the nutrient feed rate was increased to 90 L/day (increasing the wetting cycle to 60 sec/min), but the efficiency dropped from 69% and ranged by day 67 from about 22% to 65%. On day 67, the temperature was raised from 52°F to 60° F and the efficiency increased to 66%. By day 75, the efficiency was 87% and this level was maintained to day 83. After day 83, the temperature was raised further, in 10° F steps, to 90° F, but the efficiency did not improve. In fact, for the rest of the run, at 90° F and 60 L/day, the efficiency ranged between about 58% and 83%. The run was terminated on day 152.

Biofilter C: The first biofilter run on the synthetic pelletized medium was made to evaluate the effect of pressure drop, and then temperature, on toluene removal efficiency. The biofilter was charged with pellets used in the previous run. These pellets were washed by hand in hot water (150° F) until the accumulated surface biomass had been removed and the pellets were free flowing. Figure 5 presents the biofilter performance as a function of time. The biofilter was started up at 52° F, 50 ppmv toluene, and 2 minutes EBCT. By day 21, the removal efficiency was 99% and, by day 27, it approached 100% and remained at this level until day 50. From day 51 to day 57, the EBCT was gradually reduced to 1 minute, causing the efficiency to drop to 84%. Subsequently, the toluene removal efficiency rapidly increased to the low 90's and remained in that range until day 81. On day 82, the temperature was raised to 60° F and the efficiency steadily rose until complete biodegradation of the toluene was reached on day 89. This essentially 100% efficiency in toluene removal was maintained through day 97. During the period between day 54 and day 97, pressure drop across the system increased from 0.2 to 5.5 in. water. From day 97 to day 111, the efficiency dropped steadily from 100% to 86%, while the pressure drop increased from 5.5 to 6.0 in. water. On day 112, the temperature was increased to 70° F, and the efficiency rebounded by day 113 to a peak value of 97%, after which it dropped to 85% by day 188. On day 119, the temperature was raised to 80° F, and the efficiency rose to about 89% by day 120. During the period from day 112 to 120, the pressure drop increased from 6 to 18 in. water. By day 128, the efficiency had steadily dropped from 89% to 77% as the pressure drop increased from

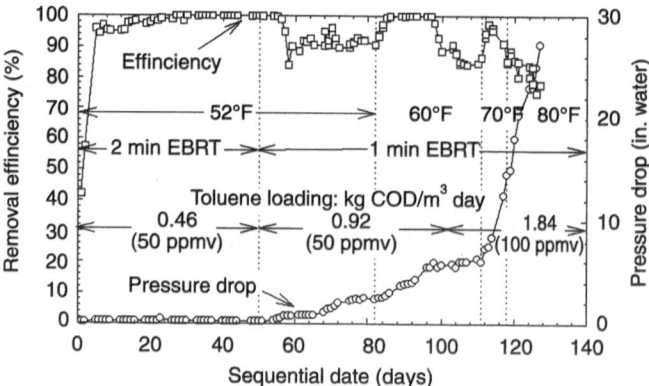

Fig. 5. Performance of pelletized biofilter with respect to toluene removal at 1 and 2 minutes EBRT without backwashing

18 to 27 in. water. This pattern of a steady loss of efficiency with a concomitant increase in pressure drop suggests the development of short circuiting within the biofilter medium due to biomass accumulation, which results in a significant reduction in actual contact time. The run was terminated on day 128.

The second biofilter run on this medium was conducted to evaluate routine biomass control by backwashing. The biofilter was charged with a 50:50 mixture of fresh pellets and pellets from the previous run. The used pellets were thoroughly washed by hand in tepid water (90° F) until the accumulated surface biomass had been removed and the pellets were free flowing. Figure 6 shows the biofilter performance as a function of time. The filter was started up at 90° F, 50 ppmv toluene, and 2 min EBCT. By day 4, the removal efficiency approached 100%. (Note: this second run, started up with pellets washed in tepid water, contrasts with the slower start-up in the first run, where the pellets were washed with hot water.) On day 8, the feed was increased to 250 ppmv toluene and the efficiency dropped to 97%, ranged between 92% and 98% until day 25, when it again reached 99%. Subsequently, the efficiency dropped as low as 86% before regaining 99% on day 81, after which the efficiency was nearly always 99+%.

Initially, backwashing was performed once per week by using 100 L of fresh water at a rate of 6 gpm. After day 28, the frequency was increased to twice per week and, after day 38, the volume was increased to 200 L. These changes were made because measurable pressure drop was observed between backwashings. On day 73, the backwash rate was increased to 15 gpm in order to induce full fluidization. The backwash system consisted of clean water supply tank, a recycle tank, and a rinse water receiving tank (Fig. 7). The backwash cycle consisted of an initial period of fluidization whereby water is recycled between the biofilter and the recycle tank. The recycle period was followed by a brief flushing period where clean water was pumped from the clean water supply tank into the biofilter in order to remove the suspended solids in the recycle flow. Although the pressure drop increase was minimal, the efficiency did not improve, suggesting some form of channelizing within the bed. Therefore, on day 80,

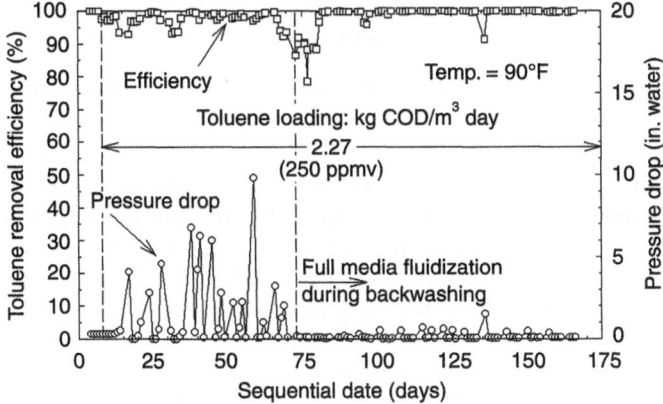

Fig. 6. Performance of pelletized biofilter with respect to toluene removal at 2 minutes EBRT with backwashing

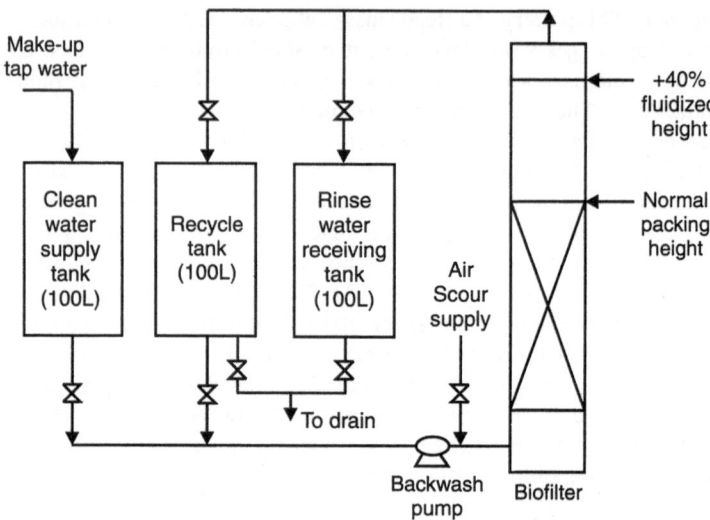

Fig. 7. Schematic of backwash system

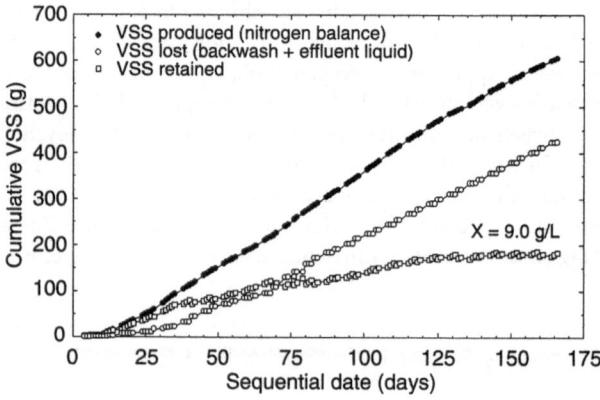

Fig. 8. Development of pelletized biofilter with time (VSS closure) with ammonia-N nutrient feed

the length of the backwash period was increased to 1 hour by recirculating the backwash water. After this final adjustment, the toluene removal efficiency, as mentioned above, achieved and sustained 99+%. During this latter period, the total volume of water used per backwash was optimized to 120 L. Of this volume, 70 L were used for the 1 hour backwash recycle, while the remaining 50 L were used to flush the released solids from the reactor. Figure 8 shows the development of biomass with time. After day 38, the rate of biomass accumulation declined with the increase in the wash volume. After day 73, the accumulation rate became nearly zero with the implementation of full fluidization. No further change in the backwash procedure was made and the accumulation of biomass within the biofilter leveled off at about 180 g with the pressure drop between backwashings typically under 0.2 in. of water.

3.2.4
Conclusions and Future Work

A marked improvement in toluene removal efficiency with increasing temperature was demonstrated in this study for the compost mixture, the channelized medium, and the pelletized medium. The direct consequence of this finding: much less medium would be needed for a biofilter operating at 90° F than at 52° F, resulting in a proportional reduction in capital cost. The economic tradeoff with the cost of heating the incoming air should usually favor operation at these warmer conditions.

The modest performance of the compost mixture with respect to increased loading complimented our earlier findings with respect to decreasing EBCT (Sorial et al., 1993). Unfortunately, implicit limitations of the experimental apparatus may have resulted in reduced performance. Specifically, the manufacturer recommended using a width-to-depth ratio of 1/1, rather than 1/8. They also stated that, from their experience, the only effective means of controlling bed moisture content was to weigh the entire biofilter. This was impossible with the heavy stainless steel unit used here, which was bolted to a support frame. Several moisture measurement and control strategies were attempted, but it was never possible to be certain that the bed moisture content was consistently at the reported optimum range, i.e., between about 50% and 60% (Bohn, 1993; Van Lith et al., 1990). The sometimes erratic performance may have been influenced by variations in bed moisture content. However, it can be seen that the best removal efficiencies achieved by the compost mixture were better than shown by the channelized media, but worse than shown by the pelletized media.

The performance of the channelized medium also confirmed our earlier findings that this medium was distinctly inferior to the pelletized medium (Sorial et al., 1993). The best performance was achieved during the use of new medium blocks. After biomass accumulation within the channels and subsequent removal by hosing, the performance never regained the previous, still modest, levels. Attempts to adjust nutrient flow as a means of testing the effect of the duration of the wetting in the nutrient application cycle did not overcome the previously demonstrated efficiency limitations. The more erratic performance of this medium after removal of the biomass suggests that this medium may be unsuitable for sustained efficiency after periodic cycles of biomass removal. This erratic performance, due to suspected random uneven plugging of channels by biomass, coupled with its relatively low overall removal efficiency, difficulty in biomass removal, and intrinsically high medium cost suggests that this medium may not be a viable option for this application.

The pelletized medium exhibited the best and most consistent performance of the three media tested. It rapidly achieved high removal efficiencies at high toluene loadings. As the first run demonstrated, however, an excessive accumulation of biomass, shown by a rise in the pressure drop across the medium, results in a substantial loss in efficiency, followed by a very rapid rise in pressure drop. This suggested that efficient, sustained performance might be achieved through early and periodic control of biomass accumulation by backwashing. In the second run, the implementation of a suitable backwashing strategy for biomass control was achieved by using full medium fluidization. This strategy permitted sustained operation of the biofilter at high loadings with efficiencies consistently at 99 + %. According to mass balance calculations, the biomass retained within the biofilter stabilized at a nearly constant level.

Future work will concentrate on further optimizing the use of the pelletized medium, with the objective of minimizing the medium volume required for a selected APC technology application.

3.2.5
References

Bohn HL (1993) Biofiltration: Design Principles and Pitfalls. Paper No 93-TP-52A.01 presented at the 86th Annual Meeting & Exhibition of Air & Waste Management Association, Denver, Colorado, June 13

Lee B (1991) Highlights of the Clean Air Act Amendments of 1990. J Air Waste Manage. Assoc, 41 (1): 16–24

Ottengraf SPP (1986) Exhaust Gas Purification. Biotechnology, 8, Rehn HJ, Reed G, (eds.), VCH Verlagsgesellschaft, Weinheim

Sorial GA, Smith FL, Smith PJ, Suidan MT, Biswas P, Brenner RC (1993) Evaluation of Biofilter Media for Treatment of Air Streams Containing VOCs. Paper No. AC93-070-002 Proceedings of the Water Environment Federation 66th Annual Conference & Exposition, Anaheim, California. October 15

Sorial GA, Smith FL, Smith PJ, Suidan MT, Biswas P, Brenner RC (1993) Development of Aerobic Bio-filter Design Criteria for Treating VOCs. Paper No. 93-TP-52A.04 presented at the 86th Annual Meeting & Exhibition of Air & Waste Management Association, Denver, Co, June 13

Van Lith C, David SL, March R (1990) Design Criteria for Biofilters. Trans Inst Chem Eng, 68 B : 127–132

3.2.6
Further Readings

Recent Advances in Biofiltration
Rakesh Govind, Univ of Cincinnati, OH and Dolloff F, Bishop EPA, Cincinnati OH
EPA Risk Reduction Eng Lab 20th Annu Res Symp, Cincinnati, OH, Mar 15–17, 94, p 115 (7) conf paper
Abstract: Limitations of current biofiltration systems for VOC degradation have spurred research on operational dynamics, media cleaning and scale-up requirements, and alternate media to sustain different microbial assemblages. Related studies conducted by EPA and the Univ. of Cincinnati focus on packed bed peat and compost, activated carbon, ceramic celite packed bed, and ceramic celite plate-straight passage biofilters. The demonstrated advantages of biofilters with porous pellet or structured media and recycled nutrient and buffer solutions relative to conventional soil or peat/ compost biofilters for biodegradable VOC removal from air are cited. (7 references, 1 tables)

Biotechniques for Air Pollution Control
Johan W van Groenestijn and Paul GM Hesselink, TNO Inst of Environ Sciences, Delft, Netherlands;
Biodegradation, 1993–1994, v4, n4, p 283 (19) journal article
Abstract: The benefits, limitations, and economics of biological techniques under development for off-gas treatment are examined. This treatment option focuses on volatile contaminant absorption in the aqueous phase or biofilm, followed by oxidation by microbial action. Control of odors, VOCs, and inorganic compounds is facilita-

ted by use of bioscrubbers, biotrickling filters, and biofilters. Low investment and operating costs relative to physicochemical methods favor biocontrol in many applications. The use of membranes and activated carbon is being pursued for enhanced elimination of hydrophobic compounds and buffering of fluctuating loads. (2 diagrams, 3 graphs, 81 references, 2 tables)

Biofiltration: an Air Pollution Control Technology for Hydrogen Sulfide Emissions
Eric R Allen and Yonghua Yang, Univ of Florida, Gainesville;
Proc Xth Annu Symp on Ind Environ Chemistry, College Station, TX (Plenum), Mar 24–26, 92, p 273 (15) conf paper
Abstract: Biofiltration, originally developed to control odorous emissions, is applicable to the treatment of dilute waste-gas streams. Naturally occurring microorganisms are supported on a stationary bed filter to continuously treat pollutants in a flowing waste-gas stream. Optimal operating conditions for hydrogen sulfide emissions control by compost-based biofilters in packed-tower systems were determined in the laboratory. Field studies were also performed at a wastewater-treatment facility to design and operate a full-scale on-ground compost biofilter bed odor-control system. (2 diagrams, 1 graph, 37 references, 4 tables)

3.3
Deodorization in Wastewater Treatment Plants by Wet-Scrubbing on Packed Column and Chlorine Oxidation

CHRISTOPHE BONNIN · ALAIN LAPLANCHE

3.3.1
Introduction

Urban wastewater treatment are frequently a source of olfactory nuisances caused by a multitude of compounds. The main pollutants are sulphur and nitrogen containing molecules and organic compounds such as aldehydes, ketones and volatile fatty acids. The odours vary widely (faecal, rancid, etc.) depending on the chemical type. Moreover, the olfactory perceptibility thresholds for some of these compounds are extremely low, that for hydrogen sulphide being, for example, 1.10^{-1} mg/m^3 (Van Gemert, 1977; Loenardos, 1958).

The main sources of odour problems in a wastewater treatment plant are the mercaptans, organic sulphides and bisulphides and, above all, hydrogen sulphide. These may be accompanied by nitrogen containing molecules (NH$_3$, amines, etc.) while aldehydes, ketones and organic acids can form a secondary source (Bowker, 1985; Kienow, 1982; Kangas, 1986; Nishigushi, 1990; Frechen, 1987).

All these compounds result from the anaerobic degradation of sulphates, proteins, sulphur and nitrogen amino acids, urea, etc. present in the urban effluent. The formation mechanisms have been accurately described (Harkness, 1980; Pommeroy, 1972; Thislethwayte, 1972).

They are frequently present in the effluent as it enters the plant and the main sources of olfactory pollution are the initial stages in treatment (pumping, pretreatment, etc.), which account for 30% of the odoriferous pollution and the sludge treatment line (thickener, dehydration, etc.) which generate roughly 65%. The pollutant concentrations vary enormously depending on the quality of the effluent, the design of the treatment line and the size of the structures. Very general typical values are 5 to 15 mg/m^3 for H$_2$S, 2 to 4 mg/m^3 for mercaptans and 0.1 to 1 mg/m^3 for NH$_3$ rising to maximum values of between 5 and 10 mg/m^3 when the sludge is finally stabilized by lime (Bonnin, 1991).

Current deodorization techniques involve trapping and/or destroying the malodorous molecules. The techniques used include:

- masking (Parthum, 1985; Le Cloirec, 1984)
- thermic oxidation with or without a catalyst (Pantel, 1987; Valentin, 1989)
- adsorption (Le Cloirec, 1984)
- biological oxidation (Martin, 1987)
- wet scrubbing with or without a chemical reaction (Walltrip, 1985).

Due to the flows to be treated ($\geq$ 20,000 m^3/h), wastewater treatment stations generally use the last of these methods. An oxidizing agent (mainly chlorine) is added to the washing solution with the objective of destroying the absorbed molecules and in certain cases, improving performance by accelerating the mass transfer from the gaseous to the liquid phase.

Nevertheless, to attain a reliable removal efficiency (higher than 99.9%) requires a strict operating protocol as well as a good knowledge of reaction mechanisms.

3.3.2
Physico-Chemical Properties and Reactivity of Malodorous Sulphur compounds

The absorption of sulphur compounds by an aqueous solution obeys the general gas/liquid transfer law:

$$\Delta Na = k_L \cdot a \cdot E (C^* - C_L) \Delta V$$

The quantity of product transferred "ΔNa" in a unit volume "ΔV" is directly proportional to the transfer coefficient "k_L", the interface area "a", a concentration gradient ($C^* - C_L$) and an enhancement factor "E" which depends on the chemical reaction. To be more precise, in the concentration gradient, "C^*" represents the concentration at the interface, generally obtained by Henry's law: $yP_T = HC^*$, and „C_L" is the concentration in the heart of the liquid.

The value of "E" depends on the reactivity of the compound transferred. With sulphur compounds, two reactions can occur simultaneously:

1 – The dissociation reaction

$$H_2S \text{ or } RSH \xrightleftharpoons{H_2O\,(K_1)} HS^- \text{ or } RS^-$$

$$HS^- \xrightleftharpoons{H_2O\,(K_2)} S^{2-} + H_3O^+$$

2 – The oxidation by chlorine

$$H_2S \text{ or } RSH \xrightarrow[kHClO]{HClO} \text{ oxidized sulphur compounds}$$

The value of "E" is determined by calculation of two dimensionless numbers:

$$Ha^2 = \frac{a \cdot k \cdot C_{\text{reagent} \cdot \text{sulphur} \cdot \text{compounds} \cdot \text{diffusivity}}}{k_L^2}$$

where "a" is the stoechometric coefficient of the reaction and "k" the kinetic reaction.

$$Z = \frac{C_{\text{reagent} \cdot \text{reagent} \cdot \text{diffusivity}}}{a \cdot C^*_{\cdot \text{sulphur} \cdot \text{compounds} \cdot \text{diffusivity}}}$$

When these two reactions occur simultaneously, "E" is calculated from the "overall Ha" and "overall Z" representing each reaction:

$$Ha_0^2 = Ha_1^2 + Ha_2^2$$

where: $Z_0 = \max (Z_1; Z_2)$

Consequently, to design an absorption column, it is necessary to know at each point of the reactor:

- the transfer coefficient "k_L" which can be semi-empirically determined from the ONDA equations (1968),
- the interface area, calculated form data provided by the packing supplier,
- the concentration at the interface, which depends on Henry's constant, i. e. on the solubility of the pollutant,
- the chemical reaction rates and the reagent concentrations.

The pH is the main parameter in the dissociation reaction; but, the type of oxidant, its concentration and the pH must all be known to determine the oxidation reactions.

Product Solubility

The solubility of reduced sulphur compounds results from their ionization in the water and depends on several factors, particularly the temperature and pH.

For H_2S: $H = exp^{(-2\,035.2/T° + 13.094)}$.
At 25 °C, the value of H is 525 atm/molar fraction (Perry, 1984).

For CH_3SH, $H = 225$ atm/molar fraction.

The Sulphur Compound Dissociation Reaction

In water, published literature gives the following values (Millero, 1986; Waltrip, 1985):

	H_2S	CH_3SH
pK_1	7.04	9.7
pK_2	12 to 15	–

Analysis of these values leads to two important conclusions: sulphur compounds will only dissociate in a basic medium and mercaptans will require even higher pH values. H_2S will dissociate virtually completely at pH = 9 while a value of 11 is required for methylmercaptan.

Oxidizing Agents in an Aqueous Solution

The chlorine compounds present in an aqueous solution depend on the pH.

$$Cl_2 + 2H_2O \overset{K}{\rightleftharpoons} HClO + H_3O^+ + Cl^-$$
$K = 1.5$ to 4.10^{-4} depending on temperature (Merlet, 1986)

$$HClO + H_2O \overset{K}{\rightleftharpoons} ClO^- + H_3O^+$$
$K = 1.5$ to $3.2 \cdot 10^{-8}$ depending on temperature (Merlet, 1986)

Oxidation Reactions

There is limited literature on the reactivity and compounds formed by the action of chlorine on hydrogen sulphide and mercaptans. Table 1 below summarizes the available data.

Table 1. Hydrogen sulphide and methylmercaptan oxidation parameters

Product to be oxidized	H_2S		H_2S		CH_3SH
Oxidant	HClO ClO^-		HClO ClO^-		HClO ClO^-
Conditions	pH < 7	pH > 7	pH < 9	pH > 9	pH = 11
Products formed	S*	SO_4^{-2}	$S* + SO_4^{-2}$	SO_4^{-2}	CH_3SO_3H
Steochiometry – chlorine – soda	? ?		? ?		? ?
Rate constant $M^{-1} \cdot s^{-1}$	?		?		?
Reference	Cadena (1988)		Waltrip (1988)		Capozzi (1974)

Due to the inadequacy of published data, it was necessary to study oxidation reactions, in a batch reactor and a continuous reactor, to fill in some of the blanks in the data and, second, to determine the parameters required to design and operate industrial deodorization units.

3.3.3
Equipment and Methods

Pilot

Experiments have been carried out in a batch reactor to study the mechanisms involved in the chlorination of hydrogen sulphide and in a pilot unit (see Fig. 1) to study the various parameters which affect the process.

Pilot operating conditions were as follows:

- counter flow absorption
- column diameter: $D = 0.2$ m
- packing height: 1.5 m
- type of packing: Pall rings d = $1,6 \cdot 10^{-2}$ m
- air flow rate: 150 m³/h
- water flow rate: 0.45 m³/h
- Ratio D/d: 12.5
- pH: 8 to 12
- chlorination rate: 0 to 1 g Cl_2/l

Assays were performed at a working velocity equal to 70 % of the flooding and a wet rate of $1.2 \cdot 10^{-5}$ m³/s, which is a bit low but satisfactory. We will not revert to the hydrodynamic operating conditions which are now well known (Tambouze et al., 1984) as regards the problems of mass transfer, entailing interference between the fluids, air and water.

Regulation of pH and oxidant rate in the washing solution is performed by addition of soda and chlorine.

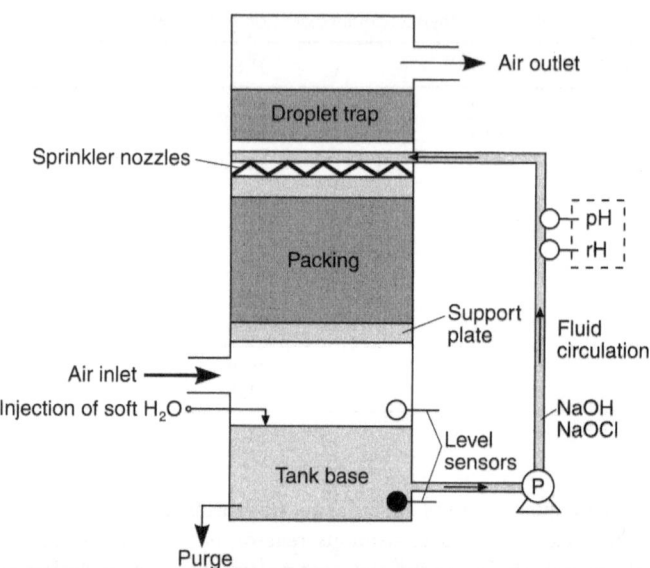

Fig. 1. Deodorization pilot unit

When chlorine was used as an oxidant, 150 g/l of sodium hypochlorite was injected directly into the washing solution recirculation circuit. The absorption therefore took place in the presence of an oxidant.

Parameters Studied

The efficiency of the process depends on mass transfer with chemical reactions: dissociation of sulphur compounds and oxidation reaction using chlorine (Copigneux, 1980; Trambouze and al., 1984). The parameters which affect design are hence the pH, the chlorination rate and the evolution of the scrubbing solution.

Choice of pollutants. Measurements at industrial sites showed that hydrogen sulfide was the most pervasive pollutant. Sulphur pollutants were injected in the form of a gaseous solution already diluted by inert nitrogen. The working concentration varied between 5 and 10 mg S/m^3 (0.1 to $0.2 \cdot 10^{-3}$ mole/m^3).

These concentrations are representative of those measured in the foul gases emanating from the sewage works (Bonnin and al., 1990).

Chemical parameters. As the process consists of a mass transfer with a chemical reaction, the reagents used take on special importance. We therefore varied the following factors:

- pH between 8 and 11 for H_2S (10 to 12 for CH_3SH). The pH levels were chosen in term of the sulphur compound dissociation equilibrium, with the knowledge that it is difficult to work in industrial conditions at pH measurements greater than 11.5, because of corrosion.

- chlorination rates varied between 0 and 16 moles of oxidant per mole of H_2S. In certain operations, we worked with an excess of chlorine of 1 g Cl_2/liter ($14 \cdot 10^{-3}$ mole/l). The residual chlorine after scrubbing is an important parameter in regulating industrial plant.

Operating parameters. Operating parameters were those determined during the scrubbing study, when the scrubbing liquid was recirculated non stop in order to simulate a continuous process, and aging of the scrubbing solution.

Method of Analysis

During the experiments, the following measurements were executed:

- transfer efficiency by measuring the sulphur compounds at the inlet and outlet of the pilot plant. The concentration was measured by gas chromatography using a column packed with butyl phthalate, and electro-chemical detection using chromic acid (Medor chromatograph, RN Electronic Society).
- residual chlorine dosage by the colorimetric method with DPD (AFNOR standard n° 90038),
- sulphide dosage in aqueous solution using a specific Orion electrode or by iodometry,
- dosage of sulphates, sulphites and thiosulphates by ionic chromatography (DIONEX, column AS₃).

3.3.4
Experimental Results

Products from Hydrogen Sulphide Oxidation Reaction

The batch experiments were designed to determine the products formed by oxidation since published literature does not contain these data.

The curves in Figs. 2 and 3 show variations in the sulphur compounds produced during chlorination at pH 8.3 and 11.5. It can be seen that, at chlorination rates below

Fig. 2. Oxidation of S(-II) by Cl_2 (pH = 11,5; Time = 5 min)

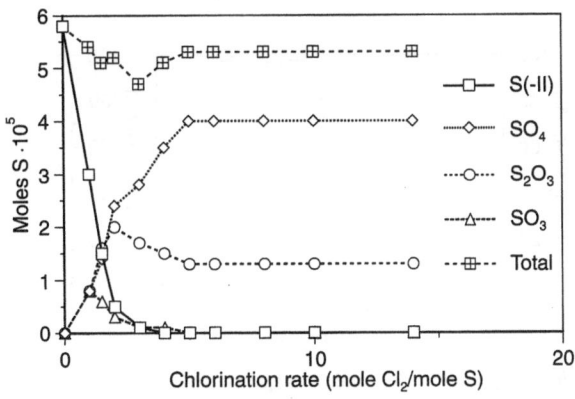

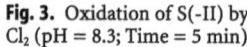

Fig. 3. Oxidation of S(-II) by
Cl_2 (pH = 8.3; Time = 5 min)

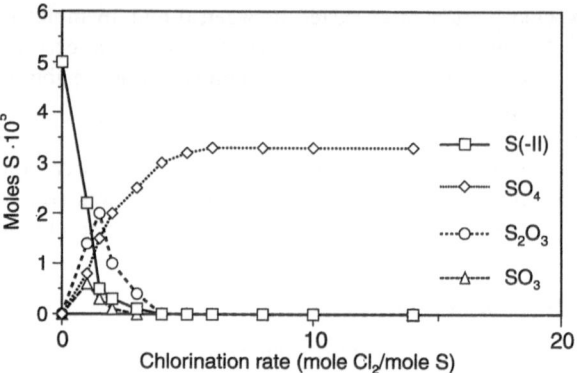

the theoretical stoichiometric rate of 4 moles/mole, sulphites and thiosulphates are present at the same time as the initial product, sulphide and the only terminal oxidation product identified, sulphates.

When the rate is higher than the stoichiometric rate and at pH = 11.5, virtually all compounds are transformed into sulphates. At pH = 8.3, sulphates represent only 75% of the initial reduced sulphur with an experimental stoichiometry of up to 16 moles/mole. In this case, the chlorine consumption is 3.1 moles/mole and it seems reasonable to assume there is an accumulation of sulphur ("S").

The Main Process Parameters

Determination studies of the main parameters have been carried out on the wet scrubber pilot plant.

With no oxidant. In this case, absorption is controlled by the dissociation equilibria and efficiency increases with pH. Up to 99.2% hydrogen sulfide is eliminated at pH = 11 but the limit for methylmercaptan is 48% (pH = 12) (see Fig. 4). The reason for this lower efficiency is the quantity of non-dissociated CH_3SH in the solution.

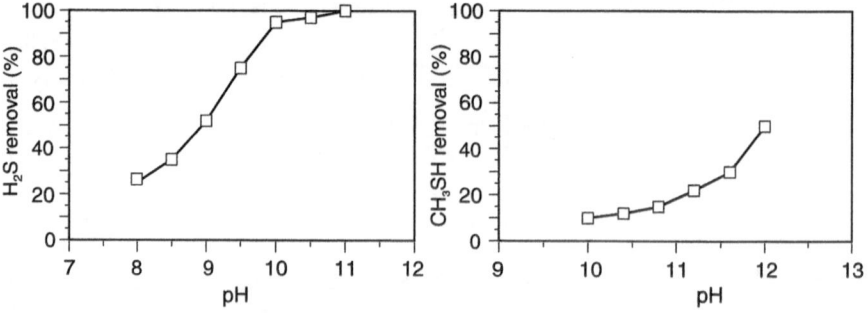

Fig. 4. H_2S and CH_3SH eliminated against pH

Calculation shows that, within the limits of experimental error, the quantity of pollutant transferred is limited by the theoretical non-dissociated CH_3SH concentration which can be obtained at equilibrium. The transfer is not limited by the dissociation of methylmercaptan but by the CH_3SH concentration at the interface. The maximum methylmercaptan concentration is achieved, and consequently the transfer is inhibited (no transfer gradient), when the interface concentration is $1.66 \cdot 10^{-7}$ mole/mole (0.008 mg/l).

In the presence of chlorine. The effect of the two chemical reactions (dissociation and oxidation) is clearly marked regardless of the chlorination rate.

More than 99% H_2S is eliminated at any pH > 11 (see Fig. 5). At this level, the addition of chlorine has no effect on the instantaneous efficiency of the process. The transfer is governed by dissociation. The chlorine constantly regenerates the washing solution by oxidizing sulphides into sulphates. At pH < 10, chlorination enhances the transfer and allows efficiencies exceeding 99% to be achieved at pH ≥ 9 and chlorination rates above 4 moles/mole. This level is essential to achieve virtually total conversion into sulphates.

Fig. 5. H_2S eliminated against pH and chlorination rate

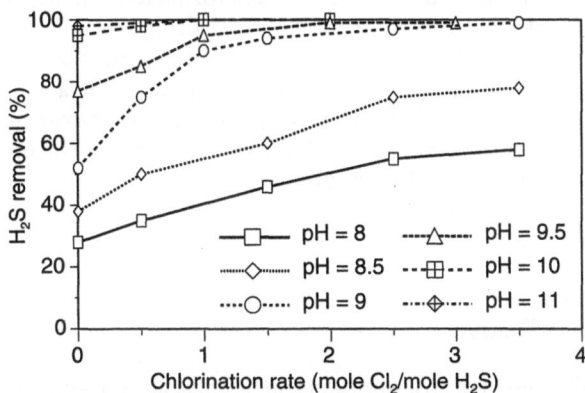

Fig. 6. CH_3SH eliminated against pH and chlorination rate

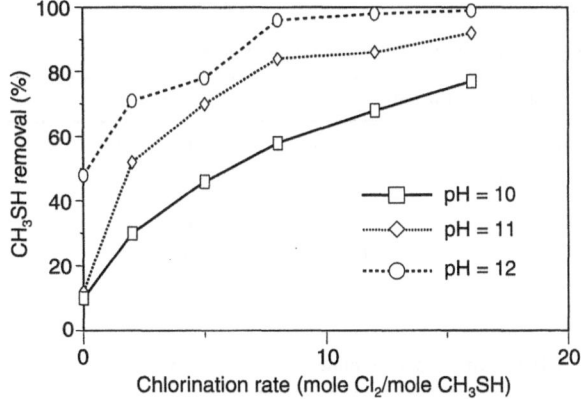

In the case of methylmercaptan removal (99.99% at pH = 12, see Fig. 6), increasing the chlorination rate improves removal. As with hydrogen sulphide, the methyl-mercaptan concentration in the washing liquid drops when the chlorination rate is increased. This gives a sharper transfer gradient, implying that residual chlorine should be present to optimize elimination.

Although published work (Millero, 1986) suggests that the theoretical reaction requires 3 to 6 moles/mole, the chlorination rate required to ensure the presence of residual chlorine is 9 moles of chlorine per mole of methylmercaptan. The final oxida-tion product is methylsulphonic acid.

At this stage in progress, it can be stated that the formation of disulphide is, almost certainly, the first intermediate stage in the reaction and that sulphonic acid is a stable final product. However, the intermediate sulphur compounds formed, and how they are oxidized by the chlorine, are unknown. The high chlorine consumption (9 moles Cl_2/mole CH_3SH) suggests that there are many complex chlorination reactions.

An aging study on the bath showed that the H_2S and CH_3SH purification efficiencies always exceeded 99% throughout 45 hours in operation and that the sulphates are definitely formed by oxidizing H_2S.

For hydrogen sulphide, the chlorine consumption was calculated at 5.2 moles/mole H_2S and the soda consumption approximately 5 moles/mole. This exceeds the theo-retical 4 moles required, the difference being explained by water losses (approximately 0.5% of the liquid flow) and secondary reactions between the chlorine and organic compounds dissolved in the water. The excess soda consumption is due to reaction with CO_2 in the air. Methylmercaptan requires approximately 9 moles/mole of oxidant. In either case, the salinity of the washing water can be up to 100 g/l of dry extract per liter without causing any problem.

From known results and those obtained from our experiments, we can draft the Table 2.

3.5.5
Modelization

The design of an absorption reactor must answer a certain number of questions: given a gas flow rate at the inlet and a maximum concentration to be obtained at the outlet,

Table 2. H_2S and CH_3SH removal operating parameters

Product to be oxidized	H_2S	CH_3SH
Oxidant	$HClO + ClO^-$	$HClO + ClO^-$
Conditions	pH ≈ 9	pH ≈ 11.5
Products formed	$SO_4^{-2} + e$ SO_3^{-2} $S_2O_3^{-2}$	CH_3SO_3H
Experimental stoichiometry	5.2	9
Packing height	1.5 m > 99.9%	1.5 m > 99.9%
NaOH consumption mole/mole	4.4	7.6

what should the diameter of the column be and what type and height of packing media should be used?

In simple cases of gas-liquid transfer, suing air and water as the main fluids (mass flow Gas and Liquid rate), the hydrodynamic definition of the scrubbing columns has been adequately described in the various publications on the subject. Starting with a type of packing defined by two diameter "d" and its packing factor "Fp", according to the studies performed by Sherwood and al. (1952) applied in the form of mathematical equations by Trambouze (1984) or Copigneux (1979), the flooding velocity U_E is determined, then the pressure drop per meter of packing height. We also examine the diameter of the column "D".

Two ratios must be respected to ensure satisfactory operating.

$8 < D/dp < 60$, in which "D" is the diameter of the column and "dp" the diameter of the packing.

Wet rate: $2.5 \cdot 10^{-5} < a\varrho_L < 2 \cdot 10^{-4}$

in which "a" is the specific packing surface (Morris and Jackson, 1953; Copigneux, 1979), "L" is the liquid mass flow rate, and "ϱ_L" is the density.

The height of packing required for a given treatment efficiency is calculated according to the transfer equation. If we consider that a small packing element, with a volume dV corresponding to the height dh, and assuming that the concentrations in the fluids on the small element are constant, it becomes possible to establish a transfer balance by following the absorption process governed by the general transfer law.

Transfer is extremely easy to calculate if enhancement factor (E), which depends upon kinetics constants, is known.

In order to determine the value of the constants k_1 (dissociation reaction) and k_2 (oxidation reaction) required to obtain enhancement factors compatible with the absorption efficiencies found with or without chlorine, experimental results were input to a computer calculation, based on the theory of amass transfer with chemical reaction.

The program operated on column height increments dh = 0.01 m, assuming the parameters remain constant over this small elementary volume. The transfer was calculated form the general formula:

ΔH_2S absorbed $= k_l \cdot a_w E\left[(H_2S)^* \cdot (H_2S)_l\right] dV$ (a_w is the wet area)

The calculation started at the top of the column (gas outlet) and is initiated at the following values:

$[H_2S]_{gas\ outlet}$: experimental value
k_1:   arbitrarily selected value
pH: experimental value
$[H_2S]_L$:   0

For each column element, the model determines Henry's constant, k_L, a_w, Ha, Z, E and ΔH_2S, it memorizes the new values $[H_2S]_{Gas}$ and compares the values of $[H_2S]_{Gas\ calculated\ inlet}$ and $[H_2S]_{Gas\ experimented\ inlet}$. The required column height is then compared to the real column height (H = 1.5 m). If the two values are different, the model iterates on k_1.

When k_1 value is obtained, constant k_2 is calculated on the same principle.

Table 3. Comparison of experimental and computed values without chlorination

pH of water	Theoretical	Experimental
8.25	17	27
8.65	35.5	37
9.00	50.2	52
9.50	81.7	77.5
10.00	97.1	95.5
11.00	99.5	98.50

Table 4. Comparison of computed and experimental removal (at Antibes sewage plant)

pH	Chorine residual [(mole/l) · 1000]	Removal %	
		experimental	theoretical
8.97	4.92	97.7	95.6
8.98	18.17	99.6	99.7
9.06	11.19	99	99.1
9.11	11.55	99.8	99.3
9.28	18.45	99.8	99.6
11.26	1	99.8	99.8
11.3	3.24	99.8	99.8
11.48	3.24	99.8	99.8

Since the kinetic constants of dissociation and oxidation reactions are known data, it is now possible to introduce them in a computerized tool based on mass transfer law. which calculates, whatever the working conditions, the enhancement as well as the corresponding absorption efficiency.

For example, a comparison between calculated and experimental value is given in Table 3. A good correlation between these two values is observed.

In order to check whether our calculation model could be applied to an industrial unit, the computed efficiency rates were compared with those obtained on the Antibes, sewage plant treating 50,000 m³ of air per hour, with a water flow rate of 125 m³/h and a packing height of 2 meters (see Table 4).

In this case, simulation again corroborates the measurements performed on site.

3.3.6
Sizing Procedure And Industrial Units

In the current state of the art, chlorine is widely used due to the simplicity of the process and the fact that the transfer of methylmercaptan is improved by the effects of the oxidation reaction in the absorption phase.

In an industrial plant, it is not sufficient only to eliminate hydrogen sulphide and methylmercaptans, the overall process must also eliminate nitrogen compounds.

Various studies have been carried out to determine the maximum concentrations of the main malodorous compounds acceptable at the outlet of the odour treatment unit. Atmospheric dilution (wind, velocity, stack, etc.) is taken into account in these calculations, the aim being to ensure that there are no perceptible odours at the limit of the treatment plant site (see Table 5).

Table 5. Maximum acceptable concentrations at outlet form the odour treatment unit (Bonnin, 1991)

Malodorous compound	Max acceptable concentration (mg/Nm^3)
NH_3	≤ 1
organic N	≤ 0.1
H_2S	≤ 0.1
CH_3SH	≤ 0.07

On this basis, a chemical washing deodorization unit must possess two or three columns:

- In the first column, known as the „acid" column, sulphuric acid (H_2SO_4) is injected to achieve a suitable pH. This column eliminates all nitrogen compounds. Ozone can be used to achieve equivalent results at a higher pH (pH = 2 with no ozone, pH = 4 to 5 with ozone).
- The second and third columns are basic and oxidizer columns. They eliminate all sulphur compounds (H_2S, RSH, etc.) to a greater or a lesser extent depending on the pH and the oxidant concentration. An oxidant increases the efficiency against relatively insoluble compounds such as mercaptans. The oxidant may be injected under the form of sodium hypochlorite or active chlorine solution produced by electrolysis.

3.3.7
Sizing of Adsorption Columns

Large air flow rates must be purified in wastewater treatment plants. This implies a low pressure drop and quick transfer. A packing column then comprises (see Fig. 1):

- one/several column elements consisting in a platform to support the packing material,
- a gas distribution system,
- a liquid distribution and, possibly, redistribution, system,
- a vesicant remover to trap any droplets entrained by the gas,
- a tank bottom section,
- air ducting,
- a pH and oxidant control system.

Optimum sizing of the transfer columns and strict respect of operating conditions (pH, residual oxidant, consumption, etc.) are vital to effective deodorization. The work described here was used to prepare treatment unit design software which initially calculates the air and pollutant flows extracted from each structure in the wastewater treatment plant then, for each column in the deodorization system:

- uses a hydrodynamic model to determine the column diameter, type of packing material and pressure drops,
- uses a gas/transfer liquid model to determine the height of packing necessary to obtain the required purification efficiency for each pollutant (nitrogen and sulphur compounds) at the appropriate pH and reagent concentration,

- determines the purge rates required to reduce the concentration of salts in the washing solution,
- calculates the consumption of reagents (H_2SO_4, NaOH, NaOCl) and the quantities of these reagents the operator must supply.

This second point is important since many authors or manufacturers omit an important factor in calculating the soda consumption: it must be remembered that the soda will react with CO_2 as follows:

$$NaOH + CO_2 \rightarrow NaHCO_3 \qquad 2\,NaOH + CO_2 \rightarrow Na_2CO_3$$

Air contains 0.033% by volume of CO_2, i.e. 640 mg CO_2 m³ air. Depending on the pH and transfer laws, 5 to 10% of the CO_2 is absorbed in the liquid phase. With the high air flows involved, the quantity of soda consumed by CO_2 cannot be neglected and can, in fact, easily represent half the total consumption.

Table 6. A few examples of chemical washing deodorization plants enacted by O.T.V.

Treatment plant	A	B	C	D	E
Size (Eq/inhbt)	60,000	200,000	75,000	1,000,000	550,000
Treated air flow rate (Nm³/h)	58,000	120,000	19,600	180,000	80,000
Number of columns	4	2 × 3	2	3 × 3	3
Column area (m²)	10	7	9	19	12
Packing height (m)	2	2	1.5	2	2
Type of packing	Hackett	Levapack	Etapack	Levapack	Multi-cellular
Oxidant	Electrolytic chlorine	Electrolytic chlorine	Ozone NaOCl 48°	Electrolytic chlorine	Electrolytic chlorine
Treatment capacity (H_2S)					
Average (g/h)	290	600	190	900	400
Maximum (g/h)	870	1,800	280	2,700	1,200
pH 1st column	2.3	3	4	3	2.7
2nd column	9.4	8.5 to 9	11	9	9
3rd column	11	10.5 to 11	–	10.5–11	10.5–11
4th column	9*				
Outlet concentration (average 24 h)					
H_2S (mg/m³)	0.006	0.008	0.003	0.1	0.01
	<0.01	<0.01	0.02	<0.01	<0.01
Ammoniac mgN/m³	0.04	0.05	0.14	0.03	0.08
Organic nitrogen total mg N/m³	<0.01	<0.01	<0.01	<0.01	<0.01

* Washed in the presence of bisulfite.

3.3.8
Examples of Applications

Table 6 summarizes several applications. In general, three columns are connected in series. However, A is extremely sensitive to malodorous effluents since the wastewater treatment plant is in a building completely surrounded by a built-up area. Consequently, the deodorization treatment plant has four columns in series.

3.3.9
Conclusions

Wastewater treatment plants have now become veritable factories employing increasingly sophisticated technology. Moreover, they are frequently installed in sensitive, densely populated areas.

In the existing context of increasing sensitivity to the environment and living conditions, particularly the problem of olfactory pollution, it is important to develop the concept of the "zero pollution" plant which new, faster treatment methods (lamella settlers, biofilters) have now made a real possibility. Consequently, a new generation of wastewater treatment stations – compact, completely enclosed, invisible and perfectly integrated into the environment – has been born.

There is no lack of examples! Who'd believe, when visiting the Saint-Tropez citadel or passing in front of the Louis II stadium at Monaco or the Salice port at Antibes, that a wastewater treatment plant is operating less than 100 yards away?

However, if we didn't have high-performance, reliable deodorization methods, all this would be pure illusion. The study described here allowed us to achieve better control of industrial plant sizing and operating parameters.

Finally, the models prepared form this work provide a computerized tool capable of scaling a chemical washing deodorization plant, predicting the pollutants remaining in the treated air and the reagents required permanently to ensure an absence of olfactory pollution.

3.3.10
References

Bailay PS (1982) Ozonation in organic chemistry: monoléfinic compounds. Academic Press, vol 2
Bonnin C (1991) Les sources de nuisances olfactives dans les stations de traitement des eaux résiduaires, et leur traitement par lavage à l'eau chlorée en milieu basique. Doctorate thesis, Ecole Nationale Supérieure de Chimie de Rennes
Bonnin C, Laborie A, Paillard H (1990) Odors nuisances created by sludge treatment: problems and solution. Wat Sci Tech 22(12), 65
Bowker R, Smith J, Webstern O (1985) Odor and corrosion in sanitary sewerage system and treatment plant. EPA design manual, EPA 625/1, 85/018
Cadena F, Peters RW (1988) Evaluation of chemical oxidizers for hydrogen sulfide control. JWPCF, 60(7), 1259
Copigneux P (1979) Distillation-Absorption: colonnes garnies. Engineer Technics, France, p j/2626–j/2626.23
Frechen FB (1987) Stoffübergänge Wasser, Luft am Beispiel von Geruchsemissionen aus Kläranlagen. Gewässerschutz, Wasser-Abwasser, Aachen
Gordon G (1987) The very slow decomposition of aqueous ozone in highly basic solutions. Proc 8th ozone world congress, IOA, Zürich, Switzerland
Harkness N (1980) Chemistry of septicity. Effluent and water treatment journal, 1, 16

Hoigne J, Bader H, Haag WR, Staehelin J (1985) Rate constants of reactions of ozone with organic and inorganic compounds in water. Part III, Wat Res 19(8), 993

Kangas J, Nevalainen A, Hamninen A (1986) Ammonia, hydrogen sulfide and methylmercaptides in Finish municipal sewage plants and pumping stations. The science of the total environment, 57, 49

Le Cloirec C (1984) Analyse et évolution de la micropollution organique azotée dans les stations d'eaux potables. Effet de la chloration sur les acides aminés. Doctoral thesis

Leonardos AD (1958) Odor threshold for 53 commercial chemicals. Washington manufactory chemists association

Martin G (1984) La biodésodorisation, cas d'usine de traitement des sous-produits d'origine animale. TSM l'Eau, 6, 338

Martin G, Besson G, Lebeault JM (1987) Traitement biologique des odeurs. Conférence IIGGE, club odeur, Lyon

Merlet N (1986) Contribution à l'étude du mécanisme de formation des trihalométhanes et des composés organohalogénés non volatils lors de la chloration de molécules modèles. Doctorate Thesis, Université de Poitiers

Millero FT (1986) The thermodynamics and kinetics of the hydrogen sulfide system in natural waters. Marine Chemistry, 18, 121

Nishiguchi I (1990) Development and evolution of odor control technology for sewerage facilities. 4th WPCF/JSWA conference, Tokyo, pp 300–311

Onda K, H Takeuchi, Okumoto Y (1968) J Chem Eng Jap, 1 56

Ottengraf SPP, Van der Oever AHC, Kempenaars FJCM (1984) Waste gas purification in a biological filter bed. Innovations in biotechnology. Elsevier Science Publishers BV

Pantel L (1987) Réduction et neutralisation des odeurs: L'incinèration. Conférence IGGEE club odeur, Lyon

Parthum CA, Leffel RE (1985) Odor control for wastewater facilities. WPCF manual of practice n° 22, p 80

Perry RH (1984) Perry's Chemical Engineers' Handbook, 6th edition, chap 3 et 8, Mc Graw-Hill book company

Pomeroy RD, Parkurst JD (1972) Self purification in sewers. 6th EWPCF conference, Jerusalem

Sherwood TK, Pigford RL (1952) Absorption and extraction. 2nd Edition, Mc Graw Hill book company

Staehelin J, Buhler H, Hoigne J (1984) Ozone decomposition in water studied by pulse radiolysis. Part II, J Physical Chemistry 88, 5999

Thislethwayte D (1972) The control of sulfides in sewerages systems. Ann Arbor Science editors

Tomiyasu H, Fukutomi H, Gordon G (1985) Kinetics and mechanisms of ozone decomposition in basic aqueous solution. Inorg Chem. 24, 2962

Trambouze P, Van Landehem H, Wauquier JP (1984) Les réacteurs chimiques: conception, calcul, mise en oeuvre. Chap 8, editions Technip

Valentin FHH (1989) Recent development towards cheaper effective odor control. Filtech conference, Karlsruhe, pp 137–146

Van Gemert LJ, Nettenbreijer AH (1977) Compilation of odor threshold values in air and water. Central Institute for Water Supply – Central Institute for Nutrition and Food Research, Netherlands

Waltrip GD, Snyder EG (1985) Elimination of odor at six major wastewater treatment plants. JWPCF, 57(10), 1027

3.3.11
Further Readings

Gaseous Emissions form Wastewater Facilities.
David S, Dawson, Dow Chemical Co, Midland, MI and Manjunath A. Gokare, Piedmont Olsen Hensley;
Water Environ Res, Jun 94, v66, n4, p 3754(4) journal article
Abstract: Research studies conducted in 1993 on gaseous emissions emitted form wastewater-treatment facilities are reviewed. Subjects of the Studies included the characterization and measurement of odor and VOC emissions from municipal waste-water-collection and -treatment facilities. In terms of control technologies, research was conducted on biofiltration, thermal oxidation, catalytic oxidation, regenerative adsorption, caustic scrubbing, and condensation. General results are presented from

extensive studies on the effectiveness of various physicochemical and biological control technologies. (49 references)

Title: Dewatered Sludge Storage Emissions Control Using Multistage Wet Scrubbing
Author: Sereno, DJ: McGinley, CM: Harrison, DS; Haug RT
Auth. Address: J. M. Montgomery Construction Engineers 300 N. Continental Blvd., Suite 650, El Segundo, CA 90245.
Source: Water Environment Research; Vol. 65, No. 1, p 66–72, January/February 1993. 3 fig. 4 tab, 6 ref.
Abstract: Los Angeles' Hyperion Treatment Plant is a municipal wastewater treatment plant with secondary treatment capacity. Digested sludge is stored as 20% solids wet cake in large storage bins. A pilot test was conducted to test multistage packed towers and various combinations of treatment chemicals. Results were encouraging for controlling wet cake storage odors with wet scrubbing. Removal efficiencies of up to 97% were encountered using a caustic/bleach scrubbing solution. The greatest Reactive organic gases (ROGs) removal achieved was 89% using acid/bleach and peroxide. Volatile organic compound removal was poor for most combinations. It is expected that odors and ROGs can be controlled to acceptable levels using this multistage wet scrubbing approach. (Author's abstract)

Title: Odour Control for the 1990s – Hit or Miss.
Author: Toogood, S.J.
Auth. Address: WRc Engineering Swindon (England).
Source: Journal of the Institution of Water and Environmental Management JIWMEZ; Vol. 4, No. 3, p 268–275, June 1990. 8 ref.
Abstract: A general strategy for dealing with odor problems at a sewage works, once the use of preventative measures alone has been ruled out is presented. A strategy for dealing with nuisance emphasizes a fully integrated approach, involving covering, ventilation design, and odor treatment. Some of the options for treatment are compared, including adsorption, dry oxidizing packings, wet chemical scrubbing and biological odor control processes (i.e., biofilters and bioscrubbers). The objective of any odor control scheme is to eliminate nuisance, not eliminate odors. The use of dispersion models can be used to determine the extent of remedial works, and acting early in the complaint process can result in a cost effective solution. It is concluded that biological treatment is the most generally applicable method to the sewage works. (VerNooy-PTT).

3.4
Regeneration by Induction Heating of Granular Activated Carbon Loaded with Volatile Organic Compounds

Pierre Mocho · Pierre Le Cloirec

Abstract

Induction heating is used to regenerate granular activated carbon (GAC) for the purpose of recycling volatile organic compounds (especially solvents). As the technological possibilities offered by induction on an industrial scale have to be taken into account, the carbon has to be selected according to its origin and its granulometry.

Coconut charcoal (Picactif NC 60) with a median diameter of 3.8 mm was selected to maximize energizing yield. The incorporation of susceptors into the carbon (10 % weight) significantly improved heating efficiency. For a current with frequency equal to 263 kHz, heating efficiency primarily depended on the granulometry of activated carbon and of suceptors. This current frequency also permitted a homogeneous heating of the GAC.

In addition, the adsorption capacity of activated carbon for removing ethyl acetate from waste water was evaluated. Once the parameters of the granular environment were maximized for both induction and adsorption in batch, cyclic adsorption-desorption experiments were conducted to evaluate regeneration adsorption capacity. An adsorption capacity value of 0.12 g/g was reached after the third adsorption-desorption cycle which represented a global loss in adsorption capacity of 32 % compared to initial adsorption.

Experimental data indicate that industrial development of the induction heating process for GAC regeneration may be feasible.

3.4.1
Introduction

Granular activated carbon (GAC) is one of the most adsorbent materials used in the treatment of pollution (Le Cloirec, 1991). Industrial activated carbon adsorption units include two steps:

- an adsorption step – conducted up to the breaktpoint which represents a selected effluent concentration limit beyond which is unacceptable.
- a regeneration step – conducted in order to clean the activated carbon (adsorbent).

Saturation by adsorption limits the life-time of activated carbon and regeneration must be conducted for continued use.

Conventional regeneration

Conventional regeneration is achieved thermally by two processes: (1) heating the rector walls, or (2) preheating the fluid entering the bed (Cocheo, 1987). These conventional processes are founded on the circulation of hot fluid through the activated carbon filter (Shork, 1988; Jedrzejak, 1983). In regeneration, the absorbed compound (waste material) is removed form the solid carbon to the fluid (solvent). The waste material is then collected by distillation or condensation and the solvent is recycled for future use in the industrial process. Recently the high unit cost of activated carbon has led to the development of new regeneration processes (Chartier, 1993) including induction heating.

Induction heating regeneration

The induction process (Orfeuil, 1981) generates in-situ heat necessary to desorb the molecules fixed on the activated carbon (Mioduszewski, 1982). Induction heating is based on two well-known physical phenomena:

- electromagnetic induction (discovered by Faraday) and
- the Joule effect.

Activated carbon is a semi-conductor and therefore generates resistive heating when a current is passed through it – Joule effect (Le Cloirec, 1990). The major advantages of induction heating include high energy density, lower energy consumption, and good environment conditions (Reboux, 1992). In addition, induction heating avoids safety concerns regarding the contact between material and electrodes.

This inductive heating effect depends of the resistivity of the material. Currently, research workers use high frequencies, in the order of megahertz (MHz), to compensate for the resistivity of porous carbon (10^{-2} m). However, these frequencies are not readily available on an industrial scale. Nevertheless, it is possible to work at lower frequencies (near 300 kHz) with the appropriate choice of activated carbon (Mocho, 1993).

Project objectives

Induction heating is used to regenerate activated carbon for the purpose of recycling volatile organic compounds (especially solvents). The objective of this study is to apply the induction process to the regeneration of activated carbon.

Section 2 evaluates the influence of different parameters (operating conditions and material characteristics) to maximize the induction heating performance and to minimize energy cost. Specifically, the influence of physical characteristics (granulometry) of activated carbon, current frequencies, and addition of susceptors are evaluated regarding induction heating efficiency. Conclusions are presented regarding the appropriate selection of a granular environment for induction heating.

Section 3 evaluates the adsorption capacity of Picactif NC 60 activated carbon (recommended in Section 2) for removing ethyl acetate form waste water. Adsorption parameters of the ethyl acetate in the fixed bed are evaluated. After the parameters of the granular environment have been maximized for both induction and adsorption in batch, cyclic adsorption-desorption experiments in the fixed bed ware conducted with ethyl acetate. Conclusions are presented regarding the regeneration adsorption capacity.

Induction heating of granular activated carbon

The application of the induction process to granular materials is relatively recent. To our knowledge, there have been no extensive studies in this subject area. Therefore the methodology of experimental research has been applied to the study of activated carbon heating by induction (Sado, 1991). Two level factorial designs were chosen to determine the influence of some well-known factors on this type of heating. This pattern takes into account the interactions between these factors.

3.4.2
Material and methods

Figure 1 shows the activated carbon heating equipment. The temperature measurements are carried out with thermocouples dispersed in the activated carbon and connected to a computer (Schlumberger Solartron 3430). Temperature measurements are made after cutting the high frequency field, because the field disturbs the functioning of the thermocouples (Schlosser, 1972). The coil measures 15 centimeters (cm) in diameter and 18 cm in height. The frequency varies with the number of spirals per unit length. This frequency depends also on the capacitor values. The rector, made of glass, contained inside the solenoid, has an internal volume of 2 liters (L) with 12 cm internal diameter and 18 cm height (see Fig. 1).

3.4.3
Results and discussion

Electrical characteristics of activated carbon influence induction heating. The activated carbon grain size distribution, activated carbon physical characteristics, and

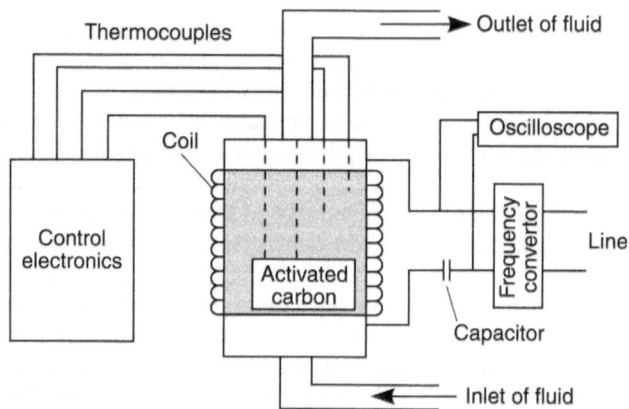

Fig. 1

the applied current frequency influence induction heating and the core temperature. These influencing factors are discussed below.

Grain size distribution

An experimental approach was necessary in selecting the appropriate activated carbon grain size distribution. Activated carbon with the following granular distributions were selected:

- diameter range of 0.5 to 1.25 millimeters (mm) with a corresponding average diameter (D_p) of 0.8 mm and
- diameter range of 2.1 to 3.2 mm with corresponding D_p equal to 2.7 mm.

Influence of carbon physical characteristics

Carbon physical characteristics such as porosity and specific area influence induction heating. Several physical characteristics of activated carbon are presented in Table 1. The influence of carbon physical characteristics was evaluated considering temperature response. The temperature response (Y) is the temperature value in the center of the granular medium after 15 minutes.

Table 2 presents the temperature response experimental field parameters. The coil power used for the induction heating was 1.35 kilowatts (kW). The effect of current frequency was considered and the selected heating rate of 5 °C per minute allowed for convection losses to be disregarded. The expression of the temperature response mathematical pattern is:

$$Y = X_0 + B_1 \cdot X_1 + B_2 \cdot X_2 + B_3 \cdot X_3 + B_{12} \cdot X_1 \cdot X_2 + B_{13} \cdot X_1 \cdot X_3 + B_{23} \cdot X_2 \cdot X_3$$
$$+ B_{123} \cdot X_1 \cdot X_2 \cdot X_3$$

The value of the coefficients B_i represent the effects of X_i on the response Y in the experimental field. The results are presented in Fig. 2. Figure 2 shows that the physical characteristics of activated carbon (carbon resistivity) are of primary importance

Table 1. Some physical characteristics of carbons

Activated Carbons	Picactif NC 60	Picabiol
Origin	Coconut	Wood
Porosity	Microporous	Macroporous
Bulk apparent ($g \cdot cm^{-3}$)	0.5	0.3
Specific area ($m^2 \cdot g^{-1}$)	1240	1750
Weight (g)	635	355

Table 2. Experimental field

X_1	grain size Dp (median diameter mm)	0.8	2.7
X_2	origin	coconut	wood
X_3	frequency (kHz)	140	220

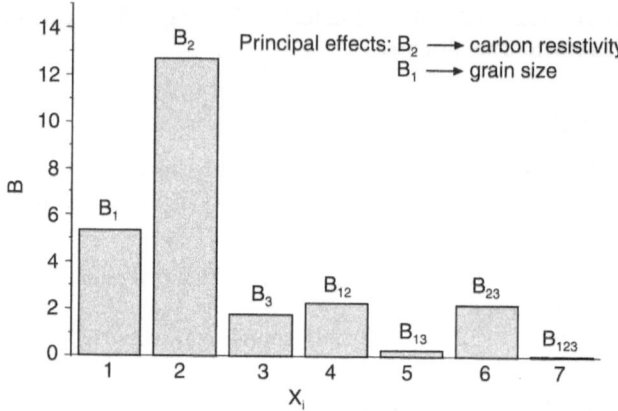

Fig. 2

regarding heating efficiency. Granular size also has an effect on the available surface for Foucault currents flowing through each grain. In addition, total electrical resistance of the bed increases as granular size is reduced.

Influence of current frequency

The heating rate of activated carbon is based on the electrical power used and on the current frequency in the coil. Table 3 presents the experimental field parameters. The effect of grain size was considered and coconut charcoal was selected for its capacity to be heated by induction. Experimental results are presented in Fig. 3. Figure 3 shows that electrical power is the principal parameter in this type of heating. The influences of granular size and current frequency are less important in this experimental field.

Reactor temperature distribution

This study evaluated the volume characteristics with respect to heating the activated carbon. Figure 4 shows a slight temperature difference between the core temperature (T_c) and the peripheral temperature (T_p) of the reactor. Eddy currents are developed superficially on each grain. The reactor is assumed to be uniformly heated. Therefore, the induction heating process (Reboux, 1992; Novelect, 1992) permits the granular medium to be uniformly regenerated. In addition, the current frequency increase from 140 to 220 kilohertz (kHz) increased the induction heating yield by a factor of 1.6.

Table 3. Experimental field

X_1	grain size Dp (median diameter mm)	0.8	2.7
X_2	coil power (kW)	1.3	2.3
X_3	current frequency (kHz)	140	220

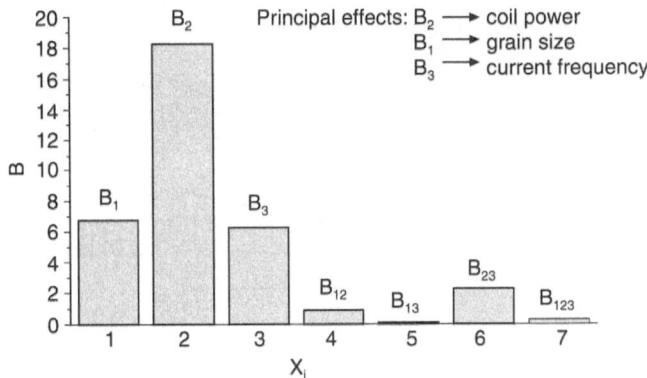

Fig. 3

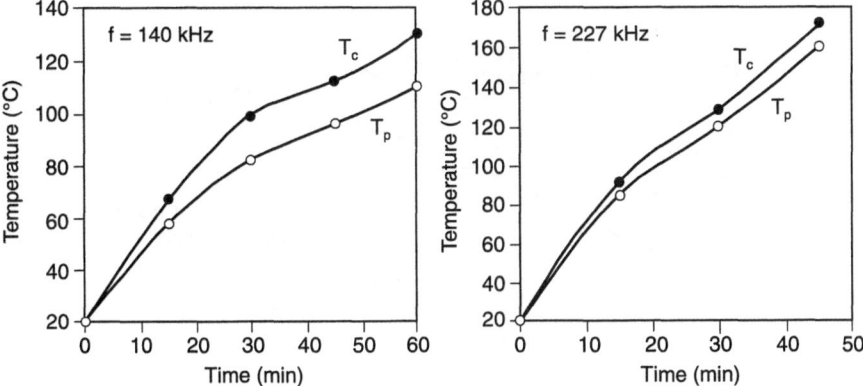

Fig. 4

Carbon-graphite composite formulation

An increase of energetical efficiency is required in a technical transfer perspective. To achieve this, the current frequency was slightly increased and the addition of susceptors were tested.

Both activated carbon and graphite have similar structures (Abram, 1969). Graphite has a low resistivity (10^{-5} $\Omega \cdot$ m) and therefore a sufficient capacity to be heated by induction (Kraus, 1984).

The low value for the coil-reactor coupling, which produced unfavorable heating conditions, was necessary in order to avoid overheating of the graphite. Figure 5 shows this induction heating phenomena by comparing single activated carbon and single graphite.

The granular size also has an influence on the heating efficiency. Improvements obtained by the addition of graphite are shown in Fig. 6. The energetical efficiency is increased by a factor 1.7 in comparison with single carbon performance.

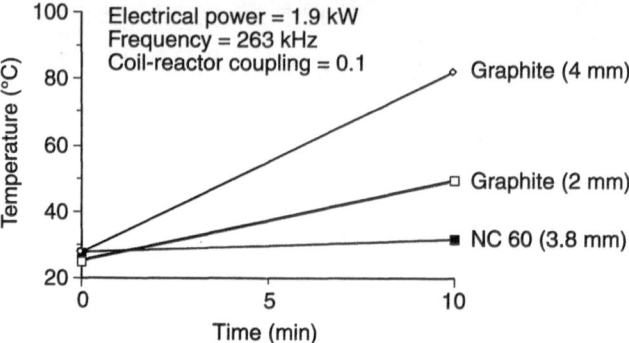

Fig. 5

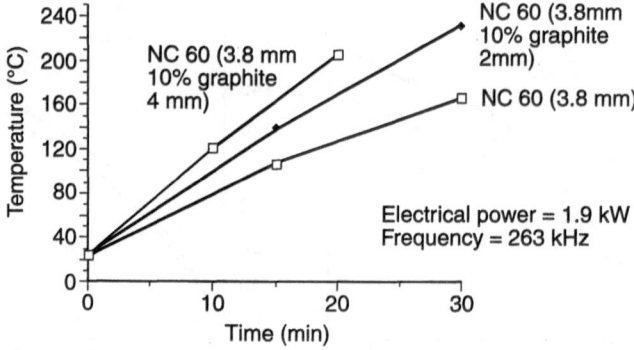

Fig. 6

Power adsorbed by the granular media

Induction heating efficiency depends on the characteristics of the carbon. Power adsorbed by the granular media can be defined by equation:

$$P = \frac{\varrho \, Ho^2 \, SF}{\delta}$$

where Ho = magnetic field intensity $(A \cdot m^{-1})$
      S = surface area of grain exposed to the magnetic field (m^2)
      F = induction transmission factor
      ϱ = resistivity of GAC $(W \cdot m)$

$$S = N \pi \, dh$$

where N = number of grains
      d = median diameter of GAC cylinder (m)
      h = height of a cylinder

$$F = \frac{1}{4} \left(\frac{d}{2\delta} \right)^3$$

where δ = penetration depth (m)

$$\delta = \sqrt{\frac{\varrho}{\pi \mu f}}$$

where f = current frequency (Hz)

μ = permeability of GAC ($H \cdot m^{-1}$)

3.4.4
Conclusion

Induction heating efficiency depends on the characteristics of the carbon. Coconut charcoal is recommended for this application.

The granular size also has an influence on the energizing yield. Ideally, the energy yield is maximized by increasing granular size while reducing adsorption capacity. A compromise was found with an average diameter of 3.8 mm and a distribution size between 3.1 mm and 4.8 mm.

The induction heating process permits homogenous heating of the granular medium which is important for regeneration quality.

Increase the current frequency improves the energizing yield, but does not contribute to heating the reactor uniformly (penetration depth effect). The addition of susceptors (graphite) may be an alternative to current frequency increase.

A simple model was presented to evaluate the induction power adsorbed by the granular media.

3.4.5
Regeneration by induction heating of granular activated carbon loaded with ethyl acetate in aqueous solution

The adsorption capacity of activated carbon for removing ethyl acetate from waste water was evaluated in this study. The selection of induction parameters including the type of activated carbon (granular environment) was based on conclusions presented in Section 2. Cyclic adsorption-desorption experiments in a fixed bed were then conducted to evaluate regeneration adsorption capacity. In general, this experimental investigation included three parts:

1. adsorption step – to define the efficiency of the separation process,
2. support heating – to evaluate the influence of operation conditions and material characteristics, and
3. cyclic sorptions – to evaluate the disposal of GAC according to experimental conditions.

3.4.6
Material and procedures

Material

Picactif NC 60 granular activated carbon (Pica Corp.) was appropriately selected as adsorbent with a grain size distribution between 3.1 and 4.8 mm and a median diameter of 3.8 mm. As discussed in Section 2, this type of activated carbon is recommended to maximize energizing yield. The carbon was washed with demineralized water 5–6 times, dried at 120 °C for 48 h, and stored in a desiccator. The specifications of the activated carbon are given in Table 4. The ethyl acetate used in the study was of industrial printing effluent.

Procedure

Adsorption isotherm and kinetics. Adsorption experiments were carried out with a known amount (100 ml) of specific initial concentration of ethyl acetate solution (13.57 grams per liter (g/L)) in 10 Erlenmeyer flasks to which different weighted carbon quantities were added. The flasks were kept on a rotary shaker Variomag (300 rpm) at room temperature (20 ± 2 °C). The samples were filtered through Whatman filter paper No. 5 and the filtrate was analyzed for the residual content in a Shimadzu TOC 5000 carbon analyzer.

Adsorption dynamics. A series of experiments to evaluate the adsorption dynamics of aqueous ethyl acetate solution were conducted. These experiments consisted of flowing ethyl acetate solution (13.57 g/L) through a fixed bed of Picactif NC 60 activated carbon at a constant rate of 2.9×10^{-5} meters per second (M/s). The experimental parameters are presented in Table 5.

These experiments were conducted with Picactif NC 60 activated carbon to which graphite was added to improve the energizing yield during the regeneration step. The specifications of the carbon-graphite composite are presented in Table 6. The concentration of the solution was determined by Total Organic Carbon Analyzer Shimadzu TOC 5000.

Table 4. Main characteristics of Picactif NC 60 activated carbon

Name	Aspect	Origin	Median diameter (mm)	Specific Area $(m^2 \cdot g^{-1})$	Apparent density
Picactif NC 60	Granular	Coconut	3.8	1240	0.52

Table 5. Experimental parameters of break-through curves

Initial adsorbate concentration Co $(g \cdot l^{-1})$	Flow rate $v \cdot 10^5$ $(m \cdot s^{-1})$	Bed Height (m)	Column Diameter $\cdot 10^2$ (m)
13.57	2.9	0.94	5

Table 6. Formulation of a carbon-graphite composite

Name	Aspect	Median diameter (mm)	Apparent density	Weight (g)
Picactif NC 60	Granular	3.8	0.52	890
Graphite	Granular	4	0.79	98.9

Table 7. Experimental parameters of desorption step

Cycle	Electrical Power (kW)	Frequency (kHz)	Nitrogen flow 10^3 $(m \cdot s^{-1})$	Heating Time (min)
1	2.4	263	0	220
2	2.4	263	0	225
3	2.4	263	3.44	246
4	2.4	263	5.89	263

Regeneration step by induction heating. The purpose of this experiment was to investigate the efficiency of the regeneration by induction heating of spent activated carbon used for waste water treatment. Wet spent activated carbon was placed into the heating equipment apparatus presented in Fig. 1.

Cyclic regeneration experiments were carried out under a nitrogen atmosphere with different convection flow. A detailed description of the experimental methods is presented in Table 7.

The desorption efficiency was deduced by comparison between break-through curves before and after the desorption step.

3.4.7
Results and discussion

Adsorption kinetics

Adsorption kinetics can be controlled by various mechanisms and conditions in adsorption phenomena. Four major rate-limiting conditions are generally cited:

1. mass transfer of solute form solution to the boundary film,
2. mass transfer of adsorbate from boundary film to carbon surface,
3. adsorption of molecules onto sites, and
4. internal diffusion of solute.

The first and second conditions consider external mass transfer resistance. The third condition is assumed to be very rapid and non-limiting in the kinetic analysis. The fourth condition considers intraparticle diffusion resistance.

The purpose of this study was to investigate the kinetic controlling mechanisms: a resistance external mass transfer model and a resistance intraparticle diffusion model. This approach was carried out by Guibal (1992).

External mass transfer resistance model

The model used to calculate the external mass transfer rate is described by the following equation:

$$\frac{dC_t}{dt} = -\beta_L S (C_t - C_s) \tag{1}$$

where β_L = external mass transfer coefficient
$\quad\quad S$ = surface specific.

According to complementary hypothesis, the initial surface concentration of solute on the sorbent is negligible ($C_s \to 0$ at time $t = 0$) and the intraparticle diffusion rate also negligible (McKay and Poots, 1980; McKay et al., 1986). Therefore, Eq. (1) can be simplified to:

$$\frac{d\frac{C_t}{C_0}}{dt} = -\beta_L S$$

and according to boundary conditions and the hypothesis formulated: $C_t \to C_0$ when $t \to 0$.

$$S = \frac{6m}{d_p \varrho (1 - \varepsilon)}$$

where m = mass of GAC (g)/volume of solution (m³)
$\quad\quad d_p$ = particle median diameter (m)
$\quad\quad \varrho$ = particle density (kg/m³)
$\quad\quad \varepsilon$ = particle void ratio

Activated carbon porosity is not taken into account in this model.

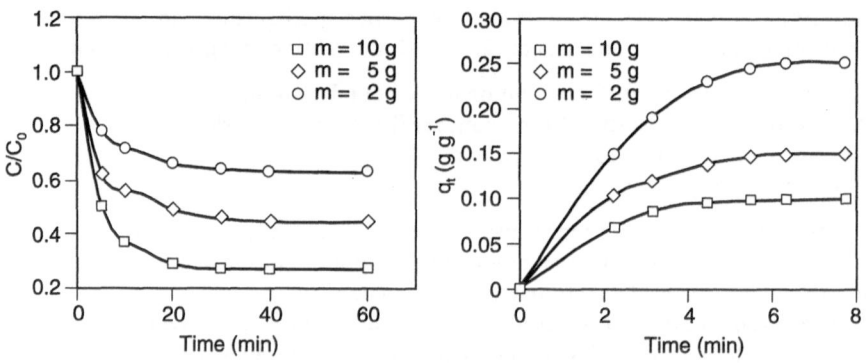

Fig. 7

Table 8. Determination of the external film transfer coefficient β_L

Carbon Weight (g)	$\beta_L S$ (min^{-1})	β_L (m $\cdot$ s^{-1})
10	$1.42 \cdot 10^{-1}$	$3.12 \cdot 10^{-6}$
5	$1.45 \cdot 10^{-1}$	$6.37 \cdot 10^{-6}$
2	$6.06 \cdot 10^{-2}$	$6.6 \cdot 10^{-6}$

Intraparticle diffusion resistance model

Morris and Weber (1962) and McKay et al. (1980) demonstrated that in diffusion studies, rate processes are usually expressed in terms of the square root of time. Therefore, q_t is plotted against $t^{0.5}$. The linearisation of the curve allows for the conclusion that the intraparticle diffusion resistance controls kinetic mechanism (see Fig. 7).

$$q_t = (C_0 - C_t) \, V/m$$

The representation of q_t versus $t^{0.5}$ shows that the intraparticle diffusion resistance is negligible. Therefore, external mass transfer is the main controlling mechanism. Table 8 presents the determination of the external film transfer coefficient (β_L).

Adsorption isotherm

The Brunauer-Emmet-Teller (BET) isotherm (1938), or multilayer Langmuir relation, was fitted to the equilibrium data.

$$q_e = \frac{Q \, K_L \, C_e}{\left(1 + K_L C_e + \dfrac{C_e}{C_s}\right)\left(1 - \dfrac{C_e}{C_s}\right)}$$

where Q = asymptotic maximum solid phase concentration (g/g)
   K_L = Langmuir equilibrium constant
   C_e = equilibrium concentration (g/L)
   C_s = saturation concentration (g/L).

Table 9 presents the BET Constants. Figure 8 presents the ethyl acetate isotherm for Picactif NC 60 GAC. This kind of isotherm is most frequently encountered when adsorption occurs on activated carbon with pore diameters larger than micropores resulting in adsorption on the external surface area. The inflection point of the isotherm usually occurs near the completion of the first adsorbed monolayer. In addition, with increasing concentration, second and higher layers are completed until at saturation, the number of adsorbed layers becomes infinite.

Table 9. BET Constants

K_L	Q (g $\cdot$ g^{-1})	C_s (g $\cdot$ l^{-1})
10.267	0.107	14

Fig. 8

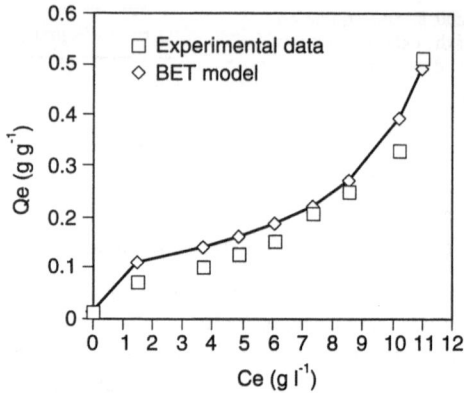

Cyclic adsorption desorption operation

The direct measure of the effectiveness of regeneration is adsorption capacity. Adsorption capacity was determined from the break-through curves integration area and compared to the difference between the amount of ethyl acetate in the initial and final solutions. The effect of cycling on the break-through curve shape is presented in Fig. 9. Experimental adsorption results for ethyl acetate are presented in Fig. 10 (adsorption capacity after regeneration versus number of cycles) and Fig. 11 (adsorption capacity after regeneration versus nitrogen flow).

If only free convection is used during the regeneration step, there is a progressive loss in adsorption capacity due to the holding back of residual ethyl acetate in the porous structure of the activated carbon. Forced convection applied to the desorption process resulted in an adsorption capacity of 0.12 grams per gram (g/g) which represents a global loss in adsorption capacity of 32% compared to initial adsorption (0.177 g/g). A nitrogen flow rate of $5.89 \cdot 10^{-3}$ m/s resulted in a recovery per cycle of 99%.

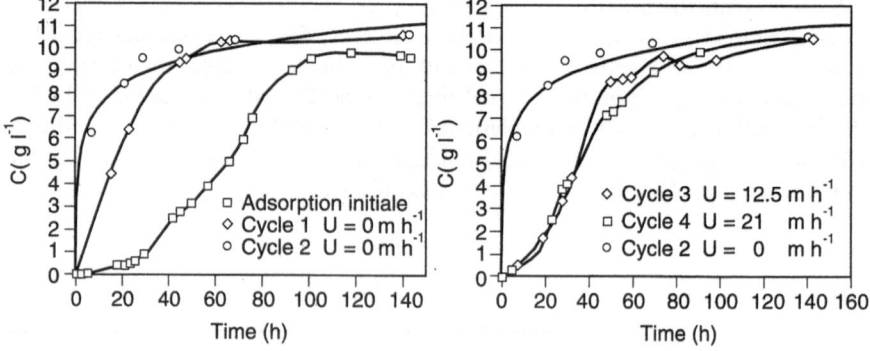

Fig. 9

Fig. 10

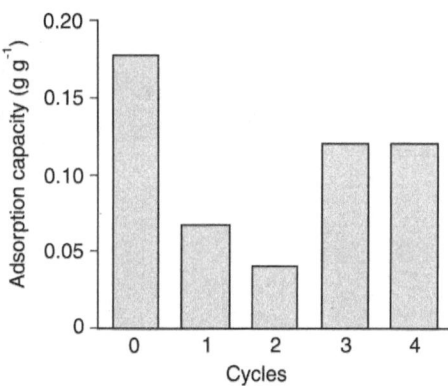

Fig. 11

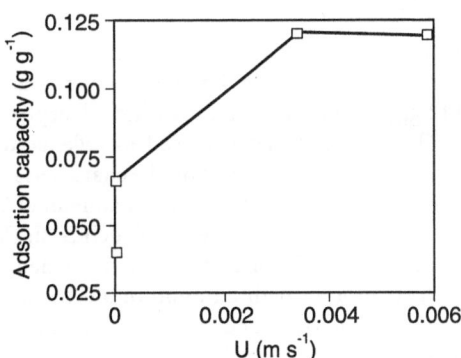

Forced convection has a positive effect on the regeneration of the material but not on the recovery of the ethyl acetate. The global concentration of the ethyl acetate in the nitrogen flow is 106 g/m³ which represents a partial pressure of 2.94 kilojoules (kJ) at a temperature of 293 °K and pressure of 1 atmosphere (Atm).

The temperature of the filter is influenced by the nature of adsorbates onto the carbon. During the desorption step, the temperature is relatively constant and corresponds to the heat of vaporization of the compounds (ethyl acetate and water). At the end of this step, the heating of the filter (Q) can be estimated by the following equation (conduction, convection, and radiation losses are neglected):

$$Q = m \, C_p \, dT$$

where m = mass of carbon in kilograms (kg)
$\quad\quad C_p$ = specific heat of carbon (kJ/kg · °K)
$\quad\quad dT$ = temperature increase (°K)
$\quad\quad Q$ = (kJ)

Figure 12 presents the heating of the GAC filter during the regeneration step.

Fig. 12

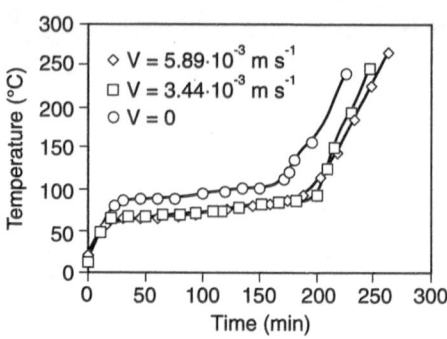

3.4.8
Conclusion

Adsorption capacity of Picactif NC 60 activated carbon for removing ethyl acetate from waste water was demonstrated in this study. An adsorption capacity value of 0.12 g/g was reached after the third adsorption-desorption cycle.

The induction heating used to regenerate the activated carbon was tested. The process efficiency depends on the characteristics of the carbon (nature and grain size). Coconut charcoal with a median diameter of 3.8 mm was recommended. This process with an appropriate frequency depends on the characteristics of the carbon (nature and grain size). Coconut charcoal with a median diameter of 3.8 mm was recommended. This process with an appropriate frequency (263 kHz) permits a homogeneous heating of the GAC. The energizing yield was improved by addition of graphite (10% weight) with a median diameter of 4 mm in the granular media.

The regeneration of the material may be carried out under the condition of forced convection with nitrogen flow ($3.44 \cdot 10^{-3}$ m/s) and the end of desorption step may be evaluated by controlling the temperature increase in the GAC filter.

These results indicate that industrial development of the induction heating process for GAC regeneration may be feasible.

3.4.9
References

Abram HC (1969) Chem Ind 1557
Brunauer S, Emmet PH, Teller E (1938) J Amer Chem Soc 60, 309
Chartier P (1993) Programme de recherche et de dévelopement technologique 1992–1996, Rapport ADEME, 27–29
Cocheo V, Bombi S (1987) Am Ind Hyg Assoc J 48(3) 189–197
Guibal E, Saucedo I, Roussy J, Le Cloirec P (1993) Water SA 19 2, 119
Jedrzejak A (1983) Modelling of activated carbon desorption by a circulated inert gaz. Chem Eng Tech 11, 352–358
Kraus JD (1984) Electromagnetics. Chap 4, 122, Ed: Mc Graw-Hill, Singapour
Le Cloirec P, Fanlo JL, Degorce-Dumas JR (1991) Etudes des odeurs et désodorisation industrielle, 114–124, Ed: Innovation 128, Paris
Le Cloirec P, Baudu M, Martin G (1990) Dispositif d'adsorption à couches superposées espacées et régéneration ar effect Joule, Brevet Francais n° 9003923
Le Cloirec P, Mocho P (1993) Procédé de régéneration d'un adsorbant granulaire, dispositif de mise en oeuvre et composite, Brevet Francais n° 93 10755

McKay G, Balir HS, Findon A (1986) Immobilization of ions by bio-sorption. Ed: Eccles H, Hunt S, Ellis Horwood Limited, Chichester

McKay G, Otterburn MS, Sweeney AG (1980), Wat Res 14, 150–20

McKay G, Poots VJP (1980) J Chem Tech Biotechnol 30, 279–292

Mioduszewski D (1982) Inductive heating of spent granular activated carbon. QED Corporation, Ann Arbor, Mi, USA

Novelect (1992), Les guides de l'innovation: Les applications innovantes de l'induction dans l'industrie. Ed: EDF

Mocho P, Le Cloirec P, Reboux J (1993) Récents Progrés en Génie des Procédés. 7, 333–338

Morris JC, Weber WJ (jr) (1962) Advances in water pollution research, Proceedings 1st In Conf on Water Pollution Res, Pergamon Press, New York, 2, 231–266

Orfeuil M (1981) Electrothermie industrielle, Ed: Dunod, Paris

Reboux J (1992) Induction heating: industrial applications. Ed: UIE, Paris

Sado G, Sado MC (1991) Plans factoriels complets à deux niveaux, Les plans d'expériences, Chap. 3, 23–41, Ed: Afnor, Paris

Schlosser WJ, Munnings RH (1972) Cryogenics 7, 302–303

Shork JM, Fair JR (1988) Ind Eng Chem Res 27, 1545–1547

3.4.10
Further Readings

Title: **Electrochemical regeneration of granular activated carbon**
Author: Narbaitz, Roberto M.; Cen, Jianqi
Corporate Source: Univ of Ottawa, Ottawa, Ont, Can
Source: Water Research v 28 n 8 Aug 1994, p 1771–1778
Publication Year: 1994
Coden: WATRAG ISSN: 0043-1354
Language: English
Document Type: JA; (Journal Article) Treatment Code: X; (Experimental)

Abstract: Laboratory experiments investigated the feasibility of a novel granular activated carbon (GAC) regeneration technique: electrochemical regeneration. GAC was loaded with phenol, via batch adsorption tests, then electrochemically regenerated and finally reloaded with phenol. Regeneration was conducted in a batch reactor filled with a 1 % NaCl solution as the electrolyte. As limited experiments showed that cathodic regeneration was 5–10 % more efficient than anodic regeneration, the investigation concentrated on the former. Although anodic regeneration was more efficient in destroying phenol residuals form the electrolyte, cathodic regeneration could also eliminate these residuals by using longer regeneration times and/or higher currents. Increasing the regeneration current to 100 mA for 5 hours can increase the regeneration efficiency (RE) to a maximum of 95 %. Lower currents applied for longer regeneration times can yield similar results. REs were also significantly affected by the electrolyte type, the electrolyte concentration, and the GAC particle size, but not by the phenol loadings. Multiple regenerations only reduce the REs and additional 2 % per cycle. Given the high regeneration efficiencies and no apparent carbon losses, electrochemical regeneration of GAC at a laboratory-scale is a feasible alternative to thermal regeneration and merits further investigation. (Author abstract) 11 Refs.

Title: Mathematical modelling of a continuous, direct resistive heating furnace for the regeneration of activated carbon
Author: Van Staden, P. J.: Bryson, A. W.
Corporate Source: Univ of the Witwatersrand, Wits, South Africa
Source: International Journal of Mineral Processing v 36 n 3 – 4 Oct 1992 p 175 – 199
Publication Year: 1992
Coden: IJMPBL ISSN: 0301-7516
Language: English

Abstract: A mathematical model of the steady-state performance of a continuous, direct resistive heating furnace (DRHF), used for regenerating granular activated carbon in gold-recovery circuits, has been developed. This mathematical model can be used as a design tool to predict the axial and radial temperature profiles, as well as the electrical potential and power requirements for a given set of carbon and team flow-rates, electrical current, and furnace dimensions. The validity of the model has been verified on a small-scale (up to about 3 kg/h) DRHF. The temperatures could mostly be predicted to within 20 K, while the electrical potential and power requirements were predicted correctly to within 10%. When the carbon's effective mean specific heat capacity was used as a fitting factor, it was concluded that the thermal decomposition of the carbon contributes about 300 J/(kg K) to the effective specific heat capacity of the sample of carbon used. Furthermore, it was concluded that carbon approaching the inlet is pre-heated by water vapour rising from the furnace, which condenses on the carbon, hence increasing its moisture content above the original moisture content of the carbon fed into the furnace hopper. (Author abstract) 15 Refs.

4 Wastewater Treatment

4A Biological Treatment

4.1
Effect of the Grease Solubilization and the Optimal Process Monitoring on the Grease Aerobic Digestion

ETIENNE PAUL · XAVIER LEFEBVRE · MICHEL MAURET · BERNARD CAPDEVILLE

Abstract

Placement of lipid residues in sanitary landfills are now prohibited by French legislation. The lipid residues used in this work were collected from a wastewater plant and restaurant traps. Aerobic degradation of these mixtures by a mixed population originated from an activated sludge was studied. First, the lipid matter underwent a saponification reaction, which solubilized partially and emulsified the substrate. An activated sludge was acclimated to saponified grease carbon source in a sequencing batch culture mode. Microbial activity was evaluated by the on line measurement of the CO_2 production rate in the exit gas. At a feed concentration of 2 gHEM/l, the substrate degradation rate was measured as high as 1.6 gHEM/l /h, the COD and HEM removals were 75% and 90%, respectively. Finally the average biomass yield was 0.3 gMLSS/gHEM.

4.1.1
Introduction

Grease residues contain a high oxygen demand. Indeed, in the domestic wastewater inflow at treatment plant, 25% of the total chemical oxygen demand (COD) comes from the lipid fraction (Bridoux, 1992). The grease matter is responsible for clogging sewer networks and disrupting waste water treatment plant operation. In the activated sludge process, oxygen transfer into biomass is limited by the buildup of a lipid film at the air/water interface and by the grease adsorption to bacterial flocs (Boutin et al., 1975; Duchêne, 1980). A decrease in waste water treatment efficiency due to lipid overload that led filamentous growth and concomitant settling problems was described by Hrudey (1980). To cope with the effects of lipids on wastewater treatment, grease removal is typically applied at the head of the plant. The process, performed mainly by air flotation, yields an average grease removal of 20%. Floated greases are made up mainly of fatty acids and glycerides (Ansenne, 1992; Bridoux, 1994). This ratio depends upon the retention time in the sanitary sewer and in the collecting tank.

Animal and vegetable substances, generally used in human cooking, are the sources of grease in wastewater. A methodology for grease characterization in the wastewater process was developed by Bridoux (1994). He showed that hexane was more efficient for extracting grease than solvents, such as chloroform or dichloromethane. Due to the high anaerobic biodegradability of the lipid residues, disposal in sanitary landfills was recently prohibited. Therefore, removal and stabilization of these agents in wastewater treatment processes has become an important component of wastewater treatment technology.

Aerobic biological oxidation is an attractive treatment alternative for lipid residues because the grease is converted to carbon dioxide, water and biomass. In addition, the process could be incorporated into existing infrastructure.

The biological reaction consists of two main stages: (i) Glycerides are hydrolyzed by extracellular lipolitic enzymes into glycerol and fatty acids. (ii) Fatty acids, like alkanes, are transformed into acetyl coenzyme A molecules by successive breaks in the carbon chain (beta-oxidation) catalyzed by four intracellular enzymes (Ratledge, 1992). These molecules are finally broken down into CO_2 and H_2O in the Krebs cycle.

Lipid biological degradation is limited by unfavourable physico-chemical properties: insolubility in water (Ralston and Hoerr, 1942) and solidification or semi-solidification at ambient temperature. The floated greases are a heterogeneous substance whose dry matter varies from 23 to 75% (Ansenne, 1992) and exists in the form of solid particles. As a consequence, slow degradation is the result of limited microbial accessibility to the substrate. In order to improve degradation, Kallel (1992) saponified a glycerides/fatty acids mixture. In saponification, glycerides are hydrolyzed into glycerol and fatty acids, and fatty acids are neutralized to form soap.

Many micro-organisms in activated sludge have the capacity to degrade grease. However, in order to enhance the aerobic degradation, the activated sludges must be acclimated to degrade grease. This can be achieved by putting activated sludges in contact with lipid compounds in a favourable environment. The resulting biomass becomes more tolerant of the fatty acids. Toxic fatty acids, especially short chain fatty acids which are intermediate metabolites of catabolism of the long carbon chain fatty acids, inhibit growth of gram negative bacteria (Fay and Farias, 1975; Sheu, 1973).

Online respirometric measurement of the biomass were used to measure the biological substrate demand during a sequencing batch culture reaction. In this paper, both saponification of the raw grease and microbial culture developed using the sequencing batch culture are discussed.

4.1.2
Materials and Methods

Inoculum and medium

The activated sludge underwent a stage of pre-adaptation in a reactor supplied each day with saponified grease (about 1 g/l). The microbial materials for the experimentation came from this reactor and were used after three days starvation to eliminate residual grease. Greases were used as the sole source of carbon and energy. The whole grease characteristics are reported in Table 1. Among the fatty acids, four major species are dominant: palmitic (C16:0, 25%), stearic (C18:0, 10%), oleic (C18:1, 40%) and linoleic (C18:2, 20%) acids.

The potassium hydroxide (0.5 N) was added in excess for the saponification reaction at 80°C during 30 minutes. The product of this reaction was diluted by tap water until the desired grease concentration was attained. The C/N ratio was adjusted to 15/1 with $(NH_4)_2PO_4$. The initial pH of the culture medium was about 8.

Pilot

The experimental assembly is illustrated in Fig. 1. The system consisted of 5 liters, continuously aerated batch reactor equipped with a dissolved oxygen meter (YELLOW SPRING INSTRUMENT MODEL 58), a pH-meter (E588 METHROM HERRISAU) and a carbon dioxide analyzer (INFRARED DETECTOR, ASC MODEL 330). It was maintained at constant temperature (25 °C) with a water jacket. The reactor was operated in SBR mode (Sequencing Batch Reactor). The output signals of these instruments were interfaced with a personal computer (EPSON ELD) equipped with a PCL-812 data acquisition card. They were used for calculation and stored during the reactor operation.

The end of the reaction was determined by means of the carbon dioxide evolution rate (RCO_2), computed from the CO_2 concentration in the exit gas. Aeration was discontinued to remove foam, 5 minutes before the mixing system was terminated. A clarification period (30 minutes) was applied to separate flocs from liquid. A draw pump purged the supernatant to eliminate the treated wastewater (2 liters). The substrate pump was switched on until the working volume (3 liters) was reached. The aeration and mixing systems were started while filling the reactor.

Analytical procedure

- Hexane extracted matter (HEM) and fatty acids
  The lipids were extracted by using hexane according to Bridoux (1992) method. After acidification of the essay samples, a mixture sample/hexane/methanol (1:1:0.5 in volume) was stirred during 30 minutes. Hexane was evaporated under vacuum at 50 °C. The dry residue was then weighed and gave the amount of HEM. To improve the gas chromatography (GC) separation of fatty acids, the dry residue was purified. First, the internal standard of GC method (heptadecanoic acid) was added to the HEM residue (1:3 w/w). With 10 ml of a solution methanol/benzene (4:1 v/v)/KOH (5% w/v), a saponification was made at 80 °C during 4 hours. Fatty acids were separated from other lipid compounds by ether extractions. Finally, they

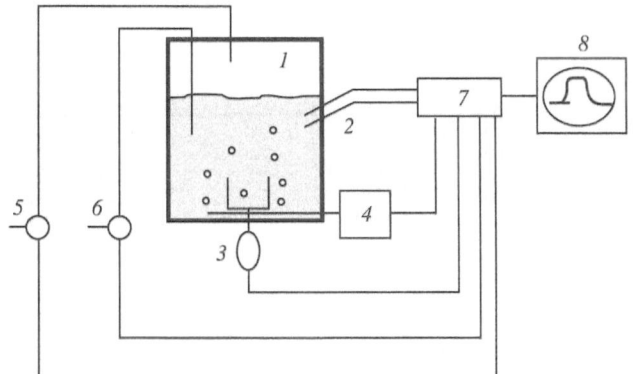

1 Reactor
2 Oxygen and pH probes
3 Mixing system
4 Aeration system
5 Substrate pump
6 Supernatant pump
7 Interface
8 Computer

Fig. 1. Experimental assembly

Table 1. Grease characteristics

Dry matter	Total carbon	hexane extracted matter	COD/HEM	free fatty acids	esterified fatty acids	other lipids
70%	70%	100%	2.4 g/g	70%	10%	20%

were esterified by diazomethane solution to turn low their evaporation temperature. The fatty acids concentration was determined by using a gas chromatograph (HEWLETT-PACKARD 5890) with a flame ionization detector. The capillary column was a DB 624 (J & W SCIENTIFIC). The carrier gas was helium (1 ml/min). The injection, detection and oven temperatures were respectively 250, 250 and 180 °C.

- Mixed liquor suspended solids (MLSS)
  A similar procedure as the lipid extraction previously described was applied. The extracting solution was hexane. The whole sample was filtered using a 8–10 micrometer membrane (SARTORIUS) that was then dried and weighed.

- Total soluble carbon (TC) and chemical oxygen demand (COD)
  Total carbon analysis was performed using a DORHMANN DC180 carbon analyzer while COD analysis was conducted using the HACH micromethod. The presence of solid grease particles made characterization difficult. In order to increase analytical accuracy, grease was first saponified using KOH to obtain a more homogenous solution. Commercial olive oil was used as a standard to check each step of the analytical methodology. The fatty acids composition (oleic acid mainly) of the olive oil was determined using gas chromatography.
  The correlation between chemical oxygen demand, total carbon and dry lipid matter are presented in Fig. 2. These data evidence by their linearity that analyses were coherent among themselves. Furthermore, this method supports the theory that small sampling volumes was accurate. Multiple testing reproduced the results for TC analysis and reduced the standard deviation among COD analyses. Moreover,

Fig. 2. Lipid COD and CT characterization

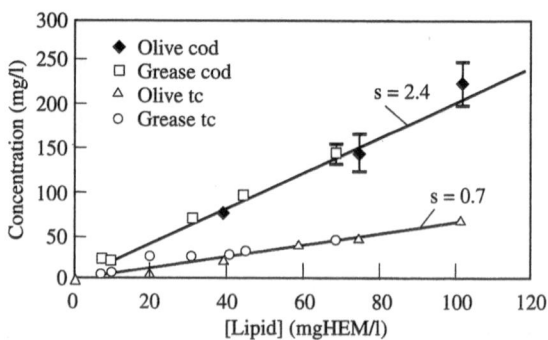

these data correctly predict the theoretical olive oil carbon percentage from its fatty acids and glycerol composition at 70+/−5%. The value was confirmed by the direct total carbon measurement (Fig. 2). Therefore, lipid materials were not lost in the analytical procedure. These data confirmed that grease residues after saponification were homogeneously dispersed in water, which allowed the sampling volume to be reduced.

4.1.3
Results and Discussion

Six sequencing batch experiments were run. CO_2 production during grease degradation was the control parameter for indicating the end of the degradation reaction and for managing the successive batch cultures efficiently. As soon as the degradation reaction was over, the aeration and mixing systems were switched off to allow the biomass to settle. Two liters of supernatant were wasted and replaced with about 2 gHEM/l fresh medium. The pH was adjusted to 8 after filling the reactor due to medium acidification from the residual volume of the previous batch. Biomass solids were not wasted at the end of the cycle.

Along the successive batches, biomass was accumulated by decantation in the reactor from 0.5 to 3.8 gMLSS/l. The average biomass yield was 0.3 gMLSS/gHEM. The initial HEM concentration (S_0) in the culture medium moved from 2 to 3.8 gHEM/l. This resulted from a grease retention, certainly due to sorption onto bacterial flocs, which increased from 0.7 g/l to 1.8 gHEM/l along the first three degradation cycles and then remained stable.

The Fig. 3 presents the evolution of the main kinetic and activity parameters: average HEM degradation rate (rs), degradation time (T), average HEM specific degradation rate (qs) and HEM removal yield (Y):

The rate rs increased to 1.6 gHEM/l/h before stabilizing, while T decreased to a minimal value of 1.2 h. This kinetic stabilization mean that, in these culture conditions, maximum biological performance was achieved from the fourth batch culture. T and rs variations depend directly upon microbial activity, that activity being the biomass concentration and the microbial specific degradation rate (qs). Specific biomass was maintained and accumulated in the reactor by settling at the end of each degradation cycle. Moreover, as described on Fig. 3, qs increased up to a maximal value of

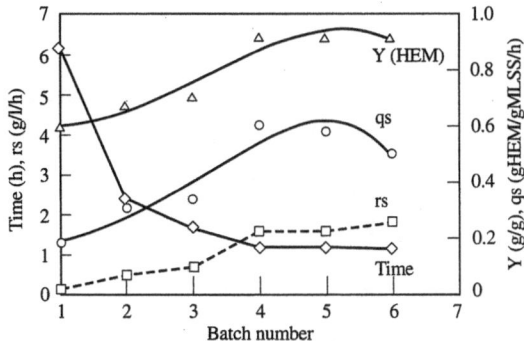

Fig. 3. Evolution of degradation time (T), degradation rate (rs), specific rate (qs) and HEM removal yield (Y)

0.6 gHEM/gMLSS/h and then decreased. Bioavailable substrate concentration, from the fifth degradation cycle, was not enough to support microbial needs. The specific activity loss might have been prevented by wasting some biomass or by increasing the substrate feed. In spite of this substrate limitation, the sorbed fraction stabilized at 1.8 gHEM/l and were not used by micro-organisms, which means that these sorbed lipids were so trapped that their transport across the cellular membrane was impossible.

The HEM removal yield (Y) calculation did not to effect the grease fraction recovered from effluent by sorption onto bacterial flocs during each previous batch. Therefore, this yield corresponded to the ratio (metabolized HEM/feed HEM). It increased along the first four batches until a maximal and constant value of 90% was reached.

Simultaneously, the COD and CT concentration of the treated effluent increased as well, and remained stable from the fourth batch (Fig. 4). 75% COD was also removed. This was calculated from the ratio COD/HEM (2.4) characterizing the feed grease. Moreover, whatever the degradation cycle, the HEM concentration of the outlet effluent was maintained a low and constant (0.15 gHEM/l) ratio.

That COD increase did not mean microbial activity loss, but a production of steadily rising metabolites, resulting from the increase of the HEM removal yield (Y) (Fig. 3). The ratio COD/HEM was 4 times as great as the ratio established for HEM characterization. The TC analysis showed, as well, that 80% of TC in the outlet effluent belonged to both inorganic carbon forms (HCO_3, CO_2) and compounds insoluble in hexane. Whereas microbial substrate demand was limited for the last batches, the constant production of these metabolites in the effluent points out their slowly biodegradable character. These compounds could be intermediate metabolites such as short carbon chain fatty acids, since the fatty acids with a carbon number less than 10 are relatively soluble in water (Bridoux, 1994). Indeed, the β oxidation of long carbon chain fatty acids could lead to an excess of short fatty acids releasing and accumulating into the bulk liquid. This result may rely on the fact that these short fatty acids could not be metabolized by the mixed population acclimated on long fatty acids.

The grease sorption onto bacterial flocs entailed lipid retention in the reactor and maintained a low HEM concentration in the outlet effluent, even when the HEM removal efficency was lower than 90%. The sorbed grease was different, by its fatty acids

Fig. 4. Characterization of effluent quality

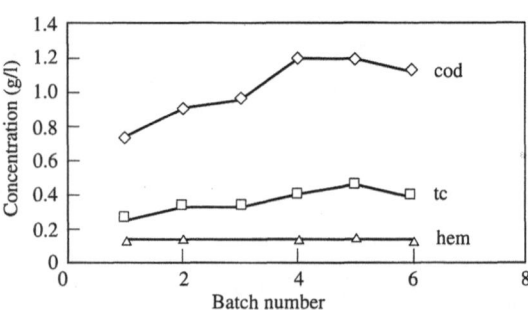

composition (stearic and palmitic acids: 70%, oleic and linoleic acids: 25%), from the feed grease (stearic and palmitic acids: 35%, oleic and linoleic acids: 60%). The greater proportion of saturated fatty acids demonstrates that they were less easily metabolized during the first batches than the unsaturated compounds. Unsaturated fatty acids owed their preferential degradation to their higher solubility in water (Loehr and Roth, 1968; Novak and Carlson, 1970). Therefore, the substrate dissolution into the bulk liquid and its diffusion into cells could be the preferential mechanism of microbial assimilation.

The increase of the specific degradation rate (q_s) and the HEM removal yield showed microbial acclimation to grease substrate. The sequencing culture mode and the lipid retention increased the contact time between biomass and refractory grease, and so promoted the improvement of consortium metabolic capabilities. As a matter of fact, the HEM removal yield of 90% pointed out that, from the fourth cycle, saturated fatty acids and unsaponified matter were degraded as well.

As a result, the maximal degradation rate (1.6 gHEM/l/h) is, for both COD and HEM removal yields, 20 times as high as the rate obtained with other processes applied to aerobic raw grease digestion reported (Kallel, 1992; Bridoux, 1992; Grulois, 1993). Many hypothesis are proposed to explain these kinetic performances:

(i)   the microbial activity inhibition by lipid, biomass or by-products was reduced because of their relatively low concentration.

(ii)   The acclimation mode was very performant. The control of microbial activity by CO_2 production limited starvation periods, which the microbial *consortium* could lose its gained capacities for. The sequencing batch culture increased the substrate and by-products tolerances. The short chain fatty acids ($n < 6$), intermediate metabolites, are gram negative bacteria inhibitors (Fay and Farias, 1975; Sheu, 1973). The ratio S_0/X_0 applied was high. The enzymatic system was also continuously induced at a high level. The active biomass accumulation led to increase degradation capacities of the biological system.

(iii) The microbial *consortium* growing on hydrolyzed grease might be different. Some micro-organisms are indeed unable to hydrolyze glycerides but are able to degrade fatty acids (Ratledge, 1992).

(iv)   Glycerides in the raw grease used by Bridoux (1992) and Grulois (1993) may account for a significant part. Their enzymatic hydrolysis controls the global degradation reaction to some extent (Kramer, 1971; Hsu et al., 1983).

(v)   As Goma (1975) showed that the kinetic of hydrocarbon degradation by *Candida lipolytica* corresponded to their water dissolution kinetic, the fatty acids degradation mechanism could proceed in the same way as their dissolution into the aqueous phase and their diffusion into cells. Therefore, fatty acids solubilization and their emulsification in micellar drops by saponification highly improved the mass transfer.

4.1.4
Conclusion

The lipid residues collected from a wastewater plant and restaurant traps were characterized (70% dry matter, 100% MEH, 80% of free and esterified fatty acids, 20% of other lipid materials, DCO/lipids = 2.4). The aerobic degradation of this lipid mixture, after saponification, by a mixed population originated from an activated sludge was studied. The control of the microbial activity by CO_2 and the sequencing batch culture mode led, in our culture conditions, to an optimization of activated sludge acclimation to saponified grease carbon source. The maximal substrate degradation rate was as high as 1.6 gHEM/l/ h and the HEM and COD removal yields were respectively 90 % and 75 %. The average biomass yield was 0.3 gMLSS/gHEM. The stabilization of kinetic performances seemed to depend upon substrate limitation.

A study about the effect of the ratio S_0/X_0 (initial substrate concentration versus initial biomass concentration) on the kinetic performances is in progress to determine the biological system limits (the maximal grease load, the by-products inhibitory concentration, the maximal biomass concentration, oxygen transfer). Finally, the characterization of the residual DCO (25 %) seems to be necessary for understanding its refractory origin.

References

Ansenne A, Destain J, Godefroid J, Thonart PH (1992) La problématique des séparateurs de graisse. Tribune de l'eau 558:33–40

Boutin P, Vachon A, Bechac JP, Lopez B (1975) Mesure de la capacité d'oxygénation dans les stations de traitement à boues activées en mélange intégral. Techniques et Sciences Municipales 11: 493–501

Bridoux G (1992) Bilan des graisses dans les stations d'épuration. Elimination des résidus graisseux par voie aérobie (Thesis). Université de Compiègne, Compiègne, France

Bridoux G, Dhulster P, Manem J (1994) Analyse des graisses dans les stations d'épuration. Techniques Sciences Méthodes 5 : 257–262

Duchene P (1980) L'efficacité des dégraisseurs en station d'épuration. La Tribune du Cebedeau 444 : 489–496

Fay JP, Farias RN (1975) The inhibitory action of fatty acids on the growth of Escherichia coli. Journal of General Microbiology 91 : 233–240

Goma G (1975) Contribution à l'étude des fermentations sur hydrocarbure. (Thesis). INSA, Toulouse, France

Grulois PH, Alric G, Bridoux G, Flyac S, Manem J (1993) Biomaster: Procédé de traitement biologique des graisses. Techniques Sciences Méthodes 5:247–251

Hrudey SE (1980) Activated sludge response to emulsified lipid loading. Water Research 15:361–373

Hsu T, Hanaki K, Matsumoto J (1983) Kinetics of hydrolysis, oxidation and adsorption during olive oil degradation by activated sludge. Biotechnology and Bioengineering 25:1829–1839

Kallel LM, Vedry B, Malesieux G, Letolle R, Saliot A (1992) Bioélimination des déchets graisseux après saponification. Techniques Sciences et Methodes 11 : 619–623

Kramer GR (1971) Hydrolysis of lipids in waste water. Journal of Sanitary Engineering Division 97:731–744

Loehr RC, Roth JC (1968) Aerobic degradation of long-chain fatty acids salts. Journal of Water Pollution Control Federation 40:385–403

Novak T, Carlson A (1970) The kinetics of anaerobic long chain fatty acid degradation. Journal of Water Pollution Control Federation 42:1932–1943

Nunn DW (1986) A molecular view of fatty acidcatabolism in E. Coli. Micobial reviews 50:179–192

Rapin F (1979) L'éfficacité des dégraisseurs dans les stations d'épuration des communes rurales: approche méthodologique. Etude n°43, CTGREF, Antony, France

Ralston AW, Hoerr CW (1942) The solubilities of the normal satured fatty acids. Journal of Organic Chemistry 7:546–554

Ratledge C (1992) Microbial oxidations of fatty alcohols and fatty acids. Journal of Chemical Technology and Biotechnology 55:397–414

Sheu WC, Freese E (1973) Lipopolyaccharide layer protection of gram negative bacteria against inhibition by long chain fatty acids. Journal of Applied Bacteriology 36 : 635–646.

Further Readings

Title: Treatment of lipid-containing wastewater using bacteria which assimilate lipids.
Author: Okuda Shin-Ichi; Ito Kazutoshi; Ozawa Hiroko; Izaki Kazuo Corporate Source: Hachinohe Inst of Technology, Aomori, Jpn Source: Journal of Fermentation and Bioengineering v 71 n 6 1991 p 424–429 Publication Year: 1991
CODEN: JFBIEX ISSN: 0922–338X
Abstract: Bacteria which grew in a medium containing olive oil as a sole source of carbon were isolated from two meat plants in the Sendai district of Japan. All of the isolates tested assimilated beef tallow, lard, olive oil and used salad oil as a sole carbon source in shaking cultures. One of the isolates, strain 351, digested lipids most efficiently, as shown by the amount of n-hexane extracts that remained. This bacterium was identified as Bacillus sp. A new and efficient laboratory-scale apparatus for the biological treatment of lipid-containing wastewater was devised using strain 351. The apparatus consisted of a water circulation system for the primary treatment of the water, in which strain 351 was inoculated, and an ordinary aeration tank using activated sludge as a secondary treatment. Lipids in the wastewater could be almost completely removed by this apparatus without physical treatment. On the other hand, an ordinary aeration system in the laboratory using an air stone and air pump resulted in the floating of lipids, and was not successful in digesting lipids even in the presence of strain 351. (Author abstract) 9 Refs.

Title: Prevention of Lipid Inhibition in Anaerobic Processes by Introducing a Two-Phase System
Author: Komatsu T; Hanaki K; Matsuo T
Auth. Address: Tokyo Univ. (Japan). Dept. of Urban Engineering.
Source: Water Science and Technology WSTED4; Vol. 23, No. 7/9, p 1189–1200, 1991. 9 fig, 5 tab, 10 ref.
Abstract: The inhibitory effect of lipids and prevention of this inhibition in a two-phase anaerobic wastewater treatment process were examined using laboratory-scale reactors and batch experiments. Lipids were satisfactorily degraded in a two-phase anaerobic filter while in a single-phase system, inhibition resulted in poor lipid degradation. Unsaturated long-chain fatty acids (LFAs) had a greater inhibitory effect than saturated LFAs. Methane production as well as beta-oxidation (degradation of saturated LFAs) were inhibited by unsaturated LFAs. The saturation of unsaturated LFAs was not inhibited, and palmitate ($C16:1$) was accumulated in the degradation of oleate ($C18:1$) or linoleate ($C18:2$). Greater inhibition was observed at low pH values. Continuous operation of a suspended-growth acidogenic reactor showed that hydraulic retention times (HRTs) of no less than 8 hours were necessary to mitigate the inhibition in a two-phase process. The fact that saturation of oleate occurred at HRTs no less than 8 hours suggests that the saturation of unsaturated LFAs in an acidogenic reactor is essential in the prevention of lipid inhibition in two-phase anaerobic processes. (Author's abstract)

4.2
Membrane Gas Liquid Contactors in Water and Wastewater Treatment

PHILIPPE APTEL · PHILIPPE MOULIN · MICHAEL CLIFTON ·
JEAN-CHRISTOPHE ROUCH · CHRISTOPHE SERRA

Abstract

The alleviation of environmental problems is one of the biggest challenges of technology today. The development and implantation of new separation processes may result in a "green" industrial revolution. Membrane technology has many advantages to offer in this respect. Membrane contactors are commonly hollow fibre devices used as substitutes for packed towers. As such, they are alternative industrial configurations for carrying out gas absorption or stripping and liquid-liquid extraction. This paper is limited to gas-liquid contactors. In the case of oxygenation of water, a resistance-in-series model with two resistances, the membrane and the liquid film resistance, was used to describe the oxygen transfer process. An induction of secondary flows is used to decrease the mass transfer resistance in the liquid phase and increase the oxygen flux. The study was based on a comparison between straight modules and coiled modules. For straight modules, the results are consistent with the Lévêque correlation. For coiled modules mass transfer coefficients were found to be two to four times higher than for straight modules and a new mass transfer correlation is presented.

4.2.1
Introduction

A membrane can be defined as a barrier between two fluids which restricts the movement of one or more components of one or both fluids across the barrier. The first commercial membranes were prepared in the late 1920's for use in bacteriology laboratories; these were symmetric microfiltration membranes. Desalination of water by reverse osmosis is one of the earliest and best-known industrial applications: the first plants were started up in the 60's. More recently, hollow-fibre ultrafiltration plants have been installed for water treatment to replace conventional physico-chemical clarification and disinfection steps with just one physical unit operation (Anselme et al., 1993), (Aptel, 1994).

Mass Transport Considerations

The primary role of a membrane is to act as a selective barrier. It should allow the passage of certain components of a mixture and retain the others. A schematic representation of a membrane separation operation is given in Fig. 1.

Fig. 1. Schematic representation of a two-phase system separated by a membrane

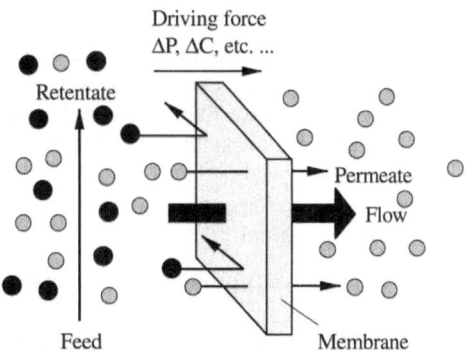

As for all transport phenomena, the transmembrane flux for each element can be described by the following simple expression:

Flux = Force · Concentration · Mobility (1)

The local driving force is the gradient of chemical potential $d\mu/dx$ of every component that can be transported.

The variation in the chemical potential of component "i" can be expressed as the sum of three terms:

$$d\mu i = RT \, d \ln a_i + V_i \, dP + z_i \, F \, d \, \Psi \tag{2}$$

where a_i is the activity (product of concentration by activity coefficient), P is the pressure and Ψ is the electrical potential.

In the case of a gas-liquid contactor, there is no gradient of pressure or electrical potential, equation (2) reduces to:

$$d\mu_i = RT \, d \ln a_i \tag{3}$$

or

$$\mu'' - \mu' = RT \ln (a''/a') \tag{4}$$

where double prime ($''$) and prime ($'$) are permeate and feed respectively.

In dilute solutions the activity may be replaced by the concentration, or the partial pressure for components in the gas phase. The chemical potential gradient must be negative for the process to be thermodynamically feasible and thus the term $\ln (a''/a')$ must be negative. This requires that $a'' < a'$.

In the case of the removal of specific contaminants from water, the values of a' is fixed and the separation can only be accomplished by providing conditions that lower the value of a''. The maximum driving force that may be attained under such circumstances corresponds to the condition under which the value of a'' approaches zero.

In applications where a membrane is being used to add a chemical substance to the water, the value of a' represents the activity of the substance on the feed side of the membrane and a'' represents its activity in the water phase. In this case the value of a'' is determined by the concentration needed in the water, and the driving force is controlled by increasing the values of a'. Clearly, in these applications, higher concentration gradients may be maintained since the large values of a' may be achieved by using high concentrations of the permeant on one side of the membrane.

Generally, for water treatment the water is present on both sides of the membrane. In this paper processes will be addressed in which the membrane is used to expose the water to a different phase to facilitate the removal of particular contaminants or the transfer of gases.

There are two differences from the other membrane processes that are specific to multiphase separation:

- the driving force is provided by maintaining a concentration gradient across the membrane,
- the membranes are largely impermeable to water.

Definitions

Multiphase membrane operations use membrane contactors for mass transfer of a gaseous component to or from a liquid. Three main operations can be distinguished (Table 1):

Pervaporation (PV): Pervaporation is a liquid/vapour separation process in which a liquid is partially vapourized through a dense membrane. The activity difference is generally maintained by creating a partial vacuum in such a way that it is situated below the vapour pressure of at least one component of the liquid in contact with the upstream side of a dense membrane.

Membrane Stripping (MS): This process differs from PV in that the membrane is porous and that air sweeping is generally used to maintain a partial pressure below the vapour pressure of the volatiles to be removed.

Membrane Absorption (MA): The membrane is used to add a substance to the water. The substance in the gas phase constitues the feed. The membrane can be porous or dense.

There are three different types of membrane that may be used for phase contact processes: dense membranes, hydrophobic porous membranes, composite membranes.

Dense membranes: Dense membranes are made of a solid non-porous polymer such as silicone rubber. Côté et al. (1989) and Hirasa et al. (1991) have used these membranes to transfer oxygen into water and numerous investigators have demonstrated the use of silicone membranes for removal of VOC's (Volatile Organic Compounds) by pervaporation (Psaume et al., 1988), (Néel, 1992). Silicone membranes have the advantage that they can be operated at high gas pressures without bubble formation and are

Table 1. Gas-liquid membranes operations

Feed	Permeate	Membrane Types	Process
water	air	porous	air stripping
water	vacuum	dense	pervaporation
gas (oxygen)	water	porous/dense	adsorption (aeration)

highly selective to many organic compounds relative to water. The ability to use high gas pressures naturally increases the concentration gradient and therefore, the mass transfer rate. However, these membranes suffer from several disadvantages: their thickness and their non-porous nature present a higher resistance to mass transfer than their microporous counterparts, silicone membranes are not available in small-diameter hollow-fibre form and lastly, they are more expensive than microporous membranes.

Hydrophobic porous membranes: Porous membranes are generally represented by a system of intersecting capillaries or microvoids that extend from one surface of the membrane to the other. Porous hydrophobic membranes are made from materials like polypropylene. They have the advantage of providing very high gas permeability, since the pores in the membrane stay dry and gas-filled. Thus, the permeant goes through the pores of the membrane by gas diffusion rather than by dissolving in the membrane itself. The disadvantage is that the transfer is non-selective; mass transfer in these membranes was evaluated by Yang et al. (1986).These membranes can be manufactured as small-diameter tubes called hollow fibres that can provide a very high surface area per unit volume (up to 10,000 m^2/m^3). Unfortunately, because of the porosity of these membranes, their use must be limited to low gas or liquid pressures to avoid bubble formation or liquid transfer by convection. Nevertheless, these membranes have found potential use in waste treatment.

Composite membranes: Composite membranes combine the advantages and selectivity of dense polymers with the faster transport kinetics of porous membranes. Composite membranes can be created by coating the surface of a microporous membrane support with a very thin dense polymer layer. This dense layer physically covers and seals the pores of the membrane so that in gas transfer, wetting is impossible even under the most adverse conditions.

Hollow-fibre contactors: The hollow-fibre membranes are assembled as a bundle into a shell and sealed at both ends of the shell with an epoxy resin (Fig. 2). Hollow-fibre contactors have three major advantages and one significant disadvantage over conventional equipment. The advantages are:

- *high surface area per volume:* Membrane contactors can supply twenty to one hundred times more surface area per volume than conventional equipment (Zhang et al., 1985).

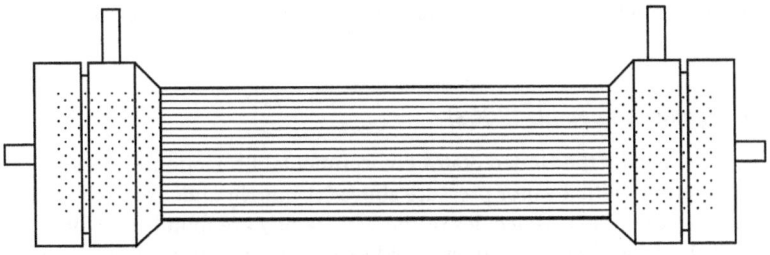

Fig. 2. Schematic representation of a hollow fiber contactor

- *complete loading:* Membrane contactors give this large area even at very small solvent flow rates where conventional packed towers do not continue to give a large liquid-liquid area per volume.
- *no flooding:* Membrane contactors can avoid flooding caused by using solvent and raffinate flow rates beyond the narrow range possible in conventional equipment.

The potential disadvantage of these contactors is also important: slower mass transfer, since both the membrane and the liquid boundary layer may offer a major resistance to mass transfer.

4.2.2
Membrane Desorption

In this paper, only two examples will be discussed: de-aeration (removal of dissolved oxygen) and elimination of VOC's.

De-aeration

Corrosion by oxygen dissolved in water is a very serious problem in industry, e.g. in the power industry where corrosion considerably reduces the life time of boilers. If dissolved oxygen is removed from water before it is fed to a boiler, corrosion can be greatly retarded and the life-time of the boiler will be increased.

Conventionally, de-oxygen resin methods are used to remove dissolved oxygen (Yinkun et al., 1993). The reduction functional groups of these resins can reduce the dissolved oxygen in water. When most of the reduction functional groups are consumed, the resin is regenerated and used for the next cycle of O_2 removal. The de-oxygenated water produced per cycle is 150 times the volume of the resin used. 5 – 10 % of the resin will normally be consumed per year. The most conventional systems have inherent drawbacks in terms of both operating costs and bulky constructions.

De-aeration can also be performed using membrane systems. Studies have been conducted by Zhang et al. (1985), Yang et al. (1986) and more recently by Tai et al. (1994): they provide data on oxygen, carbon dioxide, bromine and ammonia transfer across microporous hollow-fibre membranes. Since these membranes provide a stable and definite interface, there is no restriction on the range of operating flow rates, such as the flooding or loading point usually found in packed towers and any water- and gas-flow ratio can be selected as required. Hollow-fibre modules have been assessed for their capability to reduce dissolved oxygen to the parts per billion (ppb or µg/kg) concentration range in ultrapure water production. The best results obtained with microporous hollow-fibre modules give a final concentration around 8 ppb.

Volatile organic compounds (VOC's)

VOC's are defined by the United States Environmental Protection Agency (USEPA) as stable compounds with a vapor pressure above 0.1 mmHg under normal temperature and pressure conditions. Molecules of the VOC family are found therefore in almost every branch of the chemical industry and belong to the different groups of organic chemistry (alcohols, ketones, aromatics, chlorinated hydrocarbons, etc.). Several environmental problems are associated with the presence of VOC's in groundwater or in effluents and the emission of VOC's in the atmosphere: VOC's are sometimes toxic, carcinogenic, irritating and/or flammable compounds.

Conventionally, VOC's and other gases are removed from water by gas stripping. The industrial equipment most commonly used is the packed column; the compounds extracted from the liquid phase are stripped by a strong countercurrent of stripping gas and the initial concentration of the target compound in the stripping gas must be virtually nil. Most dissolved gases (O_2, CO_2) are only slightly soluble in water and it is the transfer in the liquid phase that determines desorption rates. In the case of highly soluble gases, such as NH_3, desorption is controlled by the gas phase.

Membrane stripping: In recent years, the use of microporous polypropylene hollow fibres for separating volatile organic contaminants from water was tested. In the hollow-fibre modules, as in packed-tower air stripping, volatile compounds are transferred from water to air through intimate contact of the two phases. In hollow fibres, contact occurs at the interface between the air-filled membrane pores and the liquid-filled fibre lumen.

Microporous membranes have been successfully employed to strip a variety of volatile species from water. Zhang et al. (1985) presented data on the separation of hydrogen sulfide and sulfur dioxide across such a membrane and other data are presented by Aurelle and Julien (chapter 2.4, Environmental Technologies and Trends, Springer 1996).

The two main advantages of separation using hollow-fibre contactors are:

- the rates of mass transfer obtained are an order of magnitude greater than the corresponding values for packed-tower aeration (Semmens et al., 1989). This means that removal can be achieved in a smaller reactor volume;
- because the air and water streams are separated by the membrane, they may be varied independently to maximize VOC separation or minimize the cost of off-gas control.

Pervaporation: The first commercial applications of pervaporation of this type concern the extraction of solvents from industrial waste water. Since the organic compound flux is proportional to its concentration, it is obvious that pervaporation alone is not an appropriate way to achieve complete decontamination of polluted water under satisfactory economic conditions. Thus, where very high purity is required, pervaporation can be combined with such conventional technologies as activated carbon adsorption, anaerobic biological treatment, steam stripping or oxidation.

4.2.3
Membrane Absorption

Field of Application

Activated sludge processes necessarily involve a phase in which the water to be purified is brought into contact with bacterial floc in the presence of oxygen (aeration), followed by a phase of separation from this floc. Thus the biodegradable organic matter is consumed by a mass of micro-organisms under aerobic conditions. The micro-organisms require oxygen to satisfy their energy demand, for their production by cellular division (synthesis of living matter) and for their endogenous respiration (auto-oxidation of their cellular mass). To illustrate these various phenomena, glucose may be used as an example of degradation of a totally biodegradable molecule. In the first stage, additional assimilable nitrogen transforms the glucose into cellular protein, whose formula can be represented as $C_5H_7NO_2$. In the second stage, the protein is degraded inside the cell itself to provide the energy required to sustain the cell. These two reactions can be expressed in the following manner.

Synthesis:

$$6\ C_6H_{12}O_6 + 4\ NH_3 \rightarrow 4\ C_5H_7NO_2 + 16\ CO_2 + 28\ H_2O$$

Auto-oxidation or endogenous respiration:

$$4\ C_5H_7NO_2 + 20\ O_2 \rightarrow 20\ CO_2 + 4\ NH_3 + 8\ H_2O$$

These two reactions clearly both occur in a purification plant, but the latter never reaches completion because the necessary retention time of the sludge would require extremely large tanks.

Even though the second reaction is not fully completed, it does occur to a varying degree, depending on the processes used. The higher the degree of completion, the less excess sludge is produced but the more oxygen is consumed. In the above example, the complete oxidation of 6 molecules of glucose required 20 molecules of oxygen.

Conventional Equipment

At the present time, there are two types of aeration system (Table 2): surface aeration (water in air) and volume aeration (air in water). With conventional bubble aeration, there are two disadvantages:

- formation of foam on the waste water,
- venting of gases to the atmosphere.

Aeration with Membranes

The immobilization of cells on solid support materials, such as hollow-fibre membranes, has opened up the possibility of much more stable and productive bioreactors.

The advantages of bubble-free aeration with hollow fibres are:

- no bubbles are formed in the aerator, so 100% of the oxygen supplied is transferred to the water phase: no oxygen is wasted.
- the aerator is highly efficient: careful design can reduce the power requirements of the aerator to very low values.

Table 2. Aeration systems

Surface Aeration (water dispersion in gas phase)	Turbine aerator Brush aerator Aerator with water jet Aearator with (air + water) jet
Volume Aeration (bubbles)	Bubbles Static's aerator Injector (or ejector)

- the membrane aerator does not vent gases to the atmosphere as is the case with conventional bubble aeration. It would be an effective means of aerating waters that are rich in volatile organic contaminants. Use of the membrane aerators can reduce volatile organic compound emissions without the need for a covered aeration tank.
- the aerator can be used to effectively supersaturate waters with oxygen. This, however, does not eliminate the mixing requirement for the biomass.

Recently a bubble-free hollow-fibre membrane aerator capable of high oxygen-transfer efficiency was developed by Semmens (1991) and Ahmed et al. (1992). The fibres are potted only at one end and this end is connected to a pressure-regulated oxygen supply tank (Fig. 3). The other end of each fibre is individually end-sealed. The module contains individual sealed hollow fibres that are free to move independently in turbulent pipe flow. Pure oxygen is supplied to the inside of the fibres and transfer occurs to the water outside the fibres without bubble formation.

Pankhania et al. (1994) studied a hollow-fibre bioreactor for waste water treatment. The aim of this study was to test the feasibility of immobilizing micro-organisms on the outside of the fibres of such an aerator and apply this to the treatment of waste water. The bioreactor was operated continuously for 164 days in five distinct phases; at two different nominal COD loadings with no backwash and after day 75, at three different nominal COD loadings with backwash (Table 3).

Results from these phases are based on operating periods of 5 successive days or more, representing a minimum Hydraulic Retention Time (HRT) of 120 h. The hollow-fibre bioreactor for treating waste water was successful in removing 86% of the COD at a high volumetric loading of 8.94 kg/m³/d and a short HRT of 36 min. The need for a daily backwash was caused by deteriorating reactor performance, as excess biomass caused channelling.

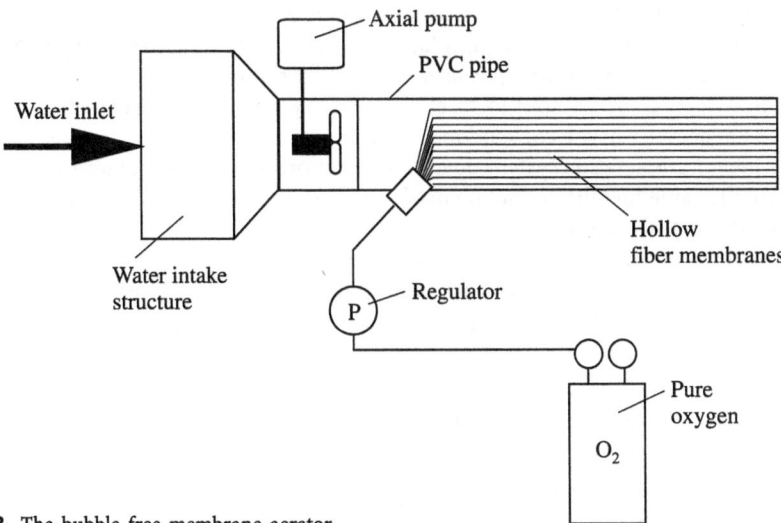

Fig. 3. The bubble free membrane aerator

Table 3. Operating conditions for hollow fiber bioreactor during 4 different organic loading periods. After Pankhania (1994)

Loading	1	2	3	4	5
Operating condition	No Backwash	No Backwash	Backwash	Backwash	Backwash
Days of operation	2–8	60–70	91–99	108–112	160–164
Total N (mg/1)	32	50	44	51	65
COD (mg/1)	185	286	253	294	369
Volumetric COD loading (kg/m³/d)	4.49	6.99	6.33	7.26	8.94

Table 4. Overall oxygen transfer coefficient. After Hirasa (1991)

Treatment process	$k_l a$ (h⁻¹)	$(C_s - C)$ (mg/l)	R_r (mg/l/d)
Trickling filtration	4.1	5.84	574
Standard activated sludge	2.1	7.76	391
Deep shaft activated sludge	4.0	10.20	979
Oxidation ditch	3.0	5.84	420
Rotary biological contactor	6.0	5.84	3644
Contact aeration	6.0	7.76	1117
Oxygen activated sludge	2.1	29.70	1496
Hollow fibre system	11.4	9.60	2626

Hirasa et al. (1991) worked on the oxygen transfer across silicone hollow-fibre membranes and compared this with other waste-water treatment technologies (Table 4). This table shows that the overall oxygen transfer coefficient K_l^a and the oxygen utilization rate (R_r) are greater with a hollow-fibre system.

The bubble-free membrane aerator has potential uses for treatment of troublesome waste waters, avoiding foaming, or preventing the stripping of volatile organic compounds. Also, it could be used to meet the high oxygenation requirements of membrane bioreactors, where a conventional aeration system would be inadequate. Finally, with a membrane aeration system, operation under pressure would not pose a problem because the same membrane can be used to deliver the oxygen and evacuate carbon dioxide (Côté et al., 1989).

In the case of aeration, today only blood oxygenation (with extracorporeal oxygenators) represents a real commercial application. In fact membrane systems have a higher cost than conventional equipment. One way of improving the economics of the membrane aeration process is to increase the mass transfer.

Use of Secondary Flows in Gas-Liquid Contactors

A severe limitation to applications of membrane processes is the occurrence of concentration polarization. The membrane's performance may be limited less by its own mass-transfer resistance than by the resistances of the adjacent fluids. This means that successful membrane module design must consider not only membrane chemistry, but also module geometry. Secondary flows can make an important contribution here (Moulin et al., 1994).

Mass-Transfer Resistance

The flux across the membrane is necessarily equal to the flux across the membrane itself and across each of the fluid boundary layers. The mass-transfer model for membrane aeration is shown in Fig. 4.

Thus the oxygen flux J in the aeration process is given by:

$$J = K (C_G - C_L) = (p/H - C_L) \tag{5}$$

K = overall mass transfer coefficient (m/s)
C_G and C_L = gas and liquid concentration (mol/m^3)
p = gas partial pressure (Pa)
H = Henry's law coefficient (Pa m^3/mol)

In the case of membrane adsorption, the mass transport involves five steps:

(i)   diffusion from the gas phase to the surface of the membrane,
(ii) gas sorption into membrane on the gas side,
(iii) selective transport through the membrane,
(iv) desorption on the liquid side,
(v)   diffusion from the surface of the membrane to the liquid phase.

The total resistance to mass transfer, 1/K, is given by:

$$1/K = 1/K_L + 1/K_M + 1/K_G \tag{6}$$

where K_L, K_M, and K_G are the individual mass-transfer coefficients in the liquid, in the membrane and in the gas, respectively.

It is possible to tell which resistance is important by varying the flow rates of gas and liquid. It is found that the influence of the boundary layer on the gas side is negligible. Depending on the permeation rate through the membrane and the mass transfer coefficient in the water, the flux is mainly controlled either by the membrane or by the boundary layer on the liquid side.

The membrane mass-transfer coefficient can be expressed as:

$$K_M = PH/\tau_e \tag{7}$$

in terms of the equivalent thickness τ_e (m) and the permeability P (mol m^{-1} s^{-1} Pa^{-1}) of the membrane material.

The mass-transfer coefficient in the liquid can be estimated using dimensionless correlations. Such correlations are available for a variety of configurations and for different ranges of operation and it is important to select a correlation that matches as closely as possible the conditions to be modelled. The correlations typically express the Sherwood number (Sh) as a function of the Reynolds number (Re) and the Schmidt number (Sc). These correlations generally take the following form:

$$Sh = a \, Re^x \, Sc^y \, (d_e/L)^z \tag{8}$$

Fig. 4. Mass transfer model for membrane
aeration

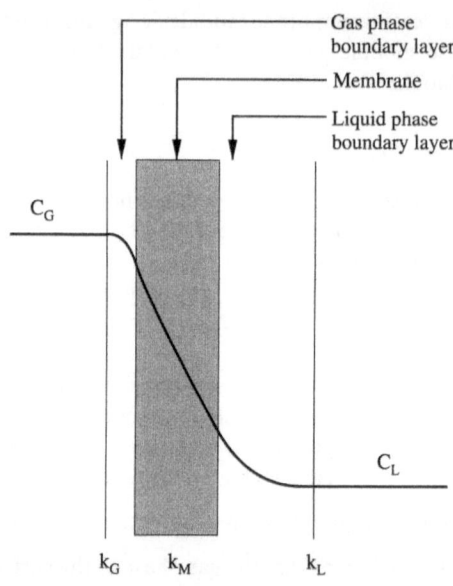

Reynolds number: $R_e = v\, d_e/v$ (–)
Sherwood number: $Sh = K_L d_e/D$ (–)
Schmidt number: $Sc = v/D$ (–)

with d_e the equivalent diameter of the channel (m) and L the channel length (m), D
the diffusion coefficient (m²/s), v the kinematic viscosity (m²/s) and v the velocity
(m/s).

The values for a, x, y and z are dependent on the operating conditions and the design
of the membrane contactor and different values can be found from the literature.
Several semi-empirical relationships have been developed to determine the mass
transfer coefficients across the fluid boundary layers next to the membrane. The
principal correlations are as follows.

Hollow-fibre or tubular contactors

(Lévêque, 1928), (Psaume et al., 1988)
   $Sh = 1.62\ (Re\ Sc\ d_e/L)^{0.33}$ Flow inside the tubes (Re < 2000)

(Yang et al., 1986)
   $Sh = 1.25\ (Re\ d_e/L)^{0.93}\ Sc^{0.33}$ Flow outside and parallel to the fibres (Re: 5 – 3500)
(Yang et al., 1986)
   $Sh = 0.90\ Re^{0.44}\ Sc^{0.33}$ Flow outside the fibres and cross flow (Re: 0 – 50)
                                                  External void fraction 0.93

Spiral-wound contactors

(Schock et al., 1987)
   $Sh = 0.065\ Re^{0.875}\ Sc^{0.25}$ (Re < 1000)

These correlations were developed to describe the fluid-film boundary-layer resistances in membrane contactors (Wickramasinghe, 1992). For the hollow-fibre modules, different correlations exist for flow inside and outside the fibres. In addition, for flow outside the fibres, different correlations exist for fluid flow parallel to the fibre axis and for flow patterns that cross the fibres, either normally or at some angle to the fibre axis.

Flow Disturbances

Concentration polarization is the reversible build-up of a non-interacting solute in the solution near the membrane-liquid interface. It is an important factor that limits the performance of membrane contactors (Psaume et al., 1988). There are different approaches to reducing the liquid-boundary resistance. These include designing membrane surfaces with organized roughness, pulsation of axial and lateral flow and the use of curvilinear flow under conditions that promote instabilities or vortices. Each of these methods is illustrated in Fig. 5 (Belfort, 1994).

Placing protuberances or corrugations directly onto the membrane surface at defined separation distances induces periodically disturbed flow in the mass-transfer boundary layer (Fig. 5a). A second approach is to replace the flat uniform membrane profile with a well-defined rough surface such as a furrowed profile (Fig. 5b). The third approach consists in the inclusion of protuberances at some defined distance away from the membrane surface in the bulk flow (Fig. 5c). It has been known for some time in the fluid-mechanics literature that superimposing an oscillating pressure gradient onto bulk axial flow in a tube produces a velocity profile with two equal maxima closer to the wall than the centre-line (Fig. 5d). Also the inner cylinder of two concentric cylinders can be rotated at a rotational Reynolds number, called a Taylor number, above the critical value for production of vortices (Fig. 5e). In order to overcome the limitations associated with rotating Taylor-vortex filter devices, Winzeler and Belfort (1993) (also Winzeler, 1990; Chung, 1993) have suggested using Dean vortices rather than Taylor vortices (Fig. 5f).

The present work deals with this last method of inducing instabilities. A new module configuration (helical module), designed to decrease the liquid-film resistance, was evaluated experimentally and compared with the conventional inside-flow configuration.

Secondary Flow in Curved Pipes

The secondary flows which appear in a curved pipe are caused by centrifugal force: the flow pattern is very complex. Dean (1927) first studied the flow in a curved pipe using a concentric toroidal co-ordinate system.

In the case of the laminar flow of a fluid through a cylindrical tube, the axis of which forms a helix of small pitch, the primary (axial) flow field is accompanied by a secondary flow field which acts in a plane perpendicular to the tube axis and which is symmetrical about the plane of curvature of the tube. Dean was the first to show that the parameter, De, now known as the Dean number, is the dynamic parameter governing fluid motion in such a coil:

$$De = Re \sqrt{(D/D_S)} \qquad (9)$$

with D the internal diameter of the tube (m) and D_S (= 2R) the coil diameter.

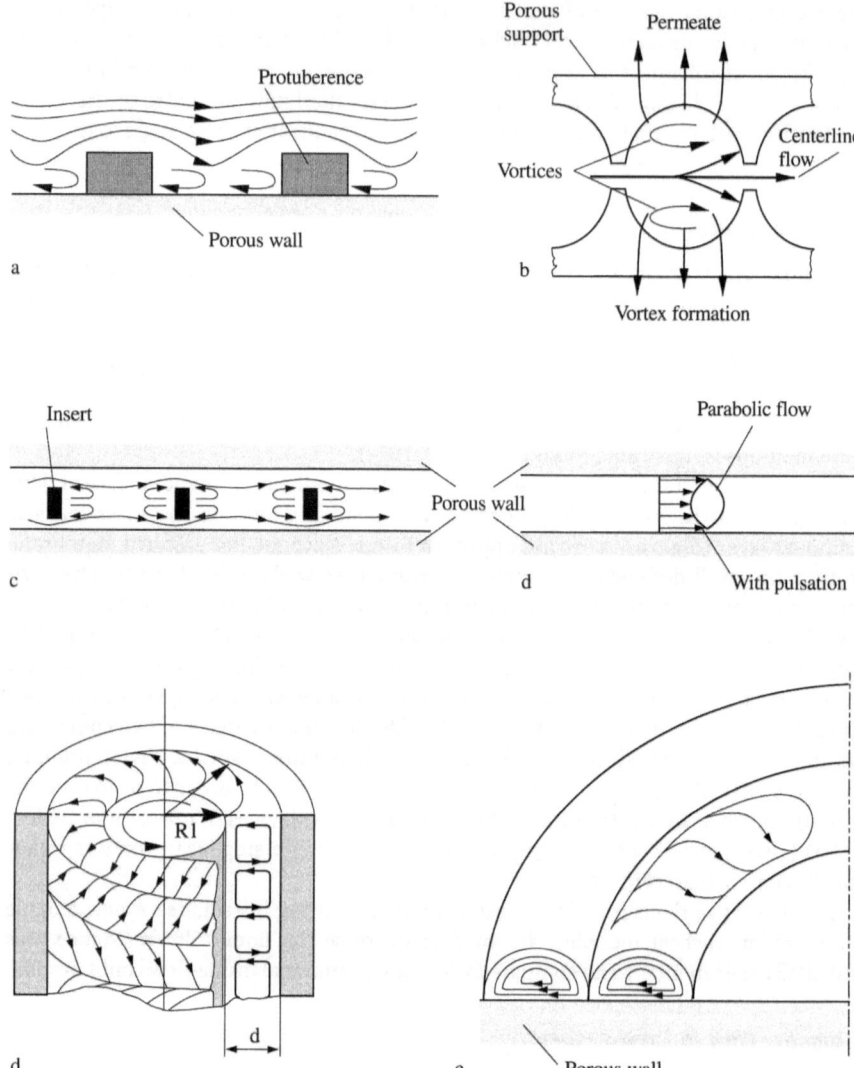

Fig. 5. Methods to induce instabilities include. **a** placing protuberances on the membrane, **b** using a corrugated membrane surface, **c** placing inserts within the flow channel, **d** superimposing pulsations on the axial or permeate flows, **e** rotating a cylinder within a stationary outer cylinder to form Taylor vortices, **f** flow in a curved channel to produce Dean vortices

The effect of the torsion on the secondary flow was discussed by Murata et al.(1981), Wang (1981), Germano (1989) and Kao (1987) and they obtained, using different orthogonal (Germano and Kao) and non-orthogonal (Murata and Wang) systems of co-ordinates, solutions for the steady flow in a helical pipe with constant curvature κ and torsion τ. These studies show that in the Dean equations extended to the case of a helical pipe flow, secondary flows depend not only on the Dean number but also on other parameters such as the characteristic numbers $\tau, \kappa, \lambda, \varepsilon$:

the curvature: $\kappa = R/(R^2 + p^2)$ (m^{-1})
the torsion: $\quad \tau = p/(R^2 + p^2)$ (m^{-1})

$\lambda = \tau/\kappa$
$\varepsilon = \kappa R$

where $2\pi p$ is the pitch of the helix (m).

Many workers have shown that secondary flow increases heat and mass transfer. Mori and Nakayama (1965, 1967) have shown that at high Dean numbers, heat-transfer coefficients in the thermally fully developed region are higher in a coiled tube than in a straight tube by a factor which varies with $\sqrt{De}$. Dorson et al. (1968) worked on a helical-tube prototype blood oxygenator to decrease the blood-channel thickness. Winzeler and Belfort (1993) worked on fluid instabilities and ways to induce them.

Mass Transfer in Gas Liquid Contactors

In this work silicone-rubber hollow fibres were used because they are dense membranes with a high oxygen permeability. With a low transfer resistance across the membrane, K_L becomes the dominant component of K and it was possible to study the influence of secondary flow on K_L.

In a closed-loop set-up containing the membrane module and a well stirred tank, a mass balance gives the following equation (Moulin et al., 1994):

$$\ln\left[\frac{C_L^* - C_{L0}}{C_L^* - C_L}\right] = \frac{Q}{V}\left[1 - \exp\left(\frac{K\,a\,L}{v}\right)\right]t \tag{10}$$

with

a $\quad$ = surface-area to volume ratio (m^{-1})
C_L $\quad$ = water-phase oxygen concentration (mg/l)
C_L^* $\quad$ = water-phase oxygen concentration in equilibrium with the gas phase (mg/l)
C_{L0} = initial water phase oxygen concentration (mg/l)
L $\quad$ = fibre length (m)
Q $\quad$ = volumetric flow rate of liquid (m^3/s)
t $\quad$ = time (s)
v $\quad$ = velocity of water (m/s)
V $\quad$ = tank volume (m^3).

Thus, the overall mass transfer coefficient can be found by plotting Ln $((C_L^* - C_{L0}))/(C_L^* - C_L))$ vs. t and measuring the slope of the line.

The membrane transfer coefficient given by Equation (7) was measured independently and the liquid film resistance was obtained by difference.

Experimental

Hollow-fibre modules were made by potting the desired number of fibres into an external shell. The hollow fibres are dense silicone rubber (Laboratorio Moderna,

Spain) and they have an internal diameter ranging from 1.0 to 3.2 mm. Two types of modules were built: straight and coiled modules. Characteristics of the hollow-fibre modules are reported in Table 5. The void fraction varied from 73 to 92%. Fibres were potted in a plug (epoxy-silicone-epoxy) for a good seal and good mechanical strength. For coiled modules, the fibres were wound around a perforated tube ($\varnothing = 2.0$ cm) with each helical fibre in contact with its neighbours. The fibres for straight modules have a total length of 0.70 m, with 0.60 m available for oxygen transfer. Straight modules and coiled modules have an equivalent surface-area to volume ratio. The liquid flows inside the fibres and the gas flows outside.

The system consisting of the module, the pipes and the tank, was completely filled with water, previously deoxygenated by adding sodium sulfite (Na_2SO_3) and a trace of cobalt chloride as catalyst. The flow rate of cold water through the heat exchanger was adjusted to maintain a tank water temperature of 25 °C. When dissolved oxygen (DO) in the tank started to increase, readings of DO and time were recorded every 10 s by a computer. The operating oxygen pressure is 200 mbar above atmospheric pressure at the inlet of the module.

4.2.4
Results and Discussion

To normalize the data and examine the dependence of the oxygen transfer coefficient on the secondary flows, correlations are conveniently expressed in terms of the dimensionless groups (Sh, Sc, Re, De). In this work, the Schmidt number was not varied, so the 0.33 power dependence in the literature was assumed.

Straight modules: Straight modules were taken as reference modules. The Sherwood numbers found from these experiments are plotted in Fig. 6 vs. Reynolds number. Results are consistent with the Lévêque correlation shown as the solid line in Fig. 6.

Coiled modules: The first observation made is that for the same Reynolds number, the Sherwood numbers for the coiled modules are two to four times greater than for the straight ones, as shown in Fig. 7. The mass transfer coefficients calculated from all experiments fall in a range from $3.05 \cdot 10^{-4}$ to $10.3 \cdot 10^{-4}$ m/s. Straight and coiled modu-

Table 5. Characteristics of straight (S) and coiled (C) hollow fibre modules

Module reference	d_i-d_o (mm)	Lenght (m)	Number of fibres	Surface area (m^2) $\cdot$ 10^2	Permeability $\cdot$ 10^{13} (mole s^{-1} Pa^{-1} m^{-1})
1S	3.2–4.8	0.60	15	9.05	2.0
1C	3.2–4.8	0.90	10	9.05	2.0
2S	2.4–4.0	0.60	18	8.14	2.0
2C	2.4–4.0	0.90	12	8.14	2.0
3S	1.6–3.2	0.60	21	6.33	2.1
3C	1.6–3.2	0.70	18	6.33	2.1
4S	1.0–3.0	0.60	25	4.71	2.2
4C	1.0–3.0	0.74	20	4.71	2.2
5S	1.0–1.5	0.60	46	8.67	2.0
5C	1.0–1.5	1.14	24	8.67	2.0

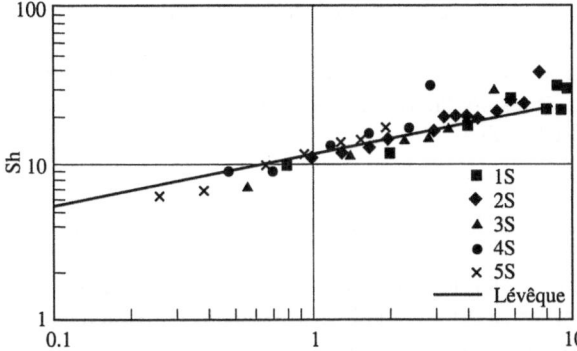

Fig. 6. Mass transfer for straight modules

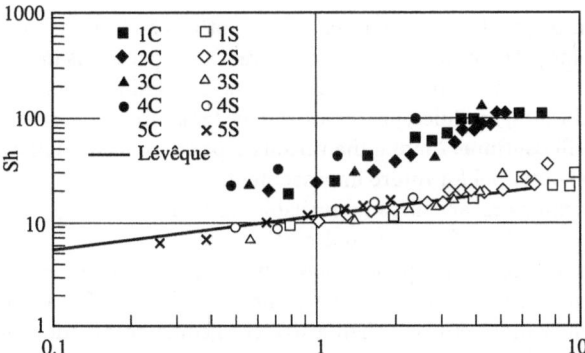

Fig. 7. Comparison between coiled and straight modules

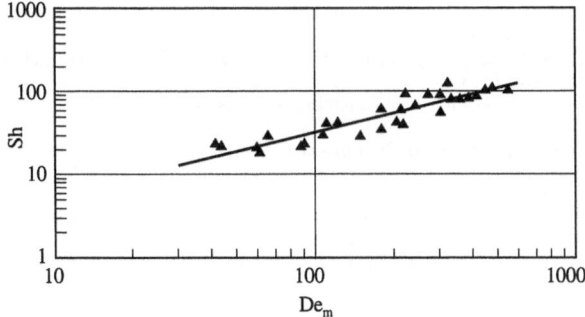

Fig. 8. The observed variation in Sherwood number with modified Dean number for oxygen transfer in coiled modules

les have the same K_M, and so a decrease of $1/K_L$ explains the greater Sherwood number for coiled modules.

In fact the curvature radius is modified by the torsion of the helix and this difference may influence results. So to obtain a correlation we use the effective coil diameter, which includes the pitch

$$Ds' = Ds \{1 + (p/\pi D_s)^2\} \tag{11}$$

This effective diameter is used to define a modified Dean number De_m ($De_m = Re \sqrt{(D/D_\S)}$).

Figure 7 shows, on a log-log plot, the Sherwood number as a function of the modified Dean number for coiled modules. A least-squares regression of the data gives the correlation shown as a solid line in Fig. 8:

$$Sh = 0.143 \, De_m^{0.75} \, Sc^{0.33} \tag{12}$$

4.2.5
Conclusion

Membrane technology offers many possibilities both as a clean technology and as a cleaning technology. The rapid evolution of membrane operations in water treatment during the last three decades has been accompanied by significant technological innovations in membranes, modules, processes and systems. Today, there is evidence that this evolution will continue. Among the various separation technologies, membranes have the greatest potentiel for future improvement.

Membrane contactors are already available, although no large-scale plants are onstream yet. Improving efficiency and cutting costs is a constant goal for all researchers and developers. This paper shows that a simple change in the design of a membrane hollow-fibre contactor can largely improve mass transfer: a coiled configuration is 200 to 400% more efficient than a straight geometry. An additional challenge will be to manufacture reliable and cheap contactors based on this principle.

4.2.6
References

Articles in Journals:

Ahmed T, Semmens MJ (1992) Use of sealed end hollow fibers for bubbleless membrane aeration: experimental studies. J Membrane Sci 69:1–10

Chung KY, Bates R, Belfort G (1993) Dean vortices with wall fllux in a curved channel membrane system. J Membrane Sci 81:38

Côté P (1989) Bubble-free aeration using membranes: mass transfer analysis. J Membrane Sci 47: 91–106

Dean WR (1927) Note on the motion of fluid in a curved pipe. Phil Mag 4(7):208–223

Dorson W, Baker E, Hull H (1968) A shell and tube oxygenator. Trans Amer Soc Artif Int Organs 15: 242–249

Germano M (1989) The Dean equations extended to a helical pipe flow. J Fluid Mech 203:289–305

Julien R and Aurelle Y (1996) Pervaporation and Membrane Stripping. 2.4. (this book)

Hirasa O, Ichio H, Yamauchi A (1991) Oxygen transfer from silicone hollow fiber membrane to water. J Ferm Bio 71(3):206–207

Kao HC. Torsion effect on fully developed flow in a helical pipe. J Fluid Mech 184:335–356

Lévêque MA (1928) Les lois de la transmission de chaleur par convection. Ann. Mines 13:201

Mori Y, Nakayama W (1965) Study on forced convective heat transfer in curved pipes. Int J Heat Mass Transfer 8:67–82

Mori Y, Nakayama W (1967) Study on forced convective heat transfer in curved pipes. Int J Heat Mass Transfer 10:37–58

Murata S, Miyake Y, Inaba T (1981) Laminar flow in a helically coiled pipe. Bulletin ISME 24:355–362

Pankhania M, Stephenson T, Semmens MJ (1994) Hollow-fibre bioreactor for wastewater treatment using bubbleless membrane aeration. Wat Res 28(10):2233–2236

Psaume R, Aptel P, Aurelle Y, Mora JC, Bersillon JL (1988) Pervaporation: importance of concentration polarization in the extraction of trace organics from water. J Membrane Sci 36:373–384

Schock G, Miquel A (1987) Mass transfer and pressure loss in spiral-wound modules. Desalination 64: 338

Semmens MJ, Qin R, Zander A (1989) Volatile organics separation from water using a microporous hollow fiber membrane. J AWWA:162–177

Semmens MJ (1991) Bubbleless aeration. Water Eng & Man 4:8–19

Tai MSL, Chua I, Li K, Ng WJ, Teo WK (1994) Removal of dissolved oxygen in ultrapure water production using microporous membrane modules. J Membrane Sci 87:99–105

Wang CY (1981) On the Reynolds-number flow in a helical pipe. J Fluid Mech 108:185–194

Winzeler HB, Belfort G (1993) Enhanced performance for pressure-driven membrane processes: the argument for fluid instabilities. J Membrane Sci 80:35–47

Winzeler HB (1990) Membran-filtration mit hoher Trennleistung und minimalen Energiebedarf. Chimia 44(9):288

Yang MC, Cussler EL (1986) Designing hollow fiber contactors. AIChE J 32 (11):1910–1916

Yang MC, Cussler EL (1989) Artificial gills. J Membrane Sci 42:273–284

Yinkun H (1993) Removal of dissolved oxygen from feed water by deoxygen resin for industrial boiler. Water Treatment 8:55–64

Zhang Q, Cussler EL (1985) Hollow fiber gas membranes. AIChE J 31(9):1548–1553

Zhang Q, Cussler EL (1985) Microporous hollow fibers for gas absorption. Mass transfer in the liquid. J Membrane Sci 23:321–332

Zhang Q, Cussler EL (1985) Microporous hollow fibers for gas absorption. Mass transfer across the membrane. J Membrane Sci 23:333–345

Zhu CL, Yuang CW, Fried JR, Greenberg DB (1983) Pervaporation membranes – a novel separation technique for trace organics. Environmental Progress 2 (2):132–143

Thesis:

Wickramasinghe SR (1992) The best hollow-fibre module. University of Minnesota, USA pp 9-30

Proceedings:

Anselme C, Mandra V, Baudin I, Mallevialle J (1993) Optimum use of membrane processes in drinking water treatment. In Proceedings of AIDE Symposium, Budapest, Hungary

Belfort G (1994) Fouling reduction through fluid mechanics and module design. In Proceedings of Seminar on Fouling in Pressure-Driven Membrane Processes. Lappeenranta, Finland

Moulin P, Rouch JC, Serra C, Aptel P (1994) Mass transfer improvement by secondary flows in gaz liquid contactors. In Proceedings of XIth Annual Summer School ESMST, Glasgow, United Kingdom. September 1994

Néel J (1992) Current Trends in Pervaporation. In Proceedings of the CEE- Brazil Workshop on membranes separation processes, Rio de Janeiro, Brasil

Book with Editors:

Aptel P. Membrane pressure driven processes in water treatment. In Membrane Processes in Separation and Purification, JG Crespo & KW Boddeker (ed), NATO ASI Series E, vol 272. Kluwer Academic Publishers. 1994, pp 263-282. ISBN 0-7923-2929-5

4.2.7
Further Readings

Title: **Hollow fibre bioreactor for wastewater treatment using bubbleless membrane aeration**
Author: Pankhania M; Stephenson T; Semmens MJ
Corporate Source: Cranfield Univ, Cranfield, Engl
Source: Water Research v 28 n 10 Oct 1994. p 2233 – 2236
CODEN: WATRAG ISSN: 0043-1354

Abstract: A laboratory scale hollow fibre membrane bioreactor which used bubbleless membrane aeration was tested over 5 months for its ability to treat synthetic sewage. A biofilm was grown on the surface of 280 µm diameter, gas permeable, sealed-end, polypropylene based fibres. The fibre lumens were pressurized with pure oxygen and synthetic sewage was pumped over the surface of the biofilm. At the highest volumetric COD (chemical oxygen demand) loading of 8.94 kg/m**3/d and a hydraulic retention time (HRT) of 36 min, 86% COD removal was achieved. A daily backwash was necessary to prevent channelling. The oxygen transfer efficiency was 100%. (Author abstract) 11 Refs.

Title: Novel closed loop air stripping process
Author: Bhowmick Madhumita; Semmens Michael J
Corporate Source: West Virginia Univ, Morgantown, WV, USA
Conference Title: Proceedings of the 25th Mid-Atlantic Industrial Waste Conference
Conference Location: College Park, MD, USA
Conference Sponsor: Bucknell University; University of Cincinnati; University
of Delaware; Drexel University; Howard University; et al.
Source: Hazardous and Industrial Wastes – Proceedings of the Mid-Atlantic
Industrial Waste Conference 1993. Publ by Technomic Publ Co Inc, Lancaster, PA,
USA. p 136–145
CODEN: HIWAEB ISSN: 1044-0631 ISBN: 1-56676-067-4
Abstract: The objective of this study was to evaluate a closed loop air stripping process that will remove the volatile organic compounds (VOCs) from water without polluting the air. The approach adopted in this study combined the use of a microporous hollow fiber membrane to separate the VOCs from water with the direct air phase photooxidation of the stripped organic compounds in the presence of ultraviolet light. The VOC-lean air was then circulated back to the membrane module for the extraction of additional volatile organics. Experiments were conducted under different operating conditions and the results were compared with those obtained from a mathematical model. (Author abstract) 7 Refs.

4.3
The Biological Treatment of High Effluent Flowrates: A Review of the Hydrodynamic Conditions and Possibilities

CHRISTIAN FONADE

Abstract

The performances of a biological treatment system for wastewater depends on both the biological reaction and on the physico-chemical micro-environment of the micoorganisms. The role of hydrodynamics is analyzed, relative to the overall behaviour of the reactor and mass transfers. A review of the different technologies which can be used to insure good mixing and mass transfer is presented. Cylindrical reactors and lagoons used for anaerobic and aerobic processes are considered.

4.3.1
Introduction

Biological treatment of organic industrial wastewaters is a currently used methodology, obviously when most of the effluent components are biodegradable. This treatment was achieved in external basins or lagoons. Lagooning was initially considered an extensive method of biological treatment, the atmosphereic oxygen supplied by a natural exchange through the free surface of the liquid (Valiron, 1985; Forster, 1985). The advantages of this process are the technological simplicity and robustness, the non consumption of energy and the possibility for the effluent to show some variations in both the flowrate and the effluent's concentration. However, very large volumes and residence times are needed to achieve the degree of biodegradation required to meet the new regulations before the waters can be directed into the rivers. More recently, new methodologies were proposed to improve the performances and kinetics of the degradation process, based either on more efficient biological reaction or on a dumping of this reaction through an external supply of energy that would improve upon the physical limitations.

What type of industrial treatment process must be used? The answer depends on the biological reaction which can be observed and on the engineering conditions. Obviously, the effluent must first be seen through its chemical composition and its characteristics of biodegradability, and especially the ratio between the BOD (biological oxygen demand) and the COD (chemical oxygen demand): these parameters are the first arguments which present a choice between an aerobic or an anaerobic treatment process. Table 1 shows, for example, the main characteristics of the effluents in the paper industry.

But engineering conditions or constraints must also be considered. Performances, economy, and the industrial achievement of the process must be looked at. This is all

Table 1. Main characteristics of some paper industry wastewaters (from Rols et al., 1994))

Type of process	Ammonium sulphite	Kraft	Recycled fibers
Product	Fluff pulp	Kraft paper	Fluting paper
Production (t/d)	400	1200	420
Effluent flowrate (m³/d)	24 000	60 000	4300
COD load (t/d)	80	25	20
BOD load (t/d)	27	10	7
Main constituents	lignosulphonate	lignine	starch
	resinic acids	hemicellulose	cellulose
	hemicellulose volatile fatty acids	methanol	volatile fatty acids

the more necessary since the flowrate of the effluent to be treated is high because the investment or daily costs have greatly increased with the volumes of the reactors.

4.3.2
General Analysis of the Hydrodynamic Conditions

It is well known that the performances of a biological reaction result in the degree of adequation between its own kinetics and the physical environment of the microorganisms. When laboratory studies are carried out, attention focuses primarily on the biological aspects of the reaction, assuming that all the cells have the same external conditions. The reactor is expected to show an ideal hydrodynamic behaviour. Such technology is available and the required energy remains low enough to be a negligible parameter of the study. However, such ideal conditions cannot be generally realized at industrial scale, either because the technological devices are not available or efficient enough, or because the required energy will lead to unacceptable costs.

This overall behaviour of the reactor, generally specifed as uniform, plug-flow or more complex combinations, is the result of what happens in the broth in the close vicinity of each cell. Ideally, the involved transfer phenomena must be observed according to the fluid particle length scale and then integrated over the whole volume of the broth. Viewing the reactor as a "black box" and using overall balance relations only allow the prediction of what transfer performances will be required if the biological reaction is ideally achieved, as for example, the total oxygen demand of an aerobic process. In fact, all the physico-chemical parameters, such as the concentrations of the various components or the dissolved oxygen, must be considered through their local value and then through their spatial fields, as they can be deduced from the computation of the fundamental equations for momentum or mass transfer:

momentum: $$\frac{\partial}{\partial t} (\varrho \vec{V}) + \vec{\nabla} \cdot (\varrho \, \vec{V} \, \vec{V}) = \varrho \, \vec{F} + \vec{\nabla} \cdot \bar{\bar{t}}$$

concentration: $\quad \dfrac{\partial}{\partial t}(\varrho \, c_\alpha) + \vec{\nabla} \cdot (\varrho \, c_\alpha \, \vec{V}) = -\vec{\nabla} \cdot \vec{j}_\alpha + \varrho \, P_\alpha$

where $\vec{V}$ is the velocity, $\overset{=}{t}$ the stress tensor, c_α the concentration of one component and $\vec{j}_\alpha$ the diffusive flux per unit area of this component.

These equations show that the transport of any scalar, vectorial or tensorial quantity results from the balance between three main phenomena: the convection, the diffusion and the production. The hydrodynamic characteristics of the flow are critical, acting both through the average velocity field for the convective transport and through the turbulence level for the diffusion. Furthermore, some energy is always required to force the movement of the fluid, and any study about the influence of the hydrodynamic behaviour must also involve the technology and cost to achieve it.

These aspects become important when high wastewater flowrates must be treated. In this case, the future of the cleaned water is essential data because the degree of treatment needed and the type of overall behavior of the reactor are dependent upon it. Fig. 1 shows schematically that, after the treatment, the water can be either directed into the river or recycled to be reused in the process. This last situation is encountered more often in industry because a lower degree of treatment can often be accepted as compared to the new water management regulations which must be applied in many countries. This is so in the case of the sugar industry, where flowrates of about 1000 m³/h are necessary to first wash the beets.

If a fed-batch operation of the reactor is chosen, the rejection will happen only when the required degree of treatment is achieved. Such a solution can be used when wastewaters are produced only during a part of the year, and generally needs to fill several successive reactors. The present trend is to use a continuous operation of the reactor, a solution which can only be practically applied when the cleaned water is recycled.

In this case, the scaling of the treatment unit mainly depends on the value of the needed hydraulic residence time T_s of the effluent in the reactor: associated to the flowrate, this value leads to the required volume. Figure 2 (Malina and Pohland, 1992) shows the range of the commonly required hydraulic residence times, as a function of the organic load of the effluent and according to the type of biological process used in the wastewater treatment. They can vary from a few hours to a few days for an aerobic treatment, to 20 or 30 days for an anaerobic digestion.

These main engineering characteristics mentioned above, the hydraulic residence time, the volume and the optimal ideal hydrodynamic behaviour of the reactor, are deduced from the fundamental kinetics of biodegradation of the pollutant and the required degradation rate. They can be obtained by preliminary experimentations, carried out with effluents and microrganism types and concentrations close to the conditions which will be encountered in the industrial reactor. However, other parameters, such as the solid residence time, the range of the organic load variations in the effluent or a casual presence of products toxic for the microorganisms, are also significant parameters which must be involved in the choice of the hydrodynamic behaviour of the reactor. At this step, other characteristics, as the gas flowrate generated during an anaerobic process or the oxygen demand in an aerobic one, can be calculated. The solid residence time will interfere when the biomass is able to leave

Fig. 1. General scheme of the water circuit

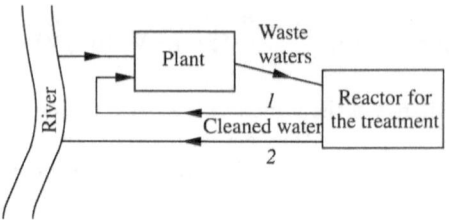

Fig. 2. Ranges of the hydraulic residence time (from Malina and Pohland, 1992)

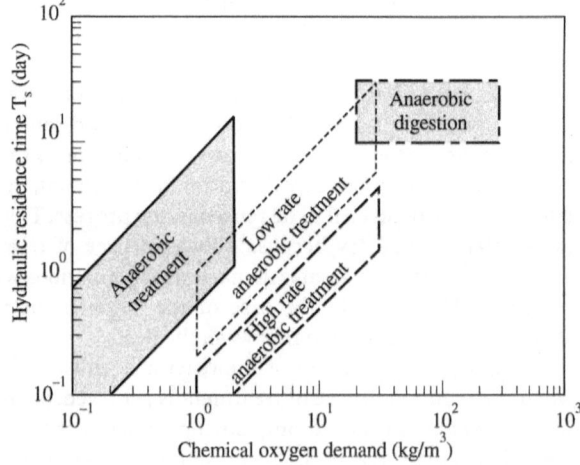

the reactor with the output flow. This residence time is directly connected to the biomass growth rate m: its minimum required value is equal to $(1/\mu_{max})$. With this condition, the washing of the biomass can be prevented.

When these ideal and overall hydrodynamic and transfer conditions are known, the next step of the study is to design the reactors and devices able to achieve them. Two questions then arise: do we know available methodologies and devices to do that, and what will be the investment and daily costs for operation and maintenance? The answer depends on the type of the involved biological reaction because the two kinds of properties of the flow, its average structure (convection) and its turbulence level (diffusion), can be either advantages or disadvantages. For example, high turbulent intensities are known to efficiently improve the transfer rates, but they are associated with high shear stresses, which is a drawback if agregates or flocs are used to fix the biomass.

As far as the structure of the flow is concerned, it is obvious that the geometry and type of the reactor will be of major influence. From this point of view, two configurations have to be distinguished: the commonly used cylindrical reactor and the large basin or lagoon. For wastewaters treatment, such configurations are generally utilized respectively in the case of an anaerobic and an aerobic biological reaction.

4.3.3
Anaerobic Treatment: the Hydrodynamics in a Cylindrical Reactor

The conversion of organic substrates to methane is a very complex anaerobic process. This complexity comes from the two main succesive steps which occur during this conversion: the acidogenesis and the methanogenesis. Several microbial populations are involved in the process and intermediate products, such as hydrogen, can be inhibitors for the following reactions. Due to these possible interactive effects, and because the used microorganisms are generally very sensitive to the temperature, such an anaerobic process can exhibit instabilities under poor physical conditions.

Because of the complexity of the biological reactions, the sizing of the reactor is generally issued from the volumetric organic loading rate B which is deduced from preliminary experimentations. It represents the biodegradation capacity per unit volume of the reactor and is expressed as kg COD/m^3d. The volume of the reactor and the hydraulic residence time are then obtained from the values of the effluent flowrate Q and COD concentration C_O:

$$V = \frac{Q\,C_O}{B} \quad \text{and} \quad T_s = \frac{V}{Q} = \frac{C_O}{B}$$

These experimentations can be carried out with different principles of reactor, and particularly with suspended biomass or fixed films. A complete analysis of their advantages/disadvantages was made by Malina and Pohland (1992). These principles are schematically represented on Fig. 3.

As for hydrodynamics, the suspended biomass reactors must show a uniform (completely mixed) behaviour and the mixing characteristics are very important for a successful process. For this reason, the ratio between the height and the diameter of the reactor must be of the order of 1. In the case of a uniform reactor, and because the biomass can flow out with the effluent, the hydraulic and solid retention times are equal. This results in larger volumes than with other principles. Some mixing of the liquid/solid medium is provoked by the upward movement of the gas bubbles, but the intensity of the vertical circulation depends on the value of the gas hold up. External energy has often been supplied to enhance the mixing. In small volume reactors, a mechanical agitator can be used but such a technology often becomes economically inadequate for large volumes, even if submerged impellers are utilized (Fig. 4). Steady or unsteady jets are then an interesting alternative method (Simon and Fonade, 1993): no moving part systems are located inside of the reactor, the jet momentum improves the convection phenomenum and the turbulence of the jet increases the diffusion. Dead zones or flow channelling can be prevented.

The obtained mixing time is well correlated to the jet momentum, so that the required pump can be easily scaled (Fig. 5). Moreover, the recycling of the fluid by the external loop decreases this mixing time and the consumption of energy. It can be noted that the bacteria are only sensitive to very high shear stresses and that they do not suffer any damage when passing through the pump or the jets.

In a fixed bed reactor, the hydrodynamic behaviour is of the plug-flow type, but some mixing is achieved by transverse diffusion and, eventually, by an external recy-

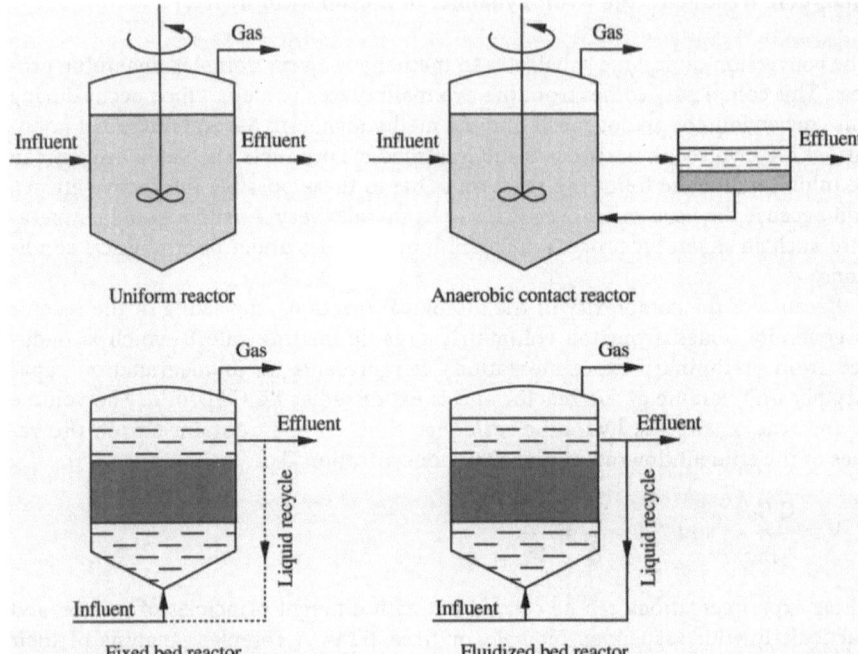

Fig. 3. Types of bioreactors

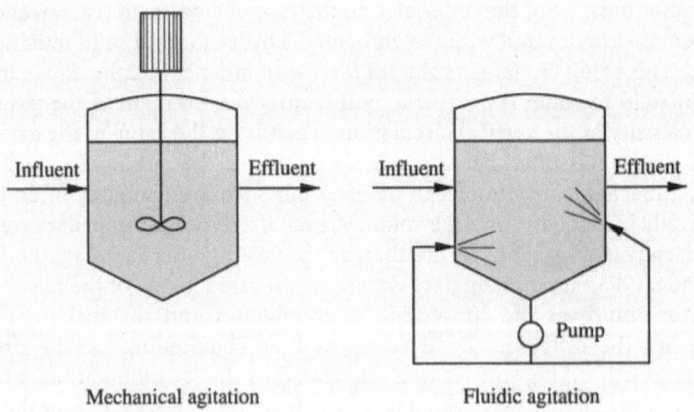

Fig. 4. Agitation principles

Fig. 5. Fluidic mixing characteristics
(from Simon and Fonade, 1993)

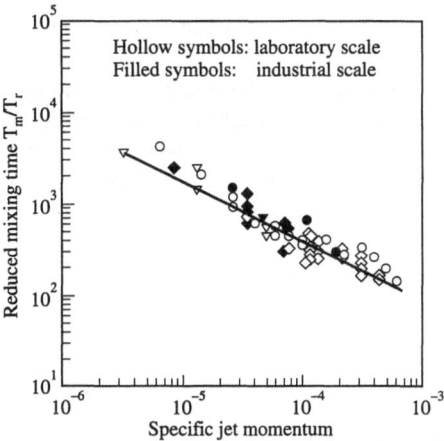

cling. However, suspended solids contained in the influent can accumulate in the pores and create flow channelling and mass transfer decrease. This disadvantage does not exist in fluidized bed reactors because of the unceasing movement of the agregates. Moreover, the biofilm is less thick near the bottom (effluent supply) so that a natural vertical circulation is induced by gravity. As in other respects, the external recycling is necessary to maintain the fluidized bed in expansion. Overall mixing is not critical for this type of reactor. However, some attention must be given to the risk of temporary flow channellings.

4.3.4
Aerobic Treatment: the Hydrodynamics in a Lagoon

As indicated in the introduction, the first wastewater treatments were achieved in naturally aerated lagoons which operated as fed-batch processes. This extensive method is technologically simple and energy-saving. The biological reactions involved in this treatment are generally stable but the physical limitations lead to high hydraulic residence times, and then the required volumes are very large. The lagoons are now managed as activated sludge basins, with a continuous mode of operation. In order that such a process is successful, the physico-chemical environment of the cells must be such that any limitation phenomena will occur. The hydrodynamic behaviour is then an important parameter but the oxygen supply is another problem to solve, especially since this process is a high energy consumer. Lastly, the lagoon geometries always lead to shallow water flows, with a small depth of liquid as compared to the horizontal dimensions. Such a geometry is very different from a cylindrical reactor because the free surface and the bottom have a great influence on the characteristics of the flows.

As for anaerobic reactions, the kinetics of biodegradation must be obtained in laboratory batch preliminary experiments. Even if many kinds of microorganisms are involved in the reaction, this kinetics does not generally appear complex and it shows clearly if the pollutants are all easily biodegradable or if some of them need a long delay to be converted. This is important information for the engineering design of the lagoon.

From this kinetics of biodegradation, the optimal hydrodynamic ideal behaviour of the reactor can be defined as uniform, plug-flow or mixed. If we note X the ratio C_R/C_O between the concentrations C_R and C_O of the product to be degraded respectively at the output and input of the reactor, and ζ the rate of degradation parenthese (ζ_R: final value), the following relations can be used:

for a uniform reactor:

$$T_S = \frac{1}{\dfrac{dX}{dt}(\zeta_R)}\,\zeta R$$

for a plug-flow reactor:

$$T_S = \int_0^{\zeta^R} \frac{1}{\dfrac{dX}{dt}(\zeta)}\,d\zeta$$

In the observation of a single behaviour, with the shape of the curve of the bio-degradabilty rate dX/dt (ζ), a plug-flow behaviour will yield a lower value of the residence time. An example can be seen on the Fig. 6 which is drawn for an effluent of the paper industry: for a conversion rate ζ of 0.7, the values of the required residence time are 48.5 h and 103 h respectively for a plug-flow and a uniform behaviour.

However, it can be seen that a uniform behaviour is better for a rate of conversion up to 0.45. In another way, it is always interesting that the reactor allows support of temporary variations of the organic load or flowrate of the wastewater without washing of the biomass. For these reasons, and in most cases, a uniform behaviour must be chosen, at least for the region of the reactor located close to the influent supply nozzles. Ideally, the total volume of the reactor must then be often separated in several elementary reactors, each of them having either a uniform or a plug-flow behaviour. For example, in the case of the aerobic treatment of the paper industry effluent, the Ricica method allows the definition of the number and volume of these elementary reactors in order to optimize the conversion rate. At this stage of the study, the main overall physical characteristics of the reactor of each compartment are known: behaviour, volume, oxygen demand. For the example of Fig. 6, the optimal configuration is found to be a uniform reactor of volume V/2 followed by a plug-flow reactor with the same volume V/2. Because the flowrate of influent is about 1000 m³/h, each of these compartments have a volume of 150,000 m³. The total oxygen demand is about 30 t/d, and 80 % of this demand must be supplied in the uniform reactor.

In the past years, engineers have essentially looked at the oxygen supply on the basis of the overall demand. Turbine aerators have been developed and are currently used. Such an equipped lagoon is schematically drawn on Fig. 7: the flow of the influent across the lagoon is only due to gravity forces. These aerators show good transfer performances and their O.E.R lies about an average value of 1.5 kg O₂/kWh. They are scaled with the volume of the lagoon, their number depending on the horizontal area. The induced flow structure is shown on Fig. 7. Measurements carried out around a 55 kW turbine have shown that the horizontal velocities decrease very quickly with the distance to the turbine. Such a turbine can be expected to have a local influence, until a distance of about 20 m from the turbine, and most of the fluid is directly recycled. In situ measurements of the hydraulic residence time distribution have shown that such a configuration gives significant dead zones and flow channelling (Fonade, 1994). On the other hand, when the reactor is expected to show a plug-flow behaviour, the turbine aerators provide an unwanted mixing.

For some years, new methodologies and technologies have appeared which can overcome these disadvantages of the turbine aerator for the hydrodynamic behaviour of the lagoon.

To improve the mixing characteristics, submerged propellers can be located inside the lagoon. However, the more obvious solution is to utilize judiciously the energy of the influent. For that, it is sufficient that one or several submerged horizontal nozzles are located at the extremity of the supply duct (Fig. 8): the momentum of the created jets can then be used for mixing purposes. Unsteady jets can also be generated, which show the great advantage to inducing an unsteady flow structure in the whole lagoon, therefore enhancing the mixing of the influent. This unsteadiness must be low frequency and electrovalves can be used. Another possibility is offered by the fluidic deviator based on a flip-flop principle (Foster and Parker, 1970): the steady influent flowrate is diverted to one of the outputs, only changing the pressure balance between the two control ports. The switching of this device is fast enough to generate very large vortices which improve the mixing process of the jets. Unlike the use of electrovalves, the switching of this device does not provoke any pressure variation in the supply duct and the pump needn't be protected against waterhammers.

Some technologies are now available which combine aeration and mixing. They use the energy of a liquid stream to entrain air and generate a two-phase jet. Under this principle a low pressure region is created where the atmospheric air is allowed to enter. The energy is provided either by a propeller or by a pump. In the first case, the whole device is immersed in the lagoon, while in the case of hydro-ejectors the pump is located on the shore. The consumed energy is of use to mixing for one part and to

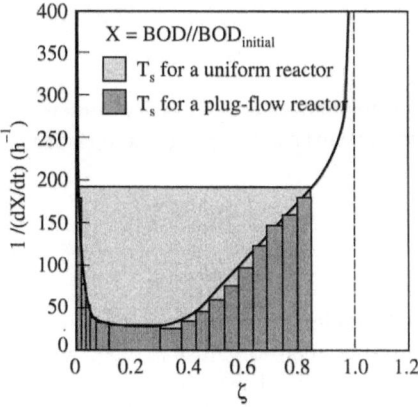
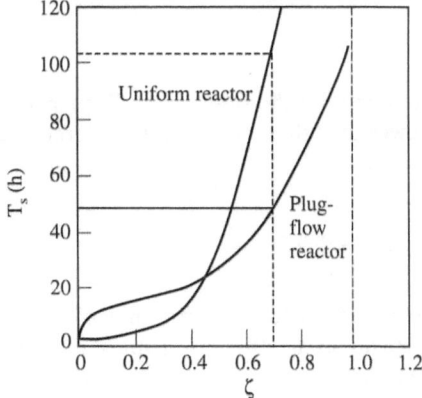

Fig. 6. Reactor behaviour and biodegradability

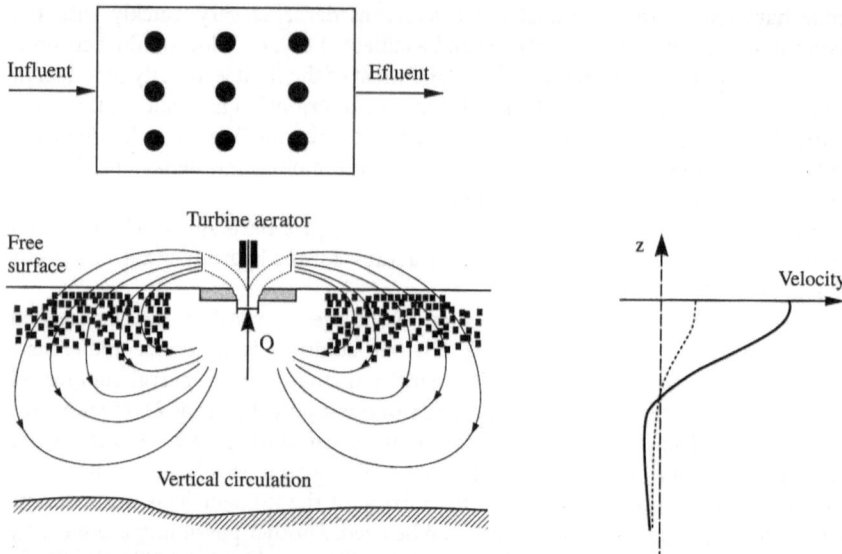

Fig. 7. Flow structure created by a turbine aerator

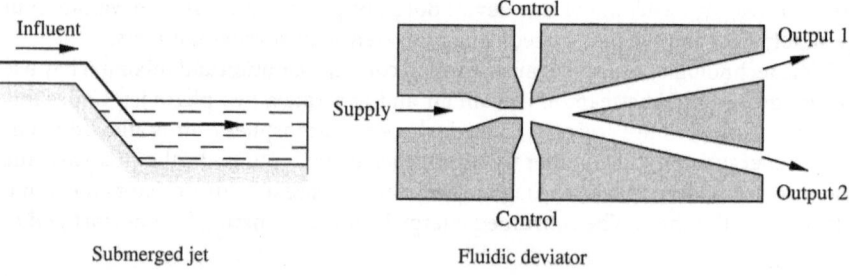

Fig. 8. Technology of the fluidic mixing

aeration for the other part. Their aeration performances are generally somewhat lower than those of the turbine aerators but many works are in progress to enhance them.

4.3.5
Conclusion

The performances of the biological treatment of high effluent flowrates strongly depend on the physico-chemical environment of the micro-organisms. When an anaerobic process is chosen, the reactor is generally cylindrical shaped. The complexity of the numerous reactions involved in the process and the problem of stability are certainly, at the moment, the main fields of interest. The hydrodynamics is only seen through the overall behaviour of the reactors. However, some works are in progress to study the influence of the local hydrodynamics on the mass transfers, essentially in the fluidized bed configurations. In aerobic treatments, more attention is given to the mixing and aeration processes, certainly because the required energy constitutes a great part of the costs. New methodologies and technologies are now available. Future work is needed in order to better control the whole process, both in the knowledge and computation of the local properties, and in the development of more efficient technologies able to insure the required mixing and aeration.

4.3.6
References

Fonade C (1994) The paper industry wastewaters treatment: engineering of aerated lagoons. Workshop on the paper industry and the environment. Toronto
Foster K, Parker GA (1970) Fluidics – Components and circuits. John Wiley & Sons Ltd
Forster CF (1985) Biotechnology and waste water treatment. Cambridge University Press
Malina JF, Pohland FG (1992) Design of anaerobic processes for the treatment of industrial and municipal wastes. Water Quality Management Library. Technomic Publishing Co Inc Lancaster, USA
Rols JL, Goma G, Fonade C (1994) Biotechnology and paper industry; aerated lagoon for waste waters treatment. Int Symposium on Environmental Biotechnology, Waterloo
Simon M, Fonade C (1993) Experimental study of mixing performances using steady and unsteady jets. The Canadian Journal of Chemical Engineering, 71
Valiron F (1985) Gestion des eaux – alimentation en eau et assainissement. Presses de l'Ecole Nationale des Ponts et Chaussées

4.3.7
Further Readings

Title: Biological treatment of kraft mill wastewater
Author: Oleszkiewicz Jan A; Trebacz Waclaw; Thompson Dave B
Corporate Source: Univ of Manitoba, Winnipeg, Manit, Can
Source: Water Environment Research v 64 n 6 Sep–Oct 1992 p 805–810
CODEN: WAERED ISSN: 1061-4303
Language: English
Document Type: JA; (Journal Article) Treatment Code: X; (Experimental)
Abstract: Wastewater from an integrated bleached kraft pulp and paper mill was treated in seven parallel treatment trains. Each train consisted of three reactors in series, with the primary reactor (denoted L1) simulating an aerated equalization/stabilization lagoon with hydraulic retention time (HRT) equal to 1.5 days. The secondary (middle) reactor was different in each train. Three reactors (R1, R2, R3) were operated at HRT equals 4.5 days and the solids retention time (SRT) equal to, respectively 10, 20, and 40 days. The next three secondary reactors (R4, R5, R6) were operated at HRT equals 1.5 days and SRT equal to 10, 20, and 40 days, respectively. The secondary reactor in the seventh train (R7) was operated at an HRT equal to SRT set at 1.5 days, that is, effectively operating as an aerated lagoon. The last, third or tertiary reactor in each of the series was operated as a settling lagoon with all reactors (denoted E1 to E7) having identical HRT equals 1.5 days. All seven treatment trains have removed 5-day biochemical oxygen demand (BOD//5) down to 10 mg/L. The removals of adsorbable organic halides (AOX) and soluble organic carbon (SOC) were found to be directly proportional to the biomass concentration and the SRT of the process and were the

highest in the R6 reactor with SRT equals 40 days and highest mixed liquor suspended solids (MLSS) of 3320 mg/L. In the activated sludge reactors R1 to R6 the removals of AOX and SOC were found to be inversely proportional to HRT, that is, the removals were larger for the reactors R4, R5, and R6 which operated at HRT equals 1.5 days. Overall AOX removal was the highest (30%) in the train L1 yields R6 yields E6, and the lowest AOX removal (21%) was in the train of lagoons L1 yields R7 yields E7. The negative effect of increased HRT on the removal of AOX and SOC was explained by biosorption since at the same SRT, the lower HRT reactors contained more than twice as much mixed liquor suspended solids. (Author abstract) 18 Refs.

Title: Interactions of Wastewater, Biomass and Reactor Configurations in Biological Treatment Plants
Author: Henze M (ed); Gujer W (ed)
Corporate Source: Technical Univ of Denmark, Dep of Environmental Engineering, Lyngby, Den
Conference Title: Proceedings of the IAWPRC Specialised Seminar
Conference Location: Copenhagen, Den Conference Date: 1991 Aug 21–23
Source: Interactions of Wastewater, Biomass and Reactor Configurations in Biological Treatment Plants Water Science and Technology v 25 n 6 1992. Publ by Pergamon Press Inc, Tarrytown, NY, USA. 320p
Publication Year: 1992
CODEN: WSTED4 ISSN: 0273-1223
Abstract: This issue of the journal contains 21 papers from the conference proceedings, grouped under the following topics: wastewater and biomass characterization; processes; modeling and simulation; filaments and solid separation; and secondary clarifiers for activated sludge. Some of the subjects discussed by the papers are here cited as examples: characterization of wastewater for modeling of activated sludge processes; transformation of wastewater in sewer systems – a review; metabolism of organic substances in anaerobic phase of biological phosphate uptake process; estimation of kinetic parameters of heterotrophic biomass under aerobic conditions and characterization of wastewater for activated sludge modeling; test of the activated sludge model's capabilities as a prognostic tool on a pilot scale wastewater treatment plant; towards a comprehensive model of activated sludge bulking and foaming; comparison of biocenoses from continuous and sequencing batch reactors; a dynamic secondary clarifier model including processes of sludge thickening; and modeling of the secondary clarifier combined with the activated sludge model No. 1. All papers are abstracted separately.

4.4
Multiphase Reactors for Biological Treatment of Urban Wastewaters

DOMINIQUE BASTOUL · MICHEL ROUSTAN · BERNARD CAPDEVILLE ·
JEAN-MARC AUDIC

Abstract

Different multiphase reactors, which can be applied to the aerobic biological treatment of urban wastewaters, are described. Solids used are new particles, patented INSA, of large diameter and low density. The hydrodynamics, mass transfer performances and, for some, the capacity of pollution removal, are presented. The adequation type of solids/type of multiphase reactor and type of reactor/type of pollution is discussed.

4.4.1
Introduction

In the last few years, studies have been carried out on different multiphase reactors with a view to apply them to the aerobic biological treatment of wastewater. It is well known that activated sludge process can be upgraded by adding a support media in order to create a three phase biofilm reactor:

- the liquid phase is the water to be treated,
- the gas phase provides the oxygen required by the biological process,
- the solid phase is the carrier of micro-organisms.

Among these three phase biofilm reactors, one can distinguish the fixed and the fluidized beds. The latter are preferred since they do not require recirculation of the bulk flow. They have emerged as one of the most promising devices for aerobic biological wastewater treatment.

In classic three phase fluidized bed, TPFB, the solid particles are fluidized by upflow liquid, which is the continuous phase, and cocurrent dispersed gas bubbles. In most previous works, solid particles (sand, glass) of relatively high density have been used as support particles for the attached microbial growth. In such three phase reactors, the operating conditions are imposed by the biological process (e.g. hydraulic retention time ~ 10–30 mn), and therefore the water phase must be recycled in order to provide the high superficial liquid velocity that is necessary to ensure fluidization. For this reason, for solid particles of high density, very small diameters (0.3–0.4 mm) must be chosen so as to limit the recycling flowrates and the energy required.

When biological growth occurs on the media, their diameter increases and the overall density is reduced. The resulting expansion is beyond that caused by fluidization of the unseeded media (Diniz Leao, 1984). Bed carry-over can occur and serious technical problems of liquid-particle and particle-biofilm separation arise.

Recent studies have led to the creation of a new kind of particles, patented INSA, and called OSBG (Optimal Support for Biological Growth) of increased diameter and reduced density (Capdeville, 1987). These particles are manufactured from plastic materials, about 900 kg/m³ in density. A mineral component is then introduced to bring it to 1030–1600 kg/m³. They are spherical and 2–4 mm in diameter. Lastly, they are submitted to an oxidation treatment to be wettable.

These particles present a double interest, first because they do not need enormous flowrates to achieve fluidization and their physical properties are not modified when covered with biofilm. Secondly, the biofilm thickness is negligible, compared to their diameter, and their density is very close to that of the biofilm, so fluidization conditions do not change during the process. Therefore bed carry-over problems are avoided.

Several studies have been performed on the application of this new kind of particles to different multiphase reactors, schematically represented in Figs. 1 a–d. They differ in:

- the type of fluidization (liquid or gas as fluidizing agent),
- the distribution of the gas and liquid phases,
- the geometry.

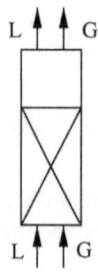

a Three phase fluidized bed (TPFB)

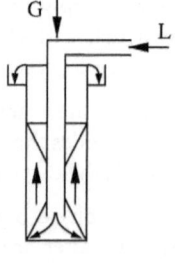

b Air predispersed reactor (PAR)

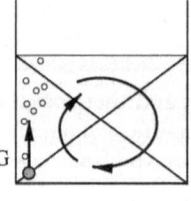

c Spiral flow reactor (SFR)

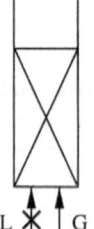

d Turbulent bed reactor (TRB)

Fig. 1a–d. Schematic diagram of the different multiphasic reactors studied

The use of multiphase reactors is not restricted to biological process.Three phase bubble columns have a wide range of application in industry (Deckwer, 1980). Solid particles are used as catalysts, products, material enhancing mass transfer (chlorination of hydrocarbons, polymerization in suspension, desulphuration of flue gas, single cell fermentation, etc.).

The control of multiphase reactors needs to characterize the hydrodynamics, the gas/liquid mass transfer and the pollution removal performances. The hydrodynamic parameters such as the minimum fluidization velocity, the bed porosity and the specific phase hold-ups, the bubble size and gas flow regimes, are among the most important information needed for the design of multiphase reactors since they govern the gas-liquid mass transfer and the removal pollution performances.

The present work synthesizes the studies performed until now.

The four reactors have been characterized by their hydrodynamics and mass transfer performances. Since the biological studies take longer to complete, only some have been tested in wastewater treatment operations

4.4.2
Three Phase Fluidized Bed (TPFB)

First, a classic three phase fluidized bed was studied. In this reactor, the solid particles are fluidized by upflow liquid, which is the continuous phase, and cocurrent dispersed gas bubbles.

Experimental Plant

The experiments were carried out in 4 PVC columns (Fig. 2). The columns 1, 2, were 0.08 m, the columns 3, 4, were 0.15 m in diameter, and 2.3 m in height. Two kinds of particles were tested:

- granular solids of small diameter and high density (glass beads, sand),
- compound granular solids of larger size and low density (OSBG1,2,3,4).

The properties of the solids, the experimental conditions (range of superficial gas and liquid velocities Ug and Ul, initial volumic fraction of solids R) are listed in Table 1.

Results

Hydrodynamics

Behaviour upon injecting gas in the liquid-solid fluidized bed
An interesting phenomenon occurs with the bed expansion or contraction when gas is injected into a liquid fluidized bed while the liquid flow is kept constant (Hatzifotiadou, 1989), (Bigot, 1990). With large particles (OSBG), the bed height increases as gas velocity increases, whereas an initial decrease of bed height exists if small and heavy particles are used. This initial bed contraction phenomena is observed in

Fig. 2. Experimental plant –
TPFB – column 1

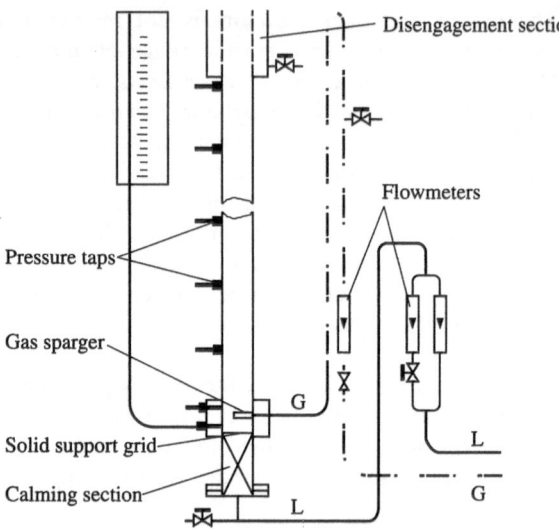

Table 1. Experimental conditions – Three phase fluidized bed

Solid	dp (mm)	ϱ_s (kg/m³)	Ug (cm/s)	Ul (cm/s)	R (%)
Glass beads	0.38	2500	0.3 – 4.5	0.5 – 3.3	10 – 40
Sand	0.32	2650	0.3 – 3	0.5 – 2.8	10 – 60
OSBG1	2.7	1180	0.3 – 4.5	0.8 – 3.3	10 – 40
OSBG2	3.5	1245	0.3 – 4.7	1.4 – 5.5	10 – 40
OSBG3	3	1600	0.3 – 4.4	1.7 – 7.2	20 – 60
OSBG4	4	1028	0.3 – 4.5	0.5 – 1.7	20 – 60

the range of low gas velocity. Bed expansion takes place when high gas velocity is used. It is believed to be caused by the wake that trails behind the bubbles.

The overall behaviour of the three phase bed is different according to the particles.

With the new particles, the two and three phase zones are easily distinguished. Except for the lighter one (OSBG4), a few particles are subjected to carry-over by air bubble. The fluidization is homogeneous. When gas bubbles in the beds of small and heavy particles coalesce, the whole bed passes into a turbulent state and packets of particles can be thrown about violently.

Gas flow regimes
Two different flow patterns can be obtained in the fluidized bed:

- the dispersed flow regime (bubbles practically uniform in size),
- the coalesced bubble flow regime (bubbles of various size and shape).

The gas flow regimes are affected by both the liquid and gas velocities, but the latter has less influence. Coalesced flow regimes are observed in the range of low liquid velocities. A critical value Ulc seems to exist beyond which there is no more bubble coalescence. Whatever the solid, coalescence appears when the solid fraction is more than 30%. General flow maps (Bigot, 1990), restricted to these experimental plants, have been established to predict the gas flow regimes accor-ding to the operating conditions.

Phase holdup

The hydrodynamic properties of the 3 phase bed have been characterized by the following parameters:

- the overall gas holdup εG (three phase fluidized bed and freeboard above the bed),
- the specific phase hold-up ε_G''', ε_L''', ε_S''',
- the bed porosity, defined as the sum of gas and liquid hold-up, $\varepsilon = \varepsilon_G''' + \varepsilon_L'''$.

Bed porosity

Experimental data of bed porosity are shown in Fig. 3. The bed porosity increa-ses significantly with liquid velocity, while the gas velocity has limited influence depending on the type of solids:

- with particles presenting bed contraction phenomena, the bed porosity decreases as Ug increases and so as contraction increases;
- on the opposite, with the OSBG particles, the bed porosity slightly increases with the gas velocity.

Three phase gas hold-up

Same tendencies are observed when considering the specific phase holdup results (Fig. 4):

- with the classic particles, addition of solids leads to a decrease of ε_G'''; whatever the solid hold-up, the three phase gas hold-up is always lower than the bubble column gas hold-up and decreases with increasing ε_S'''.

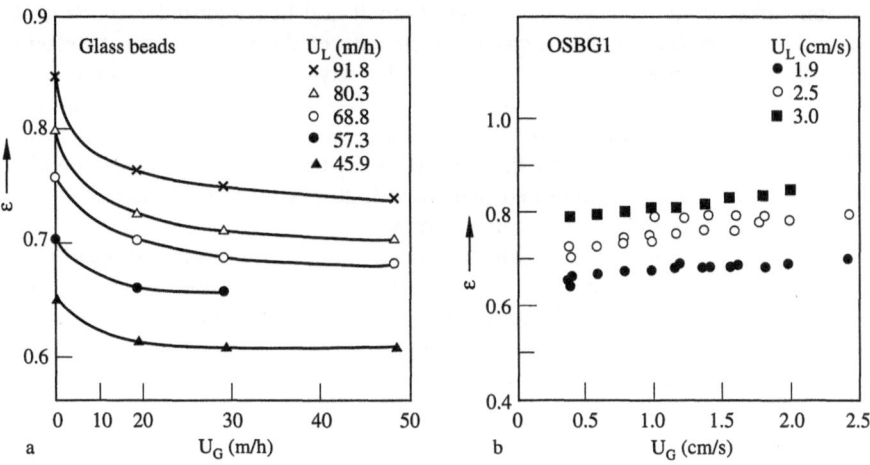

Fig. 3. Variation of the bed porosity with the gas and liquid velocity

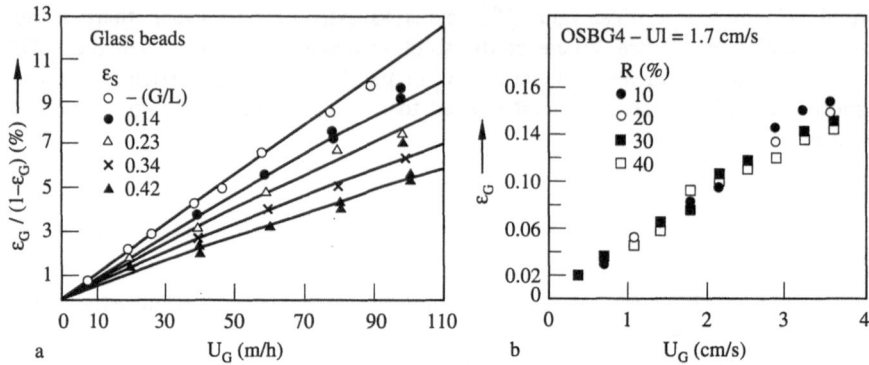

Fig. 4. Effect of gas velocity and amount of solids on the three phase gas hold-up

- with the OSBG particles, the three phase gas hold-up remains equal to the bubble column gas hold-up when the gas flow regime is dispersed, that is when $\varepsilon_S''' < 30\%$. The bubble diameter and gas liquid specific interfacial area are not affected by the presence of solids. Models have been established to predict values of ε_G''' when these new particles are employed.

Oxygen Mass Transfer

The oxygen mass transfer was characterized by the liquid side volumetric mass transfer coefficient kla'''. Some typical results for different particles are shown in Figs. 5a,b. For comparison the two phase (gas-liquid) data for the reactor operating as bubble column are also reported in the Figs.

The mass transfer results follow the gas hold-up results. With the classic particles, the higher the solid loading, the more pronounced the decrease of kla''' will be. The ratio kla'''/kla'' can decrease from 0.8 to 0.3 as the solids concentration varies from 14 to 41%. The use of small and heavy granular solids leads to a strong reduction of the oxygenation capacity of the reactor, even if its aeration device is efficient. This reduction of the oxygenation capacity may cause serious problems when such a reactor is applied in the biological wastewater treatment, since the capacity of aerobic biodegradation of the reactor depends on the oxygen transfer rate in the system. In addition, absolute values of kla''' coefficients in the three-phase bed being low, high gas flowrates must to be used to achieve the required oxygen transfer rate.

Better results are obtained when OSBG particles are used. No significant effects of the solids are observed on kla''' when the average solid fraction is less than about 30%, that is, until the gas flow regime is dispersed. For higher solids concentration, values of kla''' decrease slightly. These particles are particularly interesting for a reactor of aerobic wastewater treatment since there is no loss of oxygenation capacity in presence of solids.

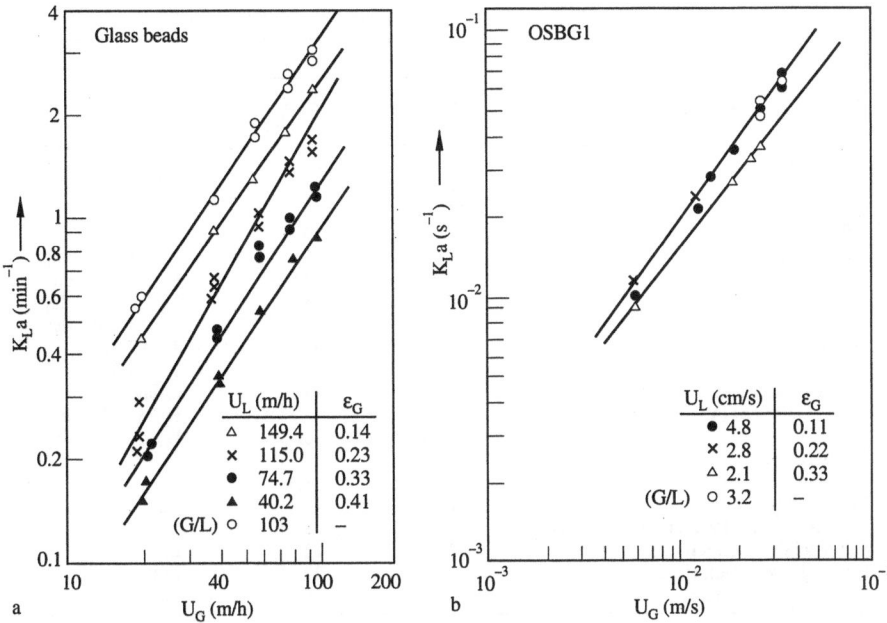

Fig. 5a, b. Effect of gas velocity and amount of solids on kla''' coefficient

Waste Water Pollution Removal

Experiments of pollution removal were carried out using the solid OSBG1 as support of biofilm growth. The particles have been first submitted to an adequate treatment to optimize the interactions between micro-organisms and particles (Lertpocasombut, 1991). The reactor was fed with a synthetic substrate composed of meat extract, glucose, peptone, yeast extract, and inorganic salts, giving a TOD of approximately 340 mg of soluble organic matter/l in the influent feed. Two groups of operating conditions, summarized in Table 2, were tested.

The results from these experiments showed a high capacity of pollution removal of this three phase fluidized bed reactor. The operating regime was stable during several dozen days. In addition, a low sludge production was observed, which can be attributed to the existence of a thin and particularly active biofilm covering the particle.

Measurements of the total deshydrogenase activity of this biofilm indicated values of activity 200 times higher than the activities for conventional activated sludge, fed with the same substrate. It can be supposed that oxidation of organic matter is more likely to occur than the synthesis of cellular components (Lertpocasombut, 1991). Values of concentration of volatiles solids attached to the particles were in the range of 0.5 g/l unit of reactor volume, which is low compared to data submitted in other papers, indicating concentrations varying from 10 to 20 g/l. These results confirm a fundamental study carried out by Belkhadir (1985).

Table 2. Operating and average performances of a three phase fluidized bed reactor

Parameter	Organic loading rate		Units
	10 kg TOD/ m³· day	15 kg TOD/ m³· day	
Influent TOD	335	340	mg/l
Effluent TOD	24	44	mg/l
Influent TOC	112	107	mg/l
Effluent TOC	7	13	mg/l
Influent organic	12	19	kg TOD/m³· day
Loading rate	3.4	4.8	kg TOC/m³· day
Removed organic	9.3	13.4	kg TOD/m³· day
Loading rate	3.2	4.2	kg TOC/m³· day
Efficiency	93	87	% of TOD
	94	88	% of TOC
Sludge production	3.5	15.7	g/day
Yield coefficient (Y)	0.05	0.15	kg/kg TOD · day
	0.14	0.45	kg/kg TOC · day
Temperature	29	28	°C
Dissolved oxygen concentration	8.3	8.0	mg/l
Influent liquid flowrate	10.1	15.4	l/h
Superficial liquid velocity in the reactor	95	95	m/h
Superficial gas velocity	12	12	m/h
Hydraulic retention time	40	20	mn
Experimentation time	70	40	days

The thin biofilms (20 to 40 µm) and their specific activity, measured by INT method, were important. The gas flowrate controls the biofilm thickness by shear forces and therefore fixes the overall biofilm activity in the reactor. The hydrodynamic behaviour of the three phase fluidized bed used for the aerobic pollution removal was very stable. The very thin biofim developed on the fluidized particles did not change the conditions of fluidization.

Conclusion

The performances of TPFB reactors largely depend on the characteristics of the employed solids (dimension, wetting properties, surface, etc.). By acting on the diameter and the density of the particles (large and light particles), it is possible to ensure a good oxygen transfer and the development of a very thin and active biofilm which induces positive efficiencies for the removal of pollutants. Solids OSBG2, OSBG3 seem to be particularly adapted to this multiphase reactor.

4.4.3
Predispersed Gas Annular Reactor (PAR)

Following the industrial development of a new ozone contactor, called the "Deep U Tube", (Brodard, 1983), (Roustan, 1992), studies have been performed on the potential application of this gas/liquid reactor to the three phase systems. The hydrodynamic behaviour and the mass transfer performances have been characterized.

Experimental Plant

It is composed of a vertical tube (0.028 m in diameter D2) inside a cylindrical PVC (0.15 m in diameter D1) closed column (Fig. 6). The vertical tube receives inlet water and gas at the top. High turbulence in the inner tube breaks the gas into bubbles. The bubble sizes are strongly dependent on the gas/liquid downward flow regime. As the liquid and gas flow downward, the pressure increases. At the bottom, the gas/liquid mixture flows upward through the external annulus to the upper part of the reactor, and fluidizes the solids. The liquid flow issued from the top of the column is partly recycled in the internal tube by a centrifugal pump. The OSBG2 particles were used as the solid phase.

Results

Results showed that the performances of this reactor (PAR) are strongly conditioned by the downward gas flow regime.

Two phase Gas/Liquid system

In the downward cocurrent flow of air and water, three types of gas flow regime can appear, depending on the gas and liquid flowrates, schematically represented in Fig. 7:

- bubbly flow: located in the range of low downward gas velocities Ugd. The gas is dispersed in the continuous liquid phase.
- slug flow: with increasing Ugd, the transition to slug flow appears, constituted by a succession of cells.
- annular flow: at the highest gas flow rates, one can reach the annular flow pattern. In the range of high gas velocity, the gas flow regime in the downward gas/liquid

Fig. 6. Experimental plant – PAR

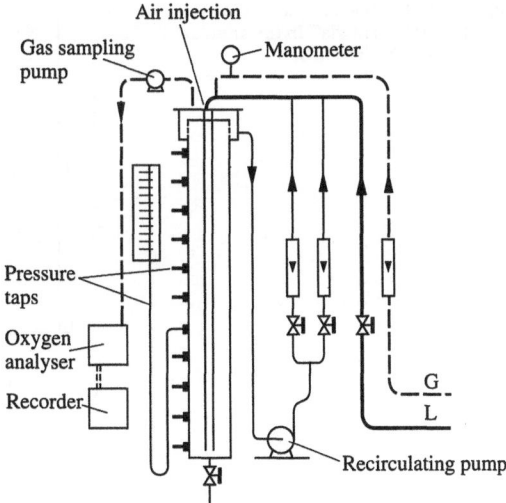

Fig. 7. Downward gas flow regimes.
(1) bubbly flow, (2) slug flow, (3) annular flow.

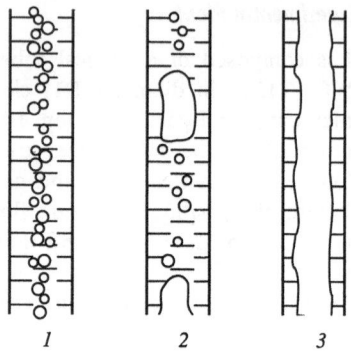

Fig. 8. Flow maps relative to the downward gas flow regimes

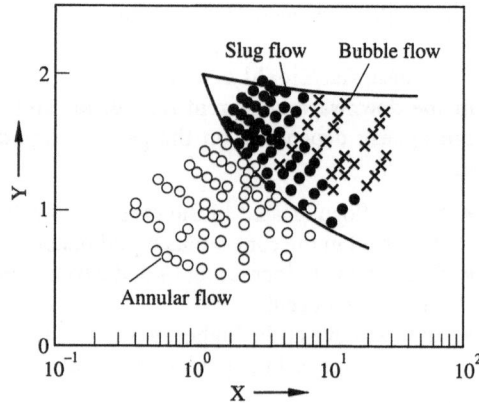

Fig. 9. Effect of the particles on the mass transfer coefficient kla′′′ in the annular three phase fluidized bed.

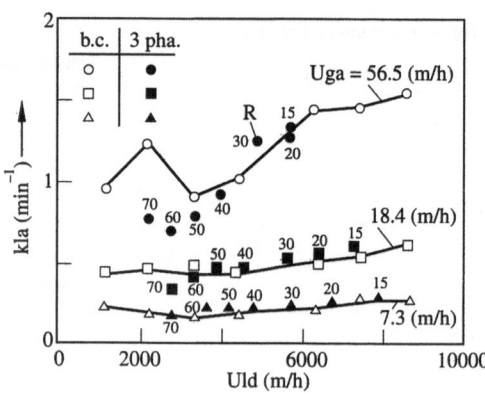

flow is affected by the both downward gas and liquid velocity and there is a cross linkage between these two factors.

The flow maps proposed by Speding and Van Thang Nguyen [1979] can be used to determine the operating conditions (downward gas and liquid velocity Ugd and Uld) required to induce the bubbly downward gas flow regime, as shown in Fig. 8, where:

$$X = L/G \quad Y = (Ugd + Uld/(g \cdot D2)^{0.5})^{0.5}$$

Three phase system – Comparison with classic TPFB

The presence of particles OSBG2 does not lead to a reduction of the gas hold-up and the volumetric mass transfer coefficient kla''' (Fig. 9), as far as the solid fraction is no more than about 30%, beyond which bubbles coalescence phenomena appear. Only the bubbly flow can ensure the performances of the reactor.

Conclusion

In this reactor, the downward cocurrent gas/liquid flow is the "key" operating condition. Only the bubbly flow guarantees good performances. In this case, the hydrodynamics and mass transfer results are close to the classic fluidization reactor characteristics.

Though this reactor must operate in given ranges of liquid and gas flowrates, and is therefore limited in the range of operating conditions, it is a potentially interesting process since it offers the same performances as the classic TPFB and

- enables a uniform distribution of the phases,
- eliminates the plugging of gas spargers problems,
- enables enhancement of the transfer by using the hydrostatic pressure.

4.4.4
Spiral Flow Reactor

In this reactor, the particles are fluidized by the liquid circulation induced by the gas bubbles, (Fig. 10).

Experimental Plant

Two geometries of reactor have been studied:

- a cubic tank of 0.125 m³ in volume,
- a cylindrical horizontal basin of 0.5 m³ in volume.

The gas phase is distributed through a horizontal perforated tube. In the first reactor, this gas distributor is set at the bottom of the vessel near a vertical wall. In the second, the tube is set longitudinally at an angle of 42°C with the axis of symmetry. OSBG1 and OSBG4 particles were used as the solid phase.

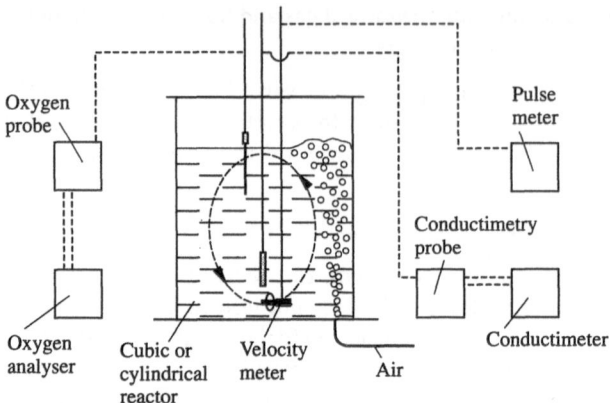

Oxygen probe

Pulse meter

Conductimetry probe

Conductimeter

Oxygen analyser

Cubic or cylindrical reactor

Velocity meter

Air

Fig. 10. Experimental plant – SFR

Results

Hydrodynamics and Mass Transfer

The most important parameter for the fluidization (keeping in motion of the solids) is the horizontal bottom velocity Vf of the liquid phase. Ghauri (1983) proposed a general correlation in order to predict this parameter. In presence of solids, this velocity can decrease when the amount of particles becomes important as shown in Figs. 11a,b where G/L is the gas flowrate defined by unit of horizontal length of the reactor and R is the initial volumic fraction of solids introduced in the reactor.

The lighter particles OSBG4, are easier to fluidize. The maximum amount of fluidized solids may reach 25% (initial volumic fraction of solid introduced in the reactor) with the OSBG4 whereas it is only 10% with the OSBG1 particles. The minimum superficial gas velocity required to fluidize the particles is lower in the cylindrical reactor. The mass transfer performances are not affected by the presence of solids as far as the overall amount of particles is fluidized as shown in Fig. 12.

Waste Water Pollution Removal

This reactor has been used for simultaneous carbonaceous and nitrogenous pollution removal (Moreau, 1993). The heterotrophic and autotrophic activities were distributed among the liquid and solid phases, phenomenon which is in favour of the nitrifying bacteria attached to the solid support.

Experiments showed that in the range of organic loading of about 0.8 kg/m³· day, the addition of solid matter leads to much higher nitrifying activity compared to classic pilot performances. It can be considered that the liquid phase is preferably colonized by the high growth rate bacteria whereas the solid favours the development of low growth rate bacteria. These results are especially interesting since they offer new perspectives for simultaneous removal of carbon and nitrogen in this reactor in the range of high organic loadings.

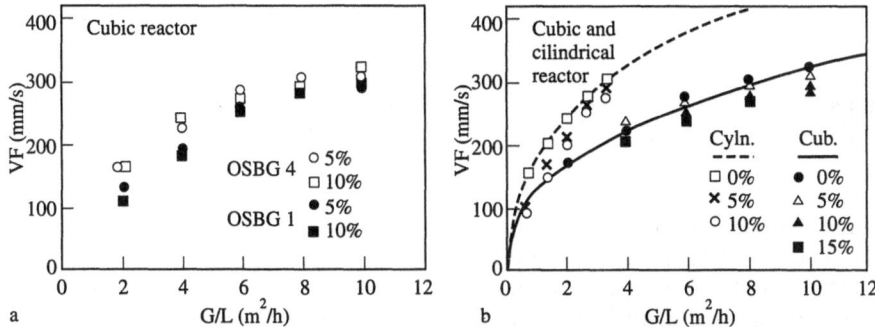

Fig. 11. Variations of Vf with the gas flowrate. Effect of the type and the amount of solids. a Cubic reactor b Cubic and cylindrical reactor – OSBG4

Fig. 12. Variation of kla‴ with the gas flowrate and the amount of solids – OSBG4

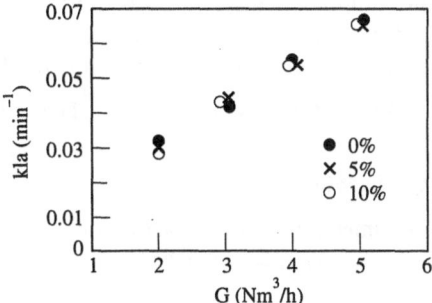

However, the OSBG particles, as raw materials, are not appropriate as support of the biofilm growth, since the fluidization gas leads to strong shear forces and consequently induces particles/particles interactions. Experiments of surface treatment have been performed on the solids OSBG4 to modify the nature of the surface by using an activated coal powder. Results are better (eliminated C, N: 0.5 kg $N-NH^4/m^3 \cdot$ aday at 20°C) but not sufficient for a total ammoniacal pollution removal in comparison to the performances of biofiltration process (1kg $N-NH^4/m^3 \cdot$ day at 20°C).

In conclusion, an important effort still remains to be made in the view to improve the micro-organisms fixation. Two prevailing parameters have to be taken into account in this new perspective of research: the nature of the surface (porosity) and the volume properties of the particles (diameter and density).

Conclusion

In this type of multiphase contactor, SFR, the best conditions are achieved when using a cylindrical reactor and solid OSBG4. Mass transfer performances are close to the gas/liquid systems performances as long as the initial volumic fraction of solids does not exceed 25%. Therefore, the spiral flow reactor presents satisfactory hydrodynamics and mass transfer characteristics. More informations on its capacity of pollution removal is still needed at the present time.

4.4.5
Turbulent Bed Reactor

Another type of multiphase column type system has been designed, to dissociate the liquid flowrate and the fluidizing agent: the turbulent bed (Fig. 13). In this reactor, the solids are fluidized by the liquid circulation induced by rising gas bubbles.

The first study (Marty, 1990) showed that particles type OSBG, of density close to that of water can be set in motion by the one gas phase in column reactors. Solids used were glass beads and OSBG particles (Table 3). Under economic considerations, the particles the more attractive, requiring the lowest fluidization gas velocity, are the OSBG4 polystyrene particles. Other sets of experiments have thus been carried out on the hydrodynamics and mass transfer performances of this multiphase system (Beck, 1994). Three kinds of polystyrene particles were used, the characteristics of which are summarized in Table 4 where Rep and Frp are the Reynolds and Froude numbers of the particles defined as:

$$\mathrm{Rep} = \mathrm{Ut} \cdot \mathrm{dp} \cdot \rho L / \mu L \quad \text{and} \quad \mathrm{Frp} = \mathrm{Ut} / (\mathrm{g} \cdot \mathrm{dp})^{0.5}$$

Experimental Plant

Experiments have been carried out in 4 vertical columns made of PVC (Fig. 13). The columns 1, 2, 3, 4 have respectively 0.08, 0.15, 0.20, 0.40 m of diameter and 2,

Fig. 13. Shematic diagramm of pilot plant – TRB

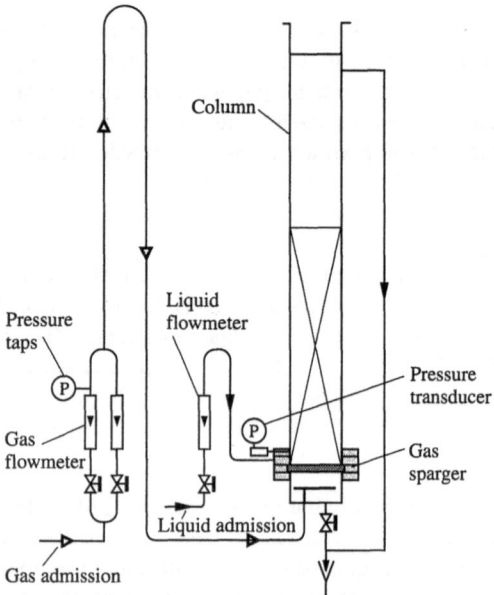

Table 3. (Marty, 1990)

Solid	dp (mm)	ρ_s (kg/m³)
Glass beads	0.375	2500
OSBG1	2.7	1180
OSBG4	4	1015
OSBG5	3.3	1140

Table 4. (Beck, 1994)

Solid	dp (mm)	ρ_s (kg/m₃)	Ut (m/s)	Rep (–)	Frp (–)
OSBG4	4.0	1028	0.037	140	0.211
OSBG6	3.6	1038	0.036	138	0.245
OSBG7	1.5	1028	0.016	18	0.120

Table 5. Characteristics of the gas spargers

Gas sparger	Sq. pitch arrangt. (mm)	Holes/cm² (–)	Orifice diameter (mm)	Thickness (mm)
Perforated plate	5	2.5	4	20
Porous plate	porosity: 0.3 20			
Membrane	5	0.5	4	2.5

2, 4, 3.5 m height (liquid height). Three different gas spargers were used: a perforated plate, a porous plate, a flexible membrane disc. Their characteristics are summarized in Table 4.3.

Results

Hydrodynamics
Critical gas velocity. The critical gas velocity required to fluidize the solids has been determined by measuring the variations of the gas static pressure ΔP with the superficial gas velocity Ug. The Fig. 14 shows that beyond a critical value of Ug, Ugc, ΔP remains unchanged. This break in the slope corresponds to the fluidization of the solids.

Unlike classical two phase fluidization process, the values of Ugc depend on the mass of particles, mp, introduced in the reactor and increase with increasing mp (Fig. 15). At low values of mp, Ugc depends only on the mass mp. Beyond a critical mass, mpc, the increase is more pronounced. Ugc vary with mp and the liquid height in the column Ho, that is mp and the solid holdup ε_s''. As in the TPFB reactors, a maximum solid hold-up must not be exceeded by 20 %, beyond which the values of Ugc are prohibitive.

The type of gas sparger, which governs the bubbles gas dispersion, has a great influence on the hydrodynamics of such a reactor, as shown in Fig. 16. Results showed: Ugc perforated plate $< =$ Ugc membrane $\ll$ Ugc porous plate. There is no direct relation between little bubbles and low Ugc but an homogeneous distribution of bubbles size seems to promote the suspension of the solids.

Fig. 14. $\Delta P = f(U_g)$

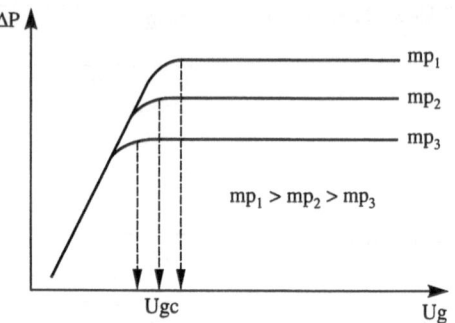

Fig. 15. Effect of mp on Ugc Col. 3 – porous plate – OSBG4

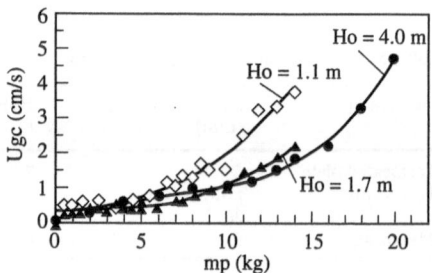

Fig. 16. Effect of the gas sparger on Ugc Col. 3 – Solid: OSBG4

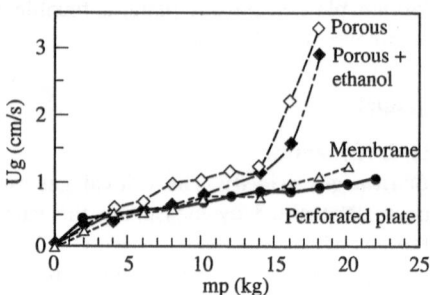

Ugc can be reduced by acting on the properties of the particles: d_p, ρ_s. The nearest the density of particles is the liquid one, the least is Ugc. Furthermore, with particles of similar density, the lowest diameter solid (OSBG7) is the first fluidized. These properties can be traduced by the modified Froude's number defined as: $Fr_p = U_t/(g \cdot d_p)^{0.5}$. The less the Frp, the less the Ugc. The use of a low upward liquid flowrate may reduce significantly (until 30 %) the critical gas velocity required.

None of the correlations of the literature (Roy, 1964), (Narayanan, 1969), (Koide, 1984), (Joshi, 1987), can give good predictions of the experimental data. The discrepancy may be explained by the fact that these expressions have been established with other types of particles and that they do not take into account of the nature of the gas sparger.

Axial solid distribution. As Ug increases, the solid moves gradually and overall solid circulation takes place. Unlike classic liquid/solid fluidization, the particles move along the entire height of the column and there is not a clear interface liquid/solid. An axial solid concentration gradient is observed, whatever the operating conditions, even if it slightly decreases with decreasing particle diameter. The solid most easily fluidized is also the easiest to distribute uniformly along the column, and is characterized by the lower Froude number.

Mass Transfer

The volumetric liquid-side mass transfer kla''' and the gas hold-up ε_G''' are affected in a similar way by the gas velocity, the gas sparger and the solids concentration. The variations of the coefficient kla''' with the gas velocity and mass of particles are presented in Fig. 17. For comparison, the 2 phase (gas-liquid) data for the reactor operating as bubble column are also reported in the figures.

The coefficient kla''' decreases with addition of particles when the perforated plate is used as gas sparger. With the two other distributors, no significant effect of the presence of solid matter is observed, as long as the average solid hold-up is less than 20%. Higher solid concentrations lead to a decrease in the gas hold-up and the oxygen mass transfer.

Conclusion

The turbulent bed reactor, a new type of multiphase reactor in which the particles are fluidized by the one upward gas flow, is now well characterized by its hydrodynamics and mass transfer performances.

Solids with the density close to that of water (polystyrene particles) must be used in such a reactor. The critical gas velocity increasing with the amount of charged particles mp, and a maximum value of mp must not be exceeded, corresponding to an average solid hold-up of about 20%. The values of Ugc decrease with decreasing particles diameter. The choice of the gas sparger prevails on the hydrodynamic and mass transfer results.

Fig. 17. Effects of Ug and mp on kla'''
Col. 3 – OSBG4 – Membrane disc diffuser

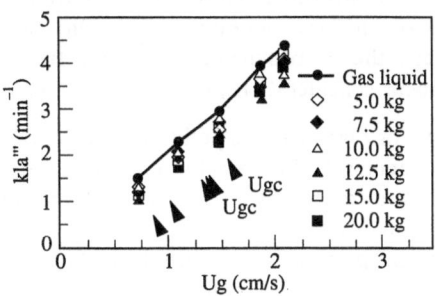

The lowest fluidization gas velocities and thus the most effective conditions are achieved with the membrane disc diffuser, when solid OSBG7 (polystyrene particle of lowest diameter) is employed. Whatever the operating conditions, an axial gradient solid is observed. This reactor remains to be tested in biological process.

4.4.6
Conclusion

In the attempts to develop three phase fluidized bed reactors for the aerobic biological waste water treatment, a new kind of particle, called OSBG, of large diameter (2.7 to 4 mm) and low density (1030–1600 kg/m³), patented INSA, has been developed and used in different multiphase reactors:

- a classic three phase fluidized bed (TPFB),
- a predispersed air reactor (PAR),
- a spiral flow reactor (SFR),
- a turbulent bed reactor (TRB).

Hydrodynamic behaviour, mass transfer performances and some capacity for pollution removal have been determined.

Whatever the reactor, the operating conditions must be defined. A restrictive condition on the nature of the gas flow regime is traduced by a maximum authorized amount of solids to fluidize, corresponding to an average solid hold-up depending on the reactor: ~30% when classic TPFB are used, ~20% for the turbulent bed reactor. This condition involves the annular predispersed air reactor to be operated in given ranges of gas and liquid velocity flowrates.

In gas bubbly flow, the presence of particles OSBG does not affect the bubble size and the gas/liquid specific interfacial area. The hydrodynamics and mass transfer parameters (ε_G''' and kla''') are very close to results obtained in bubble columns.

If all these particles lead to good results (hydrodynamics and mass transfer), an adequate type of OSBG particles/type of reactor seems to exist. When the liquid phase is used to fluidize the solids, particles OSBG1 (2.7 mm, 1180 kg/m³) and OSBG2 (3.5 mm, 1245 kg/m³) are preferred. When gas phase is the fluidizing agent, lighter particles, OSBG4, OSBG6, OSBG7 (1.5–4 mm, 1030 kg/m³), are easier to fluidize. The effect of the gas sparger is more pronounced in the latter reactors.

The hydrodynamics and mass transfer performances are now well known, at least on these laboratory pilots. One can define the operating conditions and predict the hydrodynamic and mass transfer parameters. Under waste water treatment considerations, results are less complete but it can be said that:

- for carbonaceous pollutions, the TPFB reactor shows a high capacity of pollution removal. When employed in waste water treatment bioreactors, the particles OSBG are covered by a thin and active biofilm and the removal efficiency of pollution attains 90%,
- the principle of the turbulent bed is an opportunity for simultaneous C and N pollution removal even if efforts still must be made on the nature of the surface treat-

ment of these materials to ensure a sufficient colonization of the bacteria on the support.

4.4.7
References

Beck (1994) Doctorate Thesis, INSA Toulouse, France
Belkhadir R, Capdeville B, Roques H. Etude descriptive fondamentale et modélisation de la croissance d'un film biologique. Wat Res, 22, 1, 59-70
Bigot V (1990) Doctorate Thesis, INSA Toulouse, France
Brodard E, Duguet JP, Mallevialle J, Roustan M. Appareil pour la dissolution d'ozone dans un fluide, Brevet Français N° 8307764 étendu aux pays Europeens, USA, Canada
Capdeville B. Matériaux granulaires de traitement d'eaux et procédés de fabrication, Brevet Français n° 8703611 (Mars 1987) et Europeen n° 882003841 (Mars 1988)
Deckwer WD, Alper E (1980) VDI-Ber No 321, pp 97-106
Diniz Leao MM (1984) Etude fondamentale des biomasses fixées. Description et modélisation des films biologiques anaérobies.Thesis, INSA Toulouse, France,
Ghauri B (1983) Contribution à l'étude de la circulation et du transfert d'oxygene dans les bassins d'aération. Doctorate thesis N° 91, INSA Toulouse, France
Hatzifotiadou O (1989) Doctorate Thesis, INSA Toulouse, France
Joshi et al. (1987) Int J Multiphase flow, 13, 3, 415-427
Koide et al. (1984) J of Chem Eng Japan, 17, 4, 368-374
Lertpocasombut K (1991) Doctorate Thesis, INSA Toulouse, France
Marty (1990) Doctorate Thesis, INSA Toulouse, France
Narayanan (1969) Can J Chem Eng, 24, 223-230
Roustan M, Capdeville B, Bastoul D (1993) 3rd Int Conf on bioreactor and bioprocess fluid dynamic, Cambridge UK
Roustan M, Line A, Duguet JP, Mallevialle J, Wable O (1992) Practical design of a new ozone contactor: the deep U tube. Ozone Science & Engineering, vol 14, pp 427-438
Roustan M, Beck C, Capdeville B, Bastoul D, Audic JM (1995) International Symposium of the Engineering Foundation - Fluidization VIII-Tours, May 14-19
Roy, Guha, Raho (1964) Chem Eng Sci 19, 215-225.
Spedding PL, Van Thanh Nguyen (1979) Regime maps for air water two phase flow. Chem Eng Sci, vol 35, pp 779-793

4.4.8
Further Readings

Title: Some novel aspects of multiphase reactions and reactors
Author: Sharma MM
Corporate Source: Univ Dep of Chemical Technology, Bombay, India
Source: Chemical Engineering Research & Design, Part A: Transactions of the Institute of Chemical Engineers v 71 n A6 Nov 1993. p 595-610
Publication Year: 1993
CODEN: Cerdee ISSN: 0263-8762
Abstract: Multiphase reactions and reactors, involving at least one liquid phase, are crucial in most industrial processes. Theories of mass transfer with chemical reaction have been developed for an impressive range of reaction systems. Computational methods allow complicated situations to be analysed with rigour. Mass transfer can play an important role in influencing the selectivity in simultaneous absorption with reaction of two gases, consecutive reactions, etc. The prowess of using small scale experiments to design large scale systems has been demonstrated. The benefits of deliberately converting single liquid phase reactions to two-

liquids have been delineated. The design of large scale bubble columns, mechanically agitated contactors, etc. can now be approached with much greater confidence. (Author abstract) 143 Refs.

Title: Biological treatment of hypersaline wastewater by a biofilm of halophilic bacteria
Author: Woolard CR; Irvine RL
Corporate Source: Univ of Notre Dame, Notre Dame, IN, USA
Source: Water Environment Research v 66 n 3 May–Jun 1994. p 230–235
Publication Year: 1994
CODEN: Waered ISSN: 1061-4303
Abstract: Each year, billions of gallons of wastewaters containing high concentrations of salt (greater than 3.5% w/v) and waste organics are generated by industry. Biological treatment of these hypersaline waste brines to remove organics could reduce the environmental impact and cost of waste disposal. Unfortunately, the salinity of many waste brines makes them difficult to treat with conventional waste treatment cultures. This paper demonstrates that heterotrophic halophilic organisms can be used to remove phenol from a synthetic waste brine containing 15% salt. The reactor system used in this study was a novel periodically operated biofilm reactor, the Sequencing Batch Biofilm Reactor (SBBR). The SBBR uses permeable silicone tubing to supply oxygen to the reactor. A biofilm of halophiles isolated from the Great Salt Lake, Utah, ecosystem readily developed on the tubing surface and degraded waste organics. The experimental results presented herein demonstrate that the SBBR can produce a stable, high-quality effluent despite fluctuations in influent phenol concentration. The results of periodic track studies illustrate that biomass accumulation increases oxygen demand but does not significantly improve overall reactor performance. (Author abstract) 23 Refs.

Title: Multiphase reactors. Models and experimental verification.
Author: Dudukovic MP; Devanathan N; Holub R
Corporate Source: Washington Univ, St. Louis, MO, USA
Source: Revue de l'Institut Francais du Petrole v 46 n 4 Jul–Aug 1991 p 439–465
Publication Year: 1991
CODEN: Rfptbh ISSN: 0020-2274
Abstract: This paper addresses the issue of improving our understanding of the hydrodynamics in two commonly used reactor types: bubble columns and trickle beds. We use two different approaches. For the bubble column we develop and present a technique for measurement of liquid velocities and turbulence parameters in order to provide much needed data. We then suggest that this, coupled with another experimental technique for evaluation of voidage profiles, would yield all the necessary measurements for critical evaluation of the existing two phase flow models. In trickle-beds we develop a simple phenomenological model for liquid flow and confront it with the available data for pressure drop, holdup and flow regime transition. Based on this we develop a model for liquid distribution and suggest that quantification of any such model requires the use of noninvasive imaging technology of which we give an example. 50 Refs.

Title: Study of fluidized bed dynamical behavior: A chaos perspective
Author(s): Tam SW; Devine MK
Performing Organization: Argonne National Lab., IL.
Report No: Conf-901246-3
Sponsoring Organization: Department of Energy, Washington, DC.
Contract No: W-31109-ENG-38
Notes: EPRI workshop on applications of chaos, San Francisco, CA (USA), Dec 1990. Sponsored by Department of Energy, Washington, DC.
Date: 1990 Pages: 24p NTIS Price Code: PC A03/MF A01

Abstract: Fluidized beds have long been used as multiphase reactors in the chemical processing industry and as fossil–fuel combustors in energy production. This study was undertaken to determine whether fluidization data exhibit only stochastic features, or whether they contain chaotic characteristics as well. An extensive search for chaotic behavior has been conducted on a large time series data set on fluidized bed pressure fluctuation. The non-linear dynamical techniques employed include the correlation integral and the nearest neighbor methods. Mutual information algorithm has been employed to obtain the optimal time delay necessary for the analysis of the experimental time series. A singular value decomposition technique has been used for noise reduction. The results indicate that despite the highly fluctuating appearance of the data there is a non-random component present. For the data that has been examined no low-dimensional (i.e. of dimension 2–3) attractor has been found. Any strange attractor that may be present is likely to have at least moderate dimension (i.e. above 3–4). For the analysis of possibly chaotic time series from experimental data (as in contrast to model–generated data) this study has revealed that the issues of data set size, noise and moderate-to-high dimensional attractors are critical. Careful utilization of existing and developing non–linear dynamical techniques is necessary to avoid drawing spurious conclusions. 13 refs., 5 figs.

4B Physical-Chemical Treatment

4.5
Physical Chemical Treatments for Wastewater

YVES AURELLE

4.5.1
Abstract

Over 20 years experience has led to definitive technological innovations in the physical chemical treatment of water. This discussion presents a summary of these important innovations. The first interest of this summary is to integrate these various research works and innovations into a more general context regarding the future development of wastewater treatment. The second interest is to present the various research methods applied which are based primarily on the observation of appropriate phenomena, the application of the physical chemistry theories of interfaces, and the methods of Process Engineering.

4.5.1
Introduction

In general, advances in the field of water treatment are connected with regulations. However, stricter regulations intended to improve our environment cannot be introduced unless researchers and water treatment organizations are in a position to propose new effective, efficient and, above all, economically viable techniques.

Regulations are directly dependent on technological feasibility and thus on research, which must provide efficient means of analysis and effective, innovating technical solutions at a cost that is acceptable to industry.

The fundamental ideas that have directed water treatment research in the last two decades and which will continue to direct research in the future are summarized in the following five ideas.

1. Treatment processes enabling the pollutants to be recovered will gain in importance as, through them, the industrial user can recoup his antipollution investment quickly, which acts as an incentive. In this case the water treatment is a simple consequence, an advantage for the environment, inducing no further cost. Moreover, this resolves the problem of sludge and various other forms of waste which are one of the major preoccupations of water treaters today.

   However, to attain this objective, water treatment must act closer to the source of pollution so as to avoid "mixing of wastewaters" which reduces the economic advantages. Perhaps one day, with this in mind, wastewater will be characterized by an "entropy level" that could be defined as a decrease of water quality level and would be an important parameter to be considered for optimizing water management in companies.

2. Treatment processes must, in the future, be increasingly compact as restrictions on the ground surface area of the installations will become more severe. This is already the case on off-shore oil rigs and, in the near future, stations for treating urban wastewater in particular will be situated near built-up areas where land prices are dissuasive, as was the case for the Monaco station. Similarly, we can envisage the development in Europe of treatment units in the basements of large buildings to recycle waste water on the spot; they are already in use in certain countries (Japan, Thailand, etc.).

3. Odorless, "closed container" treatment processes will also be more and more in demand. Smells must be banished from the new projects, particularly in cases where the stations are situated in built-up areas or integrated into the buildings themselves.

4. Essentially physical processes that can be easily automated will also gain importance and should oust the physical-chemical and biological processes, which are more difficult to manage, especially for fluctuating pollution loadings.

5. Super efficient processes will also gain favor as they will permit large water savings, in many cases through direct recycling, which seems to be the solution of the future.

In fact, these ideas are equally valid in the field of biological treatment as well as physical-chemical treatment. The European EUREKA project SIMBIOSE (Scientific Improvements of Biofilters and Sensors), aimed at confirming European industry's technological lead in the biological purification of wastewater, foresees the urban wastewater biological treatment station of the 21st century as being:

1. efficient,
2. of small dimensions,
3. covered and soundproofed, and
4. fitted with effective devices to deal with odors.

Figure 1 shows the characteristics of the wastewater treatment station built for Monaco State by OTV. This station represents the technological future for European stations. This station is integrated in the basement of a 10-story residential building, an arrangement that could only be envisaged thanks to the development of treatment technologies satisfying the criteria mentioned above, i.e. a close combination of efficiency, compactness, and absence of nuisance value.

This rather long but important introduction should enable us to more clearly distinguish the orientations in water treatment, particularly the physical-chemical treatments of the coming decade which will take us beyond the year 2000 and orient us towards the design of urban and industrial treatment stations of the 21st century.

4.5.2
General remarks on certain technological innovations in physical chemical treatments

This presentation is not an exhaustive study of the state of the art in the field of physical chemical water treatment. Rather, the focus of this discussion is to highlight three major technological innovations that have participated in and which follow the fundamental ideas stated in the five key points. After giving a brief view of the interest of these innovations, we shall go into the concrete contribution made by each of these

Fig. 1. Monaco wastewater treatment station (Courtesy of OTV)

innovating techniques in greater scientific detail, showing the type of methods applied, based mainly on the observation of the phenomena coming into play and the application of the physical chemistry theories of interfaces combined with the methods of Process Engineering.

Gravitational and accelerated separation techniques

The conventional physical chemical treatment is based on simple or gravitational settling which is conducted in channel-type tanks or vertical circular settlers. Settling is used as a primary and secondary treatment for urban and industrial wastewater.

The American Petroleum Institute (API) developed settling tanks several decades ago and these first generation settling tanks are still present in most of the refineries and oil installations in the world. These first generation settling tanks have proved to be great consumers of land area and, more importantly, contribute pollution in the surrounding atmosphere. These tanks are the source of the particularly distasteful odor reigning over many treatment stations. Specifically, the API petroleum installation contribute to atmospheric pollution with considerable loss through evaporation of the lightest, most volatile hydrocarbons. This material loss equates to an interesting economic loss. A study carried out at the Elf research center in Solaize showed that nearly 3000 tons of light hydrocarbons are lost each year into the atmosphere at the Feyzin refinery.

The API settling tank, which is open to the atmosphere, may be completely replaced by more efficient, compact settlers such as laminated settling tanks. Laminated settling tanks are composed of a set of parallel plates separated by distances of a few centimeters to ten centimeters which provide a larger settling surface. In addition, laminated settling tanks are easier to cover and air purification can be integrated in the separator. Although laminated settling tanks are a more costly solution, these tanks have already be installed in many stations and this trend should accelerate in the coming years.

The classic case of calculating the cut-off diameter of a settler clearly illustrates the practical interest of the laminated settlers and accelerated separation techniques. In water treatment, a settler is always characterized by its loading per unit surface area, or superficial loading (Q/S), which is the ratio of the treated water flow rate (Q), expressed in cubic meters per hour (m^3/h) to the settling area (S), which, for conventional settling tanks, is their ground surface area, expressed in square meters (m^2). This ratio, characteristic of the settler's operation, corresponds to a terminal settling velocity expressed in meters per hour (m/h).

From Stokes' law the expression for the settling velocity of a particle (see Equation 2) relates the threshold diameter for particles (dp) that are completely separated by settling with a yield of 100%.

Settler loading/unit area = $Q/S = V_{settling}$ (m/h)   (1)

Stokes' law expressing the settling velocity of a spherical particle of diameter (dp):

$V_{settling} = D_r \cdot g \cdot dp^2/18\mu_{H2O}$ (2)

Knowing the difference in density between the particles to be separated and water (D_r), the acceleration due to gravity (g), and the viscosity of water at the settler temperature (μ_{H2O}), cut-off diameter of the settler (dp_{cutoff}) can be evaluated:

$dp_{cutoff} = [Q \cdot \mu_{H2O}/S \cdot D_r \cdot g]^{-1/2}$ (3)

This simple calculation clearly illustrates the interest of a laminated settling tank composed of a set of parallel plates. For example, consider a conventional settling tank covering a ground surface of S. If N parallel plates are placed in the tank, the settling surface increases from S to (N+1) S. For equal yield, then, the laminated settling tank will cover a ground area (N+1)-fold smaller than the conventional API tank. Such equipment incorporates key point 2 concerning compactness and, if covered and provided with an air regeneration installation, satisfies key point 3 concerning closed container. In addition to the introduction of the laminated settling tank, accelerated separation techniques have been pursued in an attempt to minimize the settler tank area.

Unlike conventional settlers which rely on gravitation force to perform separation, accelerated separation techniques act directly on the different Stokes' law parameters and can thus considerably increase the separation velocity, increasing separation in a settling tank placed downstream. Accelerated separation techniques regarding the three main parameters in Stokes' relationship are discussed below:

1. Particle diameter (dp): Particle diameter is the most sensitive parameter as it is squared in Stokes' relationship. Two types of effluent can be distinguished, those containing dispersed solids in suspension and those containing dispersed liquid or emulsions, often composed of micro-droplets of hydrocarbon.

Suspended solids are typically flocculated into bulky pieces that settle out readily. This is the already well known and commonly used coagulation-flocculation technique performed by means of chemical reagents.

Micro-droplets of hydrocarbon are coalesced increasing particle diameter and separation. The technique of separation by coalescence, which was relatively unknown 20 years ago, is principally used in the petroleum industry today. Its specific field of application should develop in the next decade.

2. Gravitational force (g): The acceleration due to gravity is a constant. However, gravitational force can be increased using centrifugal acceleration. There are two specific types of separators which utilize centrifugal acceleration, centrifuges and hydrocyclones.

   It is difficult to envisage the use of centrifuges for water treatment except for the treatment of sludge, where they are widely used. Hydrocyclones, on the other hand, could play an important role in the future because of their compactness which other separation techniques can not equal.

3. Difference in density (D_r): As discussed above, it is difficult to envisage changing this parameter, the particle size. However, the particle difference in density can be changed by encouraging micro-bubbles of gas to adhere to the surface of the particles. The bubble-particle combination having a density very different from that of water separates by floating. The techniques of flotation, by either induced or dissolved air, are already widely used in the treatment of both urban and industrial wastewater. Although this technique is not completely understood, current research may lead to a better understanding. In addition, this research may lead to better modeling of the bubble-particle interactions, which will inevitably lead to the development of new, more effective flotation tank geometries directly designed using Process Engineering methodology.

Skimming techniques

Progress in skimming techniques has been particularly noticeable in the past decade. The importance of the changes is demonstrated by the recent complete fitting of surface skimmers such as oleophilic oil drum skimmers to all the API settlers in French refineries. As discussed later, this development is directly linked to research results and to physical chemical theories of interfaces in water treatment. It is also due to the introduction of process engineering methods in the development of sizing techniques for this equipment. The immediate result of this progress was the marketing of effective and selective equipment which could be fully automated and provided a solution to the light hydrocarbon evaporation from the surface of API settling tanks. Moreover, it has been possible to use this equipment to combat oil slicks and partially recover what is called "chocolate mousse". The market for these new devices, be they drums, disks or bands should be flourishing as they satisfy key point number 1 concerning the profit from recovering pollutants.

Membrane techniques

Progress during the past decade in membrane techniques has been particularly marked and is linked to the following:

a) membrane manufacturers who have made available multiple types of membranes and

b) water treatment laboratories who have conducted extensive applied and basic research.

Membranes, which were laboratory curiosities ten years ago, are now in application and provide effective solutions to numerous water treatment problems. Initially applied to the treatment of drinking water, membrane applications are continuing to develop in the field of wastewater treatment. As discussed later, the use of membranes has grown considerably from being totally absent in the marketplace ten years and are continuing to be recognized as a future treatment technique.

4.5.3
Gravity and accelerated separation techniques technological innovations

This section discusses the various innovating processes in which the Laboratoire d'Ingénierie des Procédés de l'Environnement (Laboratory) has played an important role. Specifically, the following separation techniques will be discussed in detail:

1. Spiraloil laminated settlers,
2. Oleophilic resin coalescers,
3. Guide coalescers,
4. Static or dynamic tubular fibrous coalescers,
5. Hydrocyclones, particularly the three phase type,
6. Marangoni effect induced air flotation tanks, and
7. Deep well flotation tanks.

Spiraloil Laminated Settlers (Cherid, 1993)
The Elf research center developed this specific type of laminated settler to satisfy the following requirements:

1. recover hydrocarbons emulsified in petroleum industry process water,
2. to be compact, simple from the design point of view, and easy to install in existing plants, and
3. to be able to adapt to the separation of hydrocarbon/water emulsions or dispersions over a wide range of drop diameters.

The technological solution to these specific requirements was the development of a laminated tubular settler with an inner spiral in which the turns were closely wound at a separation of 1 to 5 mm. Because of its spiral design, it was called the "Spiraloil". Figs. 2 and 3 show two possible configurations. However, it was the Fig. 3 configuration with its mixed spiral winding that was developed industrially as it was convenient to manufacture. In this configuration the distance between turns of the spiral is fixed by the height of the triangular corrugations of a corrugated sheet that is rolled up with a smooth sheet of PTFE.

For a modular Spiraloil cell 10 cm in diameter and 30 cm in length, fitted with a 0.2 mm PTFE sheet, rolled in a spiral with a corrugated iron sheet having triangle heights of 3 mm, the total settling area was 2.1 m^2 (Fig. 4).

The separation efficiency obtained was 99 % for a transit speed of 18 m/h in the case of a standard kerosene/water emulsion with a median drop diameter of 20 microns and a mean concentration of 600 milligrams per liter (mg/l).

Fig. 2. Spiraloil laminated settler: Spiral design

Fig. 3. Spiraloil laminated settler: Other spiral design

Fig. 4. Laboratory modular Spiraloil cell with integrated settler

Fig. 5. Industrial Spiraloil unit tested on an off-shore oil rig – 500 m³/h – (Courtesy of ELF)

This new type of hydrocarbon/water separator marketed by Elf is particularly efficient and is very easy to insert between two flanges into the pipes of existing units as close as possible to the sources of pollution. This modular structure (30 cm in length) can be adapted without difficulty to most separation problems, even in the presence of low concentration surfactants.

A large installation fitted with two 500 m³/day Spiraloil units has been successfully tested on an off-shore oil rig (Fig. 5).

Settlers of this type perfectly satisfy the three criteria stated above: recovery of the pollutants, compactness, and closed system.

Oleophilic Resin Coalescers (Aurelle, 1974, 1980)

These new separators, practically unknown in France in 1970, are currently relatively well established in the petroleum industry and expected to develop considerably in new sectors of industry in the future. Oleophilic resin coalescers have proved to be particularly suitable for separating secondary hydrocarbon/water dispersions that have a characteristic milky appearance and a hydrocarbon micro-droplet diameter less than approximately 20 microns. Nevertheless, it is important to note that these separators are only appropriate to the separation of dispersions or emulsions that are not stabilized by surfactants. The presence of surfactants works against the drop coalescence phenomenon which is essential to the success of this separation process.

The coalescer operating principle is very simple. The hydrocarbon/water dispersion is injected through a granular or fibrous bed which acts as a filter medium intercepting the hydrocarbon droplets. Properly designed, large drops of hydrocarbon several millimeters in diameter are collected on the output surface. These drops readily separate out in a settler of small cross sectional area placed downstream of the coalescer bed.

Fig. 6. Kerosene drops released from an oleophilic resin coalescer bed

An actual case of separation may assist in the assessment of these new separators which act directly on the most sensitive parameter of the Stokes' law, the squared parameter (dp). Let us consider a kerosene/water emulsion with a median droplet diameter of 10 microns percolating at a speed of 10 m/h through a granular coalescer bed 10 cm high having a mean grain size of 0.6 mm. The separation efficiency of the settler downstream, which has the same cross sectional area as the coalescer bed is greater than 98%. The kerosene drops released by the bed (see Fig. 6) have a diameter of 4 mm on average. Therefore, the drops leaving the bed correspond to the coalescence of 60 million of the initial 10 micron micro-droplets and the drops, coming from preferential points of the bed surface (drip points), have a rising separation velocity 10,000 times that of the micro-droplets in the initial emulsion. In fact, this magnitude increase in the settling velocity is directly related to the economics of coalescer processes. Specifically, Equation 1 relates the separation surface area (S) of a settler as inversely proportional to the settling velocity. Therefore, if the latter increases by a factor of 10,000, by passing the emulsion through an effective coalescer bed, the settler downstream of the coalescer bed will have an area 10,000 smaller.

$$Q/S = V_{settling} \quad \text{Eq. 1} \qquad \text{whence, } S = Q/V_{settling} \tag{4}$$

Therefore, the coalescer is a particularly efficient separator regarding separation efficiency, compactness, and low operating cost. In addition, for the above case, the pressure drop across the bed was only 0.25 bars.

Studies of coalescence phenomena (Aurelle, 1974, 1980) have resulted in the industrial development of a particularly efficient granular coalescer material called oleophilic resins. These are obtained very easily by ionic grafting of cationic surfactants on to strong cationic ion exchanger resins of the RSO3H type (Abadie, 1970).

Microcinematographic observation of the phenomena in play has shown that the coalescence effect is in fact the result of three consecutive elementary processes:

1. interception of the micro-droplets by the coalescer material, an essentially ballistic step the model of which can be compared with penetration filtration;

2. adhering of micro-droplets to the solid surface of the coalescer, the agglomeration leads to interactions that initiate coalescence and the formation of a coalesced phase film inside the bed which moves through the bed along preferred pathways; and
3. release of the hydrocarbon film in the form of large drops at the end of the coalescer bed.

Further studies (Aurelle, 1980) determined the parameters of these key processes. These fundamental studies, based on direct observation of the coalescence phenomena, combined with interface physical/chemistry theory and studies on penetration filtration modeling by O'Melia (Yao, 1971), concluded in the ability to develop design methodologies for these new oleophilic resin coalescers.

Patents following these studies (Abadie, 1970) (Aurelle, 1972) have also allowed the Elf Petroleum Company and fourteen (14) other companies, working under license, to market these oleophilic resin coalescers worldwide since the 1980's.

Coalescers may be applied to many domains within the petroleum industry, particularly in the treatment of water in condensates. In this case, coalescers enable the hydrocarbon phase to be recovered and the water of the condensate to be recycled directly into the boilers (see Fig. 7).

Oleophilic resin coalescers satisfy all the five key points of the introduction, providing recovery, compactness, a closed and odor-free system, ability to automate physical process, and an efficient system where both pollutants and treated water are completely recycled. These coalescence processes, still relatively unknown to users, should become important in the next decade.

Fig. 7. Industrial oleophilic resin coalescers – Up-flow process – Donges Refinery – Condensate treatment (Courtesy of ELF)

Fig. 8. Industrial oleophilic resin coalescers – Tanker terminal ballast water treatment – Down-flow process – Brest harbour Unit – 250 m³/h (Courtesy of ELF)

It should be noted that up flow coalescers are relatively sensitive to suspended solids (SS) and for this reason can only be used with water having a low SS loading or must be reserved for tertiary finishing treatments. To treat water heavily loaded with SS, the Laboratoire d'Ingénierie des Procédés de l'Environnement has developed, in association with Elf, coalescers also using resin but with down flow. The major disadvantage is that washing cycles by backflushing are necessary to remove the SS fixed in the coalescer bed. These separators are marketed for the treatment of ballast tank water at oil tanker ports. They are generally used for secondary treatment after settling in a conventional API tank. This tank can also be used for treating the water used for washing the coalescer (see Fig. 8).

Guide Coalescers (Aurelle, 1980)

Guide coalescers are enhanced up flow resin coalescers with fast through flow of the order of 100 m/h. Therefore, guide coalescers are highly compact coalescence separators.

The process in which the coalesced phase is released as a fast jet of oil dividing into small drops (see Fig. 6), which generally limits the throughput speed in conventional coalescers, is suppressed here. Instead, the film of coalesced phase formed in process 2 is, in this case, channeled by a woven oleophilic structure with a high void coefficient inside the settling tank up to the upper hydrocarbon phase. The purified water is recovered at the side surface of the guide.

A stainless steel hydrophilic grid may be added in certain cases around the guide to act as a separating capillary barrier, thus avoiding any parasite drag on the coalesced hydrocarbon phase.

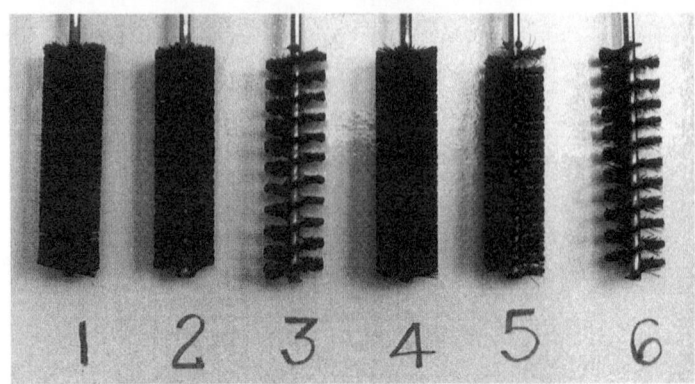

Fig. 9. Fibrous tubular coalescers: Brushes made by rolling a fibrous strip round an support axis

Static and Dynamic Fibrous Tubular Coalescers (Tapaneeyangkul, 1989)

Static and dynamic fibrous tubular coalescers have been studied most recently in the laboratory. These coalescers are only slightly sensitive, if at all, to the presence of suspended solids (SS) in the water treated. In addition, these coalescers, unlike down flow coalescers, do not need backflushing cycles.

These new types of coalescers are composed of cylindrical brushes made by rolling a fibrous strip around a support axis as shown in Fig. 9. They are characterized by very high void coefficients (e), of the order of 90 to 95%, and the diameter of their fibers is between 100 and 300 microns. The general structure of the coalescer is tubular and modular, each coalescer element comprising a 20–50 cm long tube containing the coalescing brush (diameter: 50 mm). The coalescer as a whole has the structure of a tubular temperature exchanger.

Two types of coalescer were studied, one static and the other dynamic. In the latter the brush was set in rotation to provide more efficient interception of the micro-droplets of the emulsions or dispersions treated. Given the importance of the fullness coefficient of the coalescer filling on the interception step in coalescence, rotating the brush allows the low static fullness coefficient of this type of self cleaning coalescer to be replaced by a much higher dynamic fullness coefficient that ensures better interception. The sizing analysis for this type of coalescer clearly demonstrates the key role of this dynamic void coefficient, defined by the following equation:

$$(1-e) \cdot N \cdot H/V_p = (1-e) \cdot N \cdot T_s \tag{5}$$

where N = rotational speed of the brush (rpm)
       H = height of brush
      V_p = flow rate of the emulsion treated in the coalescer bed
      T_s = residence time of the emulsion in the coalescer bed

Although the rotation has a great influence on the interception efficiency for micro-drops of diameter less than 10 microns, the mechanical complexity of manufacturing

Fig. 10. Fibrous tubular coalescer – 2 m³/h

these new devices translates to a preference for static brush coalescers which have a lower cost and simpler design.

These new coalescers, usable on water even with a high SS loading, can act as primary separators, which was not the case for the up flow coalescers discussed previously. These new coalescers (see Fig. 10), covering a wide range of possible uses, will be industrially produced by S.C.A.I., Société Commerciale d'Accessoires Industriels. For more information regarding the study of coalescence, the various theses mentioned in the references should be consulted.

Hydrocyclones (Ma, 1993)

Although well known in process engineering and particularly in mineralogy, hydrocyclones are relatively little known in water treatment and purification. However, hydrocyclones may have enormous potential in water treatment where treatment volumes are large. Indeed, hydrocyclones are the only separators whose separation efficiency increases as the flow rate of the water to be treated rises. It is difficult to find a more compact treatment processes.

The centrifugal acceleration induced in the hydrocyclone, which drives the separation, depends on the tangential velocity (V) which is directly proportional to the flow rate (Q). Helmholz's law in connection with the conservation of momentum in a hydrocyclone shows that the product of the tangential velocity (V) at a point of the cyclone and the distance (r) from this point to the axis is constant.

$$V \cdot r = \text{Constant} \tag{6}$$

We can then deduce that it is near the axis that the tangential speeds are highest and it is thus in this zone of the cyclone that the centrifugal accelerations are greatest. The area near the axis is the most active zone of the cyclone.

Historically, up to the 1980's, hydrocyclones were mainly used for separating high density suspended solids (SS) with a difference of density relative to water of 1 to 1.5. During this time it was thought that hydrocyclones were not capable of separating hydrocarbon/water dispersions because, in this case, the density difference was too small, in the 0.1 to 0.2 range. In fact it was Professor Thew (Coleman, 1988) who first modified the conventional geometry of the Rietman hydrocyclones suited to the separation of heavy SS and showed that it was possible to envisage separating hydrocarbon/water dispersions by hydrocycloning. This work was continued in a study (Ma, 1993) with the Elf Petroleum Company in order to develop a three phase hydrocyclone capable of treating wastewater containing dispersed hydrocarbons and suspended solids. This study, which will be presented in detail in the next paper, developed an original, efficient hydrocyclone that in fact includes two hydrocyclones in the same apparatus, one of the Rietman type for separating SS, the other of the Thew type for separating out the hydrocarbon phase.

The hydrocyclone offers compactness and robustness (no moving parts) which ensures reliability. Another interesting feature of the hydrocyclone is that running costs are zero as it makes use of the pressure energy at the top of the well, which is indispensable for separation. Two phase hydrocarbon/water hydrocyclones have been in industrial use for some years. However, the patented hydrocyclone (Ma, 1992) will be industrially produced by Krebs. The use of these separators in the petroleum industry will grow in the coming decade.

A particularly innovative study conducted by the Laboratory involved the coupling of a hydrocyclone with a coalescer (Ma, 1993). After all, the hydrocyclone is only a pollution concentrator, the outlet of which is in fact a concentrated oil dispersion which, treated by a coalescer, separates pure oil and pure water. This is not the case with a hydrocyclone alone.

Another study is in progress in the Laboratory to assess the potential of hydrocyclones in the general field of drinking water, urban wastewater, and rainwater treatment. However, it is the new specific domain of rainwater treatment that may offer the most promising use of hydrocyclones. Since treatment of rainwater involves enormous flows, compact separators are required. Moreover, most of the pollution in this case is in the form of SS.

In conclusion, hydrocyclones appear to be a future option regarding the water treatment and purifying where compact installations are sought at all cost.

Marangoni Effect Induced Air Flotation (Aoudjehane, 1986)

Flotation, either by induced air or by dissolved air is a basic technique in wastewater treatment. Flotation is regularly used in both urban and industrial wastewater treatment. Urban wastewater treatment involves the removal of grease from wastewater by induced air flotation followed by grease concentration using dissolved air flotation. In addition, conventional treatment processes for wastewater in the petroleum industry use dissolved air flotation tanks after an API settling tank.

Despite, the widespread use of flotation processes, new, more efficient flotation systems may be developed in the coming years. An article published in 1992 by the Western Research Center in Devon, Alberta (Canada) discusses the state of flotation technology today:

Flotation equipment design is an area of reactor engineering that has not attracted much research interest in Chemical Engineering. The commercial success of flotation remains a triumph of enlightened Know How over inadequate Know Why as key design and scale-up parameters (e.g. column height) continue to be dictated by such non-process considerations as crane rail and building heights.

The Laboratory has decided to follow this path by sweeping aside the Know How and trying to use cinematographic display to understand the basic phenomena in flotation separation and to define the limiting processes and optimize them (Aoudjehane, 1986) (Dupré, 1995). In fact, this approach is based on studying the bubble-particle interactions when a bubble of gas encounters a particle for flotation, which may be a microdrop of hydrocarbon or a hydrophilic or hydrophobic suspended solid (Aoudjehane, 1986) (Dupré, 1995).

The separation efficiency of a flotation tank is the combined result of a bubble/droplet or bubble/suspended solid collision step followed by an adhesion step. To increase the overall efficiency of a flotation tank, optimization of the collision and adhesion processes is required. Studies have shown that the adhesion process is often limiting, particularly in the case of induced air flotation of a dispersion of microdroplets of oil. In this case the limiting process is the drainage of the water film after collision between the bubble and oil droplet that determines the adhesion and thus the flotation effect.

Studies have shown that creating mass transfer from the air bubble to the continuous aqueous phase leads to bubble/drop adhesion after collision. This process was accomplished by injecting the flotation tank with an air phase, non-polluting, but very water soluble gas such as chlorine or carbon dioxide. This immediate adhesion effect depends on the drainage of the water film at the bubble/hydrocarbon drop interface, combined with the Marangoni effect. This effect, well known in process engineering and particularly in liquid-liquid extraction, is responsible for the very high coalescence effect of the dispersed liquid phase when the mass transfer occurs in the dispersed phase – continuous phase direction.

The Marangoni effect relates mass transfer where fluctuations in the concentration of the transferred compound around the bubble causes fluctuations in surface tension leading in turn to a deformation of the bubble that encourages drainage of the water film. A new type of "Marangoni effect" has been patented by Elf (Aurelle, 1985) and should have interesting applications considering its simplicity of use. This new "Marangoni effect" gives high flotation yields where induced air flotation is based on the injection of a very soluble gas into the air phase of the flotation tank before its dispersion. In fact induced air flotation tanks may readily be transformed into Marangoni effect flotation tanks.

Deep Well Dissolved Air Flotation (Dupré, 1995)

Currently, several studies are aimed at developing new, particularly efficient flotation tanks in which the two fundamental flotation processes (collision and adhesion) are abolished. In this case, the gas bubbles form directly on the particles for flotation during slow, gradual expansion of the pressurized water.

Microcinematographic laboratory studies (14) have demonstrated the technological feasibility of this heterogeneous germination of air bubbles directly on the particles to be floated when the latter present marked hydrophobic characteristics. However, in the case of hydrophilic particles, the germination of the air bubbles occurs in the homogeneous phase within the aqueous phase and does not permit flotation.

To implement the deep well dissolved air flotation tank, the deep well technique is used. In this technique, as the water level decreases, the air phase gradually dissolves in the water to be treated. When the water level increases again, the resulting gradual expansion allows controlled outgassing orientating the formation of air bubbles towards the heterogeneous phase at the very surface of the particles to be floated. A simple settling tank at ground level ensures the desired separation.

These original and theoretically particularly efficient flotation tanks are under development in cooperation with Elf.

4.5.4
Further technological innovations

Skimming techniques (Thangtongtawi, 1988)

Skimming techniques often concern the recovery of oil films or slicks on the surface of water. It was in this field that the Laboratory's contribution was decisive in allowing Elf to market its selective oleophilic oil drum skimmers. These drums are capable of continuously recovering oil floating on the surface of the API tanks of refineries without water. This precise oil recovery allows the oil to be directly recycled into the distillation units and limits the evaporation losses mentioned earlier.

The focus of this study was the development of a fluorinated macropolymer material with low surface energy and having selective wettability relative to hydrocarbons in the presence of water. The system study, which followed on the influence of the various operating parameters, enabled a reliable method to be set up for sizing these devices. This sizing method was based on the dimensional analysis methodology widely used in process engineering. The same methodology was also applied to the case of sizing the skimming disks and bands.

The photograph of Fig. 11 shows a unit of one of these oil drum skimmers placed on an API settling tank at a refinery. Figure 12 shows a STOPOL unit for recovering oil slicks with oil drum skimmers.

These very simple devices may become generalized in the short term as they provide hydrocarbon or oil recovery systems that quickly pay for themselves in industrial use.

Fig. 11. Industrial unit – Oleophilic oil drum skimmers placed on an API settling tank – Ambes Refinery. (Courtesy of ELF)

Fig. 12. Stopol unit for recovering oil slicks. (Courtesy of ELF)

Membrane techniques (Lee, 1984 – Belkacem, 1995)

Research efforts conducted at the Laboratory essentially concern the ultrafiltration treatment of used cutting oil emulsions from mechanical engineering workshops. C. Cabassud's paper will deal with this last part of our research in more detail.

The Laboratory attempted and perhaps succeeded in solving certain specific cases involving the two limiting problems inherent in all membrane processes:

1. low permeate fluxes connected with the well known phenomena of concentration polarization and concentration factor and
2. membrane clogging.

The Laboratory approach was first to understand these limitations [Lee, 1984] and then take action to avoid them (Belkacem, 1995). Aqueous cutting fluids from mechanical engineering shops are of two types:

1. Macro-emulsions with their characteristic milky aspect. They correspond to oil-in-water emulsions stabilized by surfactants that are generally anionic and by cosurfactants that are generally of the alcohol type. The diameter of the micro-droplets of oil is usually less than a micron.
2. Transparent or slightly translucent microemulsions containing higher concentrations of surfactants and cosurfactants with considerably less basic oil. They are composed of micro-droplets of oil having sizes between 100 and 500 angstroms (Å).

The technological innovations made in the cutting fluid field mainly concern the treatment of milky macro-emulsions by ultrafiltration (Belkacem, 1992; 1995).

Polarization

The Laboratory has shown that it is possible to avoid the formation of a polarization gel, which is very unfavorable to the permeate flux, by adding a salt such as calcium chloride to the cutting oil to be treated. At very low concentrations this does not lead to a breaking of the emulsion (i. e. the separation of an oily phase by settling). Adding the salt allows a part of the anionic surfactants of the macro-emulsion to form complexes, which considerably reduces the anticoagulant effect. After this partial destabilization, the emulsion can be coalesced at the membrane surface at polarization layer level. The ultrafiltration membrane thus becomes transformed into a particularly effective surface coalescer and the gel-type polarization layer is avoided. This leads to permeate fluxes four times as high as with conventional ultrafiltration. It is important to note that these considerable increases in permeate flux are only observed for very hydrophilic membranes. With less hydrophilic membranes the permeate flux falls due to the formation of a film of coalesced oil drops at the membrane surface.

Total coalescence of the oil drops at the membrane surface and their non-wettability at the surface relate to the oil drops being dragged away in the concentration circuit and separated by settling in the storage tank, where, with time, a layer of oil forms on the surface. Therefore, the concentration factor has no noticeable effect on the oil concentration in the emulsion that recirculates in the concentration circuit or on the emulsion's viscosity. Thus in this process, patented by Elf (Belkacem, 1992), the oil concentration of the emulsion in the concentration circuit remains constant over time and the permeate flow is not affected by the concentration factor. In contrast, with the conventional process, the permeate flux falls sharply when the concentration factor rises, thus leading to its limitation. This is not the case if we apply our specific ultrafiltration process, which constitutes a major technological innovation in the field of aqueous macroemulsion cutting fluid treatment.

Membrane Clogging

The Laboratory was able to solve the limiting problem of membrane clogging. The Laboratory was able to show that this is very closely linked with an impregnation of the membrane by cutting oil which gradually limits its permeability to water.

The regeneration solution proposed consisted, as in the case of assisted petroleum recovery, of performing a washing cycle using a specific micellar solution made up of a mixture of anionic surfactants and cosurfactants (Lee, 1984).

The results obtained were very impressive, the initial permeability of the membrane was restored and its hydrophilic property increased. In addition, since the micellar solution was a very good solvent it could be recycled many times.

4.5.5
Conclusion

The principal interest of this discussion was to present an overview of the various lines of research followed in the physical chemical treatment of water. In addition, this discussion presented the integration of this research into the more general context of the future developments in wastewater treatments during the coming decade.

4.5.6
References

Abadie A, Roques H (1970) Brevet français N° 70.07641
Aoudjehane M (1986) Traitement par flottation des effluents aqueux chargés en hydrocarbures. Diplôme de Docteur Ingénieur N° 167, INSA Toulouse
Aurelle Y, Abadie A, Roques H (1972) Brevet français N° 72 23768
Aurelle Y (1974) Contribution à l'étude du traitement des eaux polluées par des hydrocarbures émulsionnés par coalescence sur résines oléophiles. Diplôme de Docteur Ingénieur N°414, Université Paul Sabatier Toulouse
Aurelle Y (1980) Contribution à l'étude des mécanismes fondamentaux de la coalescence des émulsions sur lit granulaire. Thèse de Doctorat d'Etat N°954, Université Paul Sabatier Toulouse
Aurelle Y, Roques H, Angles M, Blazejczak J (1985) Brevet français N° 85.14969
Belkacem M (1995) Nouvelle méthodologie dans le traitement des huiles de coupe par ultrafiltration. Thèse de Doctorat N°336, INSA Toulouse
Belkacem M, Julien E, Hadjiev D, Cotteret J, Aurelle Y (1992) Process for ultrafiltration of stabilized emulsions. Brevet français N° 92.07262
Cherid S (1986) Conception et étude de nouveaux séparateurs lamellaires hydrocarbure-eau de type Spiraloil. Diplôme de Docteur Ingénieur N°165, INSA Toulouse
Colman DA, Thew MT (1988) Cyclone Separator. US patent N° 4,764,287
Dupré V (1995) Etude des mécanismes et de la modélisation des procédés de flottation à air dissous. Thèse de Doctorat N° 325, INSA Toulouse
Lee SB (1984. Contribution à l'étude de l'ultrafiltration des émulsions d'huile de coupe. Diplôme de Docteur Ingénieur N° 118, INSA Toulouse
Ma BF, Hadjiev D, Seureau J, Aurelle Y (1992) Séparateur triphasique à cyclone. Brevet français N° 92.04608
Ma B (1993) Epuration des eaux résiduaires de l'industrie pétrolière par hydrocyclonage. Mise au point d'un nouveau hydrocyclone triphasique. Thèse de Doctorat N° 234, INSA Toulouse

Tapaneeyangkul P (1989) Etude et modélisation d'une nouvelle génération de coalesceurs: Les coalesceurs à garnissage fibreux dynamique.Thèse de doctorat N° 93, INSA Toulouse

Thangtongtawi S (1988) Contribution à l'étude et au dimensionnement de récupérateurs de surface du
type tambours et disques déshuileurs.Thèse de Doctorat N° 72, INSA Toulouse

Yao K, Habibian MT, O'Melia CR (1971) Water and waste water filtration: Concepts and applications.
Environmental Science and Technology, 5, N° 11, 1105-12

Zhu S (1990) Etude des traitements physico-chimiques d'épuration des émulsions d'huile de coupe.
Influence de leur formulation. Thèse de Doctorat N° 115, INSA Toulouse

4.5.7
Further Readings

Title: New technology for wastewater treatment

Author: Secerov-Sokolovic, R.; Sokolovic, S.; Galesev, B.

Conference Title: Proceedings of the 16th Biennial Conference of the International Association on Water Pollution Research and Control – Water Quality International'92 Conference Location: Washington, DC, USA Conference Date: 1992 May 24–30

Source: Water Science and Technology v 26 n 9–11 1992. p 2507–2509

CODEN: WSTED4 ISSN: 0273-1223

Abstract: New technology for wastewater treatment has been designed as modular technology. For specific physical-chemical operations, a new equipment has been developed on the basis of modern principles of the processed industry. The following modules represent the main elements of the suggested technology: 1. Multipurpose column reactor (KA/00); 2. Deep bed filter for separation of suspended solids (SUM); and 3. Coalesced filter (W0/04). Pilot plant investigation of the new technology was tested on the wastewater from a car wash shop. The efficiency of the modular technology was observed using the changes of the following parameters: COD value, suspended solids content and oil content. The average decrease of COD value is 83.2%, suspended solids content 94.5% and oil content 97.8%, in comparison to the values before treatment by modular technology. (Author abstract) Refs.

Title: Exploring wastewater treatment. A treasure chest of technologies

Author: Goldman JC Jr; Bowen, Paul T

Corporate Source: Metcalf & Eddy, Atlanta, GA, USA

Source: Pollution Engineering v 24 n 15 Sep 1 1992. p 57–62

CODEN: PLENBW ISSN: 0032-3640

Abstract: Although the selection of a specific treatment technology will depend on the actual waste stream to be treated, currently available physical-chemical processes may prove to be economical and reliable alternatives for many waste streams. An examination of the range of options available to modern industry for wastewater treatment requires an understanding of the most frequently applied phys-chem processes, innovations to those common processes, and newly developed technologies. The following processes are addressed: adsorption, coagulation, precipitation, clarification, flotation, filtration, technical innovations.

An Integrated Approach to Industrial Wastewater Management
Daniel P Gallagher
Environ Sci & Engineering, Peoria, IL; Natl Environ J,
Sep – Oct 93, v 3, n 5, p 42(6) journal article
Integrated environmental regulatory action in the US is prompting industries to adopt integrated waste-wastewater management schemes. Effective wastewater treatment trains that can be incorporated into a multimedia management approach are described. Biological treatment techniques addressed include activated sludge, fixed-film systems, and upflow anaerobic sludge blanket systems. Membrane separation equipment, activated carbon systems, ion exchange, and chemical oxidation are also discussed, followed by a summary of source reduction and waste minimization benefits. (1 photo)

4.6
Hydrocyclone Based Treatment Methods for Oily Wastewaters

DIMITRE HADJIEV · IAN SMYTH · MARTIN THEW · YVES AURELLE

Abstract

The successful introduction of oil/water separating hydrocyclones into the offshore oil industry over the last ten years has provided a basis for the further development of the technology and scope for applications in the wider field of wastewater treatment where liquid contaminants are a problem; e.g. removal of fats and oils from food processing and agricultural effluents. Central to this development process has been establishing the limits of hydrocyclonic water deoiling capabilities, especially in streams contaminated with other components, and the extension of the hydrocyclone concept to achieve multi-component separations (e.g. oil and sand from water) in a single unit. These aspects will be discussed with reference to both laboratory tests and field data.

Where more rigorous treatment is required, the potential integration of hydrocyclones with other wastewater treatment technologies will be considered, notably membranes. In such systems hydrocyclones would typically function as primary separators, although they could also be employed to remove coalesced drops from recycled membrane concentrate streams.

4.6.1
Introduction

The development of the hydrocyclone as a liquid-liquid separator in the early 1980's was largely driven by the demands of the offshore oil industry for an increased produced water treatment capacity. This could not be met by extending existing plate interceptors and gas flotation systems because of space/weight limitations but the deoiling hydrocyclone provided a compact, efficient and cost effective alternative which has now become the dominant technology for this application, initially supplementing and later replacing the conventional systems. After outlining the operating characteristics of deoiling hydrocyclones, this paper focuses on how other constituents of produced water streams affect its oil-water separation function, leading on to an examination of how the hydrocyclonic principle has been extended to allow simultaneous separation of dense solids as well. The potential for some new applications of hydrocyclones for treating complex oily wastewater is also considered, together with how they might integrate with other separation technologies.

4.6.2
Characteristics of Deoiling Hydrocyclones

Hydrocyclones operate by imparting spin to a flowing liquid by tangential injection into a tapering cylindrical body to create an intense radial acceleration field such that density sorting of any dispersed material occurs very rapidly. The absence of moving parts and low flow residence times ($\geqslant 1$ sec) means they are both robust and compact separators.

The particular geometry type developed for treating oily water is illustrated in Fig. 1. Compared with the more conventional dense solids designs found in the mineral processing industry, the deoiling hydrocyclone is a relatively low shear and more elongate unit, reflecting the fragility and small differential density of the oil drops. The oil collects at the centre of the vortex and emerges in a reject or overflow stream which is typically only 1–2% of the throughflow, inlet oil concentrations usually being well below 1%. Higher oil levels can also be effectively treated if the size of the overflow is increased.

Control of the split (reject flow fraction) is readily achieved by applying back pressure. Separation efficiency increases with flow rate and where natural driving pressures are not available (as from oil-field reservoirs) the system would need to be pumped. Hydrocyclone sizes (diameters at the inlet) range from several tens to over a hundred mm, the smaller units tending to be more efficient and having lower pressure requirements but reduced capacities. Minimum operating pressures over this size range go from 2 to 6 bar, generating flows from 1–10 m³/h for indi-

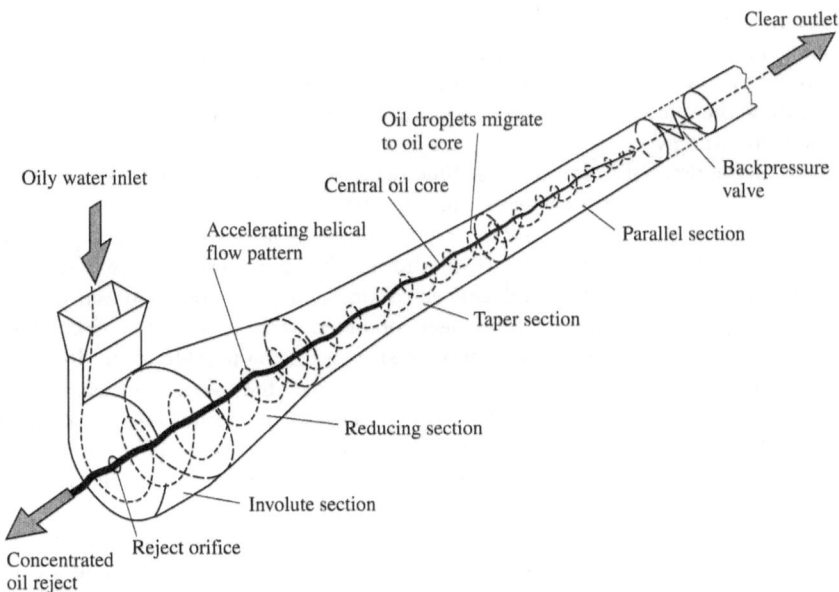

Fig. 1. Deoiling hydrocyclone schematic

vidual units with the potential to run at more than 3 times these flows if the pressure is available. Larger capacities would be achieved by parallel operation. Under favorable operating conditions, 50% of 5-10 μm oil drops could be expected to be separated in a deoiling hydrocyclone, rising to almost 90% at 20 μm (the separation efficiency is determined as the concentration ratio of oil in the water discharge to that in the inlet). More detailed background information can be found in (Smyth and Thew, 1990), (Thew and Smyth, 1992) and (Skilbeck and Thew, 1993).

4.6.3
Effect of Contaminants on the Deoiling Process
(Oilfield Produced Water Treatment)

While the deoiling hydrocyclone was initially developed in the lab using a range of stabilized oil/water systems, practically, oily wastewater can be much more complex. This section considers the influence of contaminants on the oil-from-water separation process, taking as an example the oilfield produced water application which has been the focus of much of the research effort for this type of hydrocyclone.

The treatment of produced water is typically a secondary process following the primary separation of commingled well-head fluids in large gravity separation tanks into gas, oil and water streams. The produced water tends to be hot (40-70°C) and saline and may contain between a few hundred and a few thousand mg/l oil. This would need to be reduced to below 40 mg/l for direct discharge from most offshore oil platforms, for example. In addition, the flow is likely to incorporate solids (from both the reservoir and pipeline), dissolved and possibly some free gas (light hydrocarbons) and a range of production chemicals. Determining whether or not such contaminants change the performance of the deoiling hydrocyclone is an important factor in effective prediction of field operation and evaluating the potential for other applications. This has been examined at in a number of studies.

The main area of interest regarding dense solids is relatively fine material (< 50 μm) as the coarse particles tend to drop out in the primary separator. Hydrocyclone tests with a Canadian crude (Terra Nova) and simulated produced water reported by (Simms et al., 1992) showed that the presence of a silica flour (15 μm mean diameter, 100 mg/l concentration) had little effect on the deoiling efficiency and similar results have been obtained with a fine rust simulacrum (Smyth, 1994). In both cases, solid particle sizes were comparable to oil drop sizes but, probably of more significance, was the fact that they were injected into the test rig as water wetted. Dense particulates that are oil wetted could potentially carry oil away into the water discharge stream. However, it is believed that the small scale turbulence present in hydrocyclones generates a scrubbing action which mitigates this effect.

Chemicals may be added to the produced water stream to specifically aid the treatment process, although it should be emphasized that typically they are not used in practice for deoiling hydrocyclones. In the work of (Simms et al., 1992), a reverse demulsifier (cationic) injected upstream of the hydrocyclone at a level of only 1 ppm produced a 15% increase in mean feed oil drop size (up from 23 μm) which was reflected in a modest 2 point rise in separation efficiency (up from 81%). The action of such chemicals encourages coalescence by displacing stabilizing interfacial films around droplets but in so doing reduces interfacial tension. The turbulent environment associated with hydrocyclones appears to promote rapid dispersion of the reverse demulsifier to the interfaces such that low dosages are effective but also provide a potential for droplet shearing which is increased by the lowering of interfacial tension. Testing at higher reverse demulsifier concentrations did indeed prove to be less effective. Further experimentation at the optimal low dosage in combination with other production chemicals – corrosion inhibitor, scale inhibitor and antifoaming agent – showed little change from the original test.

As the pressure progressively falls through the treatment train, it is likely that the produced water is saturated with gas at the feed to the deoiling hydrocyclones. It might be anticipated that gas bubbles would evolve as the pressure falls towards the axis of the separator and that the separation of oil could be enhanced by flotation mechanisms. However, recent research using both air and CO_2 saturated feeds (7.5 barg, 50 °C) and a Perspex model (Smyth, 1993, 1994) has shown no evidence of bubble generation within the hydrocyclone or of improved oil removal. Narrow gas cores were observed, however, discharging to the overflow. Only in the case of the considerably more soluble CO_2 (solubility at inlet conditions = 3.8 v/v cf. 0.11 v/v for air) was this to the noticeable detriment of the performance in terms of both a higher pressure requirement to achieve a typical working split and a fall off in oil separation. While the increased pressure drop to the overflow reflects a higher fluid discharge rate to the outlet, the loss of separation efficiency is presumably a result of the gas core displacing the oil core outwards, and so increasing the chance of oil being entrained in flow moving towards the water outlet. The gas most likely to be present in greatest abundance in the field is methane. This was not used in the lab tests for safety reasons but its solubility (0.18 v/v at inlet conditions) is most closely comparable with air, so it is anticipated methane would not affect deoiler operation.

Intriguingly, dissolved gas evolution in the form of dispersed bubbles does occur beyond the region of swirl in the water outlet due to the pressure let down across the hydrocyclone and with the further pressure drop across the control valve. This has been utilized by adding a low residence time (⩾1 min) skim tank (downstream enhancement vessel) after the hydrocyclone which has produced a further halving of oil levels in the water discharge in a number of installations (Schubert et al., 1992).

Free gas in the hydrocyclone inlet is potentially a more significant interference. Its impact depends on the degree of gas expansion through the hydrocyclone and the size of the overflow outlet. Tests with high pressure drops to an atmospheric overflow discharge (7.5 bar) and a conventional narrow oil stream outlet resulted in a halving of oil separation for 5% volume at the inlet (⩾40% of feed flow rate at atmospheric pressure). By 20% the deoiling function had been completely choked off (Simms et al., 1992). However, if the outlet is widened and/or back pressure applied, these amounts of free gas can be tolerated without significant performance disruption (Colman et al., 1980); (Skilbeck et al., 1993).

It appears, therefore, that the combined effect of these contaminants on the hydrocyclonic separation of produced water is limited, provided free gas levels are low, which is generally the case in the field. In other words, there appears to be only minor interaction between the various components of the flow. This implies that field separation performance can be effectively predicted from appropriate oil-water tests in the lab and this has been borne out where accurate analysis of oil drop sizes have been made on production platforms (Flanigan et al., 1989), at least for fairly fine dispersions ($\geqslant 20\,\mu m$). There is some evidence to suggest, however, that for coarser drops ($> 70\,\mu m$), which tend to be associated with higher oil concentrations ($> 1000\,mg/l$), the dispersion may be more stable and less prone to break up in real produced waters than in the test rig, with correspondingly higher separations achieved. This possibly reflects the short life (4–5 seconds typically) and, hence, relatively "fresh" interfaces of the artificially created drops (Simms et al., 1992).

4.6.4
Multi-Component Separation in a Single Hydrocyclone

As it was stated earlier, oily wastewater can be much more complex. In addition to oil, the flow is likely to contain a few hundred to over a few thousand mg/l dense solids. This would need to be reduced for direct reuse in most offshore oil platforms, for example.

The particular geometry type developed for treating three phase systems is illustrated in Fig. 2. It is similar to the more conventional dense solids designs found in the mineral processing industry with a flow reversal within the vortex finder. The latter is a relatively low shear and elongate part of the unit designed to maintain an intense radial acceleration field. In the lower part of the device the density sorting of any dispersed solid material occurs very rapidly. This underflow is recommended to be 3% of the throughflow, inlet particles concentration usually being well below 1%. Higher dense solid levels can also be effectively treated if the size of the underflow is increased. The oil collects at the centre of the vortex and emerges in a reject or underflow stream which is typically only 2.5% of the throughflow, inlet oil concentration usually being below 200 mg/l.

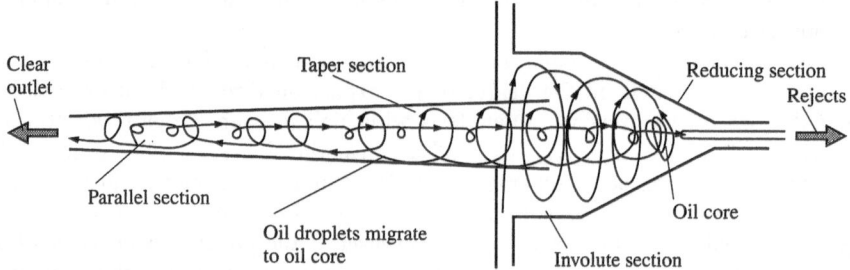

Fig. 2. Three phase hydrocyclone schematic

Control of the splits (reject flow fractions) is achieved by applying back pressure to both underflow and overflow (dense solids and water). Hydrocyclone sizes range from several tens to over a hundred mm, the smaller units tending to be more efficient and having lower pressure requirements. Minimum operating pressures over this size range go from 2 to 10 bar, generating flows from 1–10 m³/h for individual units. A 75 mm three phase cyclone can effectively match with 10 m³/h. Under favorable operating conditions 65 % of 25 μm oil drops could be expected to be separated in the three phase unit, rising to almost 90% at 50 μm oil drops and at 5 μm dense solids.

The main area of interest regarding dense solids is relatively fine material (<30 μm) as the coarse particles tend to drop out in the primary separator. Hydrocyclone tests with Pecorade crude oil and simulated produced waters showed that Durcal particles (6 to 16 μm mean diameter, 300 mg/l concentration) are effectively removed (Ma et al., 1993). Under favorable operating conditions 50% of 20 μm oil droplets could be expected to be separated, rising to almost 85% at 45 μm and in all cases 95% of 6 μm dense particles are separated.

4.6.5
New Applications for Hydrocyclones in Treating Oily Waste Waters

An application where hydrocyclones may have an impact is the treatment of used industrial cutting oil emulsions. These emulsions are very polluting, having organic oils, detergents and suspended solids, and so need to be treated and not discharged. The cost of the treatment is high due to the low level of organics (some 94–96% of water).

Conventionally, the primary separator is an ultrafiltration unit but the permeate flux is steadily decreasing due to problems related to membrane fouling and increasing oil concentration. Recently a new method was developed where a partial destabilization of the emulsion is proposed. It leads to higher permeate flux and the membrane is transformed into a surface coalescer where partial coalescence takes place. The formed drops have to be removed from the emulsion to maintain low oil concentration. As natural driving pressure is available it is interesting to include a deoiling cyclone in the process and separate the coalesced drops. Figure 3 shows a simplified separation process incorporating hydrocyclones. On-site tests have been carried out with a deoiler hydrocyclone under a range of conditions. It is of note that by operating at high recirculation rates a good oil removal rate was obtained. The oil content in the initial emulsion was 2% and the recovery rate was 90–95% of the coalesced drops, so that the oil concentration was maintained constant during ultrafiltration. The permeate flux was constant during the process and equal to 90% of that obtained for water ultrafiltration.

Another application where hydrocyclones may have an impact is the wool scouring industry. Scouring or washing is an essential first stage in the processing of wool. The scour effluent is very polluting, having a high level of organics and suspended solids, and so needs to be treated before being discharged. Also wool grease (lanolin), which represents the principal insoluble organic constituent, is commercially valuable and its recovery is, therefore, desirable. Figure 4 shows a simplified counter-current scouring process incorporating hydrocyclones. The

Fig. 3. Schematic of ultrafiltration rig incorporating hydrocyclones

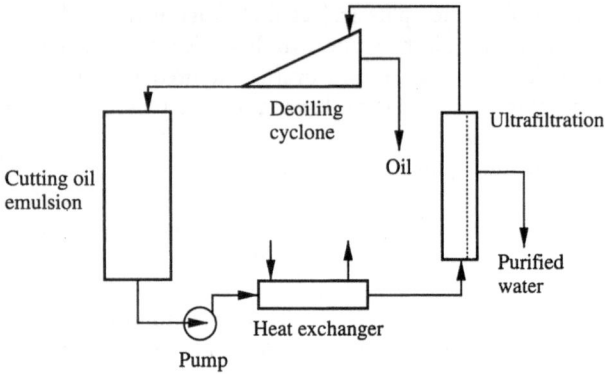

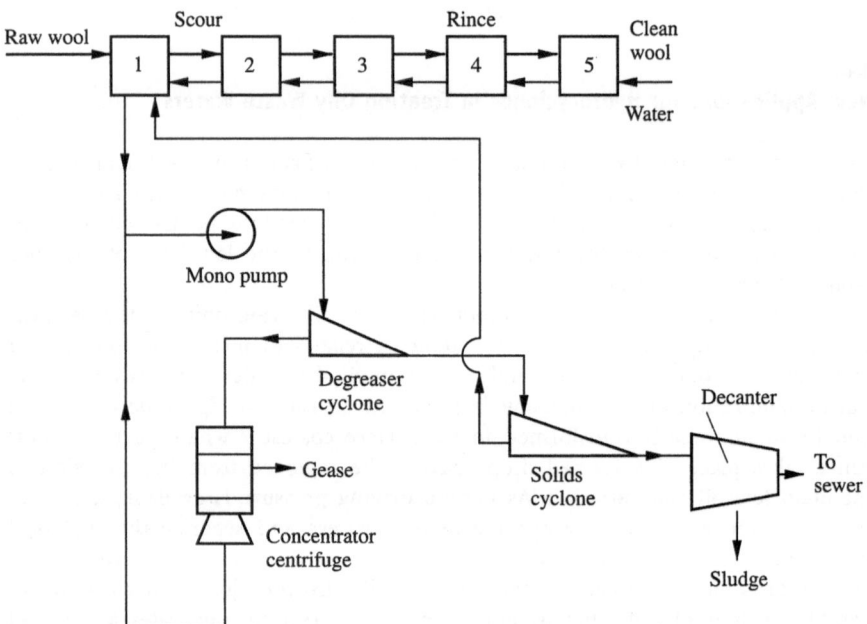

Fig. 4. Schematic of counter current wool scouring process incorporating hydrocyclones

wool passes through a series of bowls with detergent usually added in the first
few, the later ones providing a rinsing action. Hot water is added in the last bowl
and flows back to bowl 1 where it is discharged (at $\geqslant 65\,°C$), taking the contaminants with it. The primary treatment of this flow can then be dealt with by a
degreasing (effectively deoiling) hydrocyclone, the grease enriched "cream" being
concentrated in a secondary centrifuge, the bulk of the water being recycled back
to the bowl via a solids hydrocyclone. The solids-rich water stream from this

second hydrocyclone is then put through a decanter centrifuge to release a spadeable sludge from the discharge to the sewer.

Conventionally, the primary separator is a centrifuge and so it is against this that the hydrocyclone must be compared. Typical wool grease concentrations in the effluent are 2-3% and successful operation would be considered to be 35-45% recovery rates, a figure which reflects a fine grease droplet size, a density near to that of water (0.89-0.95 relative), and a close association between oxidized fractions of the grease and dirt particles. On-site tests have been carried out with a number of sizes and configurations of deoiler hydrocyclone under a range of conditions (Vortoil Separation Systems Ltd, 1992, 1993). While grease separation efficiencies achieved for the best geometry were roughly between a third and a half, those for the centrifuge, by operating at higher recirculation rates achieved comparable absolute grease removal rates at the expense of a relatively dilute overflow stream. It is of note that higher grease overflow concentrations were obtained during tests in which process changes were introduced to lower system shear, and although this might require additional processing to concentrate the grease, the potential exists to outperform the centrifuge by running hydrocyclones in parallel. Savings are obtained by the reduced operating and maintenance costs of hydrocyclones. Other potential advantages include lower noise levels and improved safety (centrifuges have been known to run dry and fail catastrophically). More significantly, increased grease recovery in itself would bring in a greater revenue from grease sales, while the lower grease levels in the scour bowls would allow reduced water consumption and also lower charges for wastewater treatment by the local sewage works.

Although not evaluated in the trials reported here, it is envisaged that the solid-liquid-liquid hydrocyclone described earlier could be also effectively utilized in this application.

4.6.6
Conclusions

The operation of a deoiling hydrocyclone is not significantly affected by the typical range of contaminants found in oil-field produced water, although the presence of free gas can prove disruptive without geometry adjustments.

A new design of hydrocyclonic separator capable of effectively treating a 3-component system (solid-liquid-liquid) has been demonstrated.

The possibility of new applications of hydrocyclones for treating complex oily/greasy waste waters are evident; e.g. in the processing of wool scouring effluent.

4.6.7
References

Colman DA, Thew MT, Corney DR (1980) Hydrocyclones for oil/water separation, 1st Int Conf on Hydrocyclones, Cambridge, Pub BHRA

Flanigan DA, Skilbeck F, Stolhand JE, Shimoda E (1989) Use of low shear pumps in conjunction with hydrocyclones for improved performance in the clean up of low pressure produced water, Paper SPE 19743, 64th Annual Technical Conf and Exhib of the Soc Petroleum Engrs, San Antonio, Texas

Ma B, Hadjiev D, Aurelle Y, Seureau J. Three phase hydrocyclone for simultaneous separation of solids from liquid-liquid mixtures. Canad J Chem Eng (Accepted for publication)

Schubert MF, Skilbeck F, Walker HJ (1992) Liquid hydrocyclone separation systems, 4th Int Conf on Hydrocyclones, Southampton (UK), Pub Kluwer

Simms KM, Zaidi SA, Hashmi KA, Thew MT, Smyth IC (1992) Testing of the Vortoil deoiling hydrocyclone using Canadian offshore crude oil, 4th Int Conf on Hydrocyclones, Southampton (UK), Pub Kluwer

Skilbeck F, Thew MT (1993) The development and application of liquid-liquid hydrocyclones for cleaning produced water, Offshore Water and Environmental Management Conf, London, Pub Business Seminars Int

Smyth IC, Thew MT (1990) Use of hydrocyclones in the treatment of oil contaminated water systems, 1st Int Symp on Oil and Gas Exploration and Production Waste Management Practices, New Orleans (USA), Pub US EPA

Smyth IC (1993, 1994) Internal reports, University of Southampton

Thew MT, Smyth IC, Hydrocylones for liquid-liquid separation, Recents Progres en Genie des Procedes, Vol. 6, N° 20 (1992)

Vortoil Separation Systems Ltd, Internal reports (1992, 1993)

4.6.8
Further Readings

Title: Concentration of oil-in-water emulsion using the air-sparged hydrocyclone
Author: Beeby JP; Nicol SK
Corporate Source: BHP Research, Newcastle Lab, Wallsend, Aust
Source: Filtration and Separation v 30 n 2 Mar–Apr 1993. p 141–146
CODEN: FSEPAA ISSN: 0015-1882

Abstract: The air-sparged hydrocyclone was originally designed for use in the mineral processing industry. Certain properties that make it very effective in that area have now been exploited in a wastewater treatment application – the removal of emulsified oil from water. Statistical design experimental procedures were used to investigate the interrelationship between several operating variables. This led to optimisation of the oil removal process with this unit and production of treated water with 95% removal of the turbidity resulting from the emulsified oil. The optimisation work also led to an understanding of the mechanism of oil removal. This is compared with the mechanism of separation of fine particles using the same device. The use of the air-sparged hydrocyclone in a one- or two-stage oil removal process was explored and efficiencies compared under various operating conditions. (Author abstract) 10 Refs.

Title: Separation of solids from produced water using hydrocyclone technology
Author: Lohne K
Corporate Source: Satoil-Facilities Engineering Division, Stavanger, Norway
Source: Chemical Engineering Research & Design, Part A: Transactions of the Institute of Chemical Engineers v 72 n A2 Mar 1994. p 169–175
CODEN: CERDEE ISSN: 0263-8762

Abstract: In addition to dissolved and dispersed hydrocarbons, produced water usually contains some solids. The solids are predominantly of reservoir origin and are usually coated with a thin film of oil. This paper discusses the origin and need for separation of such solids from produced water. The paper includes an identification of the adverse effects of the solids, which include erosion and environmental considerations. In the case of produced water reinfection, solids removal for the prevention of reservoir plugging is discussed. Basic process theory of solid-liquid hydrocyclones is presented. The paper also discusses the performance characteristics in produced water applications. Finally, results from a recent successful offshore pilot plant trial using a high pressure desanding hydrocyclone are presented. 7 Refs.

4.7
Application of Membrane Separation Processes to Oily Wastewater Treatment: Cutting Oil Emulsions

CORINNE CABASSUD · HUGO MATAMOROS · YVES AURELLE

Abstract

The problem of treating oily industrial effluents such as used cutting fluid emulsions arises in many types of industry, mainly in metalworking plants and mechanical engineering workshops. In France, these effluents represent a volume to be treated of about 300,000 m³ per year.

This document gives an overview of the state of the art in conventional treatment methods and practices, and presents more recent procedures based on membrane procedures, essentially of the ultra-filtration type. The interest and present limitations of these processes are discussed and examples of their industrial implementation presented.

There follows a summary of the most recent work performed by the authors, demonstrating three original and complementary lines of research which may lead to advances in the treatment of this type of effluent by membrane processes.

The first line of research takes the problem upstream, in the design of clean processes and products; i.e. it consists of studying and suggesting new formulations for cutting fluids which would be less polluting and easier to treat than the cutting fluids on the market at present.

The second is aimed at pushing back the economic limits of the ultrafiltration process when applied to the treatment of macro-emulsions. At the moment these economic limits are mainly imposed by the low permeate flows obtained because of concentration polarisation phenomena, which means that installations require too great a total membrane area when the processed flow exceeds 5 m³/h. A new method based on a combination of partial chemical destabilization and ultrafiltration has been studied, proposed and patented. The first results of this study show that high productivity gains can be achieved with this process under certain conditions.

Finally, the treatment of used micro-emulsion effluents with high soluble TOD loading was studied. The third line of research consisted of gathering knowledge in this domain on treatment processes using different membrane techniques, ultrafiltration, reverse osmosis and nanofiltration, so as to be able to propose a process and operating conditions resulting in an oil-free treated effluent having a reduced soluble TOD content.

4.7.1
Introduction

The legislation for industrial fluid wastes in France and Europe grows more rigorous, encouraging many industries to find a proper solution to their effluent problems. This study more specifically centers on the effluents composed of used emulsions of soluble oils encountered in the metalworking industries and mechanical engineering workshops. Cutting fluids are used for the lubrication, cooling, cleaning and protection of the cutting parts of machine tools in industries of many kinds, mainly the metalworking industries and mechanical or automobile construction workshops.

With time, these emulsions become less effective due to thermal degradation of the oils and/or bactericides and their pollution by various substances in suspension, by free lubricating oils or by the development of bacteria. They must therefore be replaced regularly and the effluents treated. Problems arise at several levels:

- as requested by the users, the emulsions are very stable products that may have a very high residual COD after clarification. The treatment must reduce the COD;
- because the water content is sometimes very high, transport and treatment costs are also high. An effluent concentration step may enable treatment costs to be reduced;
- the quality and volume loading of the oily effluents to be treated varies with time, particularly in mechanical or automobile construction shops. The effluent may arrive continuously or in batches, thus requiring a suitable treatment process of great flexibility.

Definition of Cutting Fluids

Cutting fluids must have both lubricating and cooling properties. A great variety of such products exists on the market, but they can be divided into two broad categories: straight cutting fluids, which are mineral or, more rarely, synthetic oils without water; and aqueous fluids which are diphasic oil/water solutions that spontaneously form an emulsion in water. The present study is centered on these aqueous fluids.

Composition of Aqueous Cutting Fluids

Their formulation is complex and may contain up to a dozen products, each playing a specific role. Six families of compounds can be distinguished:

1. The basic oils
   These are generally paraffinic, naphthenic or aromatic mineral oils and are the main component of most cutting oils.
2. Surfactants
   The most used surfactants are anionic, carboxylate compounds or sulphonates. Their essential function is to lower the interface tension between the oil and water of the emulsion. As the surfactants are localized at the oil/water interface they act as a mechanical barrier, through steric hindrance, and an electrical barrier. They limit the coalescence of the oil droplets and stabilize the emulsion.
3. Cosurfactants
   These compounds are usually of the alcohol type (butyldiglycol, benzyl alcohol) and are highly soluble in water. They encourage auto-emulsification by increasing the

solubility of the basic oil. In addition, they play a very important role in the stability of the emulsions as they are localized at the oil/water interface between the anionic surfactants. By limiting the repulsive forces, they allow the density of surfactants to increase at the droplet surface, thus enhancing the stability of the emulsion.

4. Corrosion inhibitors
   These are products such as fatty amines, fatty amides, calcium carbonate and sodium borate which prevent corrosion of the metal parts during machining.
5. Anti-foaming agents
   In certain cases the use of surfactants can cause foam to appear, which leads to problems further down the production line. Foaming can be avoided by incorporating microcrystalline waxes, silicone emulsions or aluminum soap into the cutting fluid.
6. Bactericides
   To check the proliferation of bacteria that make the cutting fluid rancid, products which gradually release formol and chemical products with a bacteriocidal or bacteriostatic action are added.

Classification of Cutting Fluids

Cutting fluids can be divided into two types of emulsion: micro-emulsions and macro-emulsions, differing by their composition, the size of the oil droplets dispersed in the aqueous phase, their physical aspect and their stability. Figure 1 presents an example of the typical distribution of components for a macro-emulsion and a micro-emulsion. Macro-emulsions come in the form of an opaque, milky product and contain oil droplets having diameters varying form 0.2 to over 10 μm. Micro-emulsions generally contain little oil (5 to 15% on the average) and have a higher proportion of surfactants

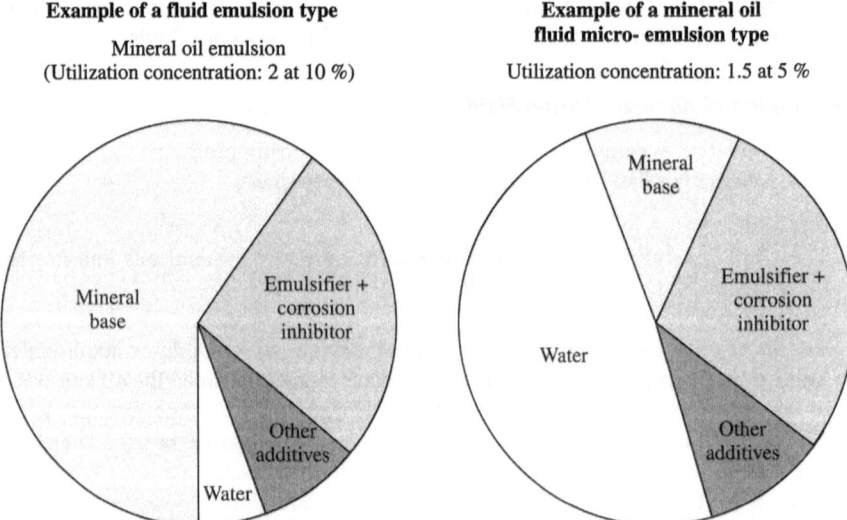

Fig. 1. Examples of standard components distribution in the emulsions

and cosurfactants relative to the oil content. Consequently, they are very stable and very easily auto-emulsifiable. These emulsions are transparent and are composed of very fine drops of oil (of the order of 10 to 150 nm) in the water. Their great stability has led to their adoption in the metal working industries in the past ten years.

4.7.2
Conventional Treatment Procedures

The most conventional and most widely used of all treatments is incineration. It has the essential advantage of being applicable to all aqueous cutting fluids without restriction, but has the great disadvantage of high expense (500 to 1000 FF per m³) because of the quantity of energy required.

The other traditional treatment process uses a preliminary step to separate the oily phase from the water with its dissolved organic substances. Each of the phases (oil and water) is then treated independently, the processes used depending on the nature and composition of the fluids.

Various methods can be used for breaking the emulsions, including:

- Thermal methods, which may involve hot breaking (65 – 80 °C) in an acid medium of pH 1 or 2, in conjunction with chemical destabilization (Degrémont, 1989) or multistage evaporation of the aqueous phase by evaproators with natural circulation, after elimination of particles and non-emulsified oils (Vandevenne, 1978). The major disadvantage of these processes is their very high energy consumption and their use of large amounts of cooling water. For the treatment of micro-emulsions, the problem is amplified because of the great quantities of water that evaporated. Solutions of this type contain 95 to 99 % water.
- Physico-chemical treatments based on destabilization of the emulsion, which is often called breaking or rupture of the emulsion, and is generally obtained by the action of chemical reagents like acids, salts or poly-electrolytes (Aurelle, 1979; Degrémont, 1989; Vandevenne, 1978). These are suitable for treating macro-emulsions but are difficult or, in certain cases, impossible to use for micro-emulsions. They may include a flocculation step for suspended or colloidal substances. The breaking procedures are specific to each effluent, work stoichiometrically can be compromised by a fluctuation in the quality of the effluent. This procedure is therefore more suitable for batch than continuous operation. Moreover, the use of chemical products entails operation and storage costs, and produces an effluent loaded with reagents which must be eliminated by further treatment.

Thermal or physical chemical treatments are followed by the separation of the two phases by settling and, more often than not, because the floating bodies have such low density, by air-or electro-flotation. The sequence of the various individual operations may vary with the composition of the fluids to be treated.

The complete method of thermal treatment combined with chemical breaking is thus relatively complex (see Fig. 2 for an example of configuration).

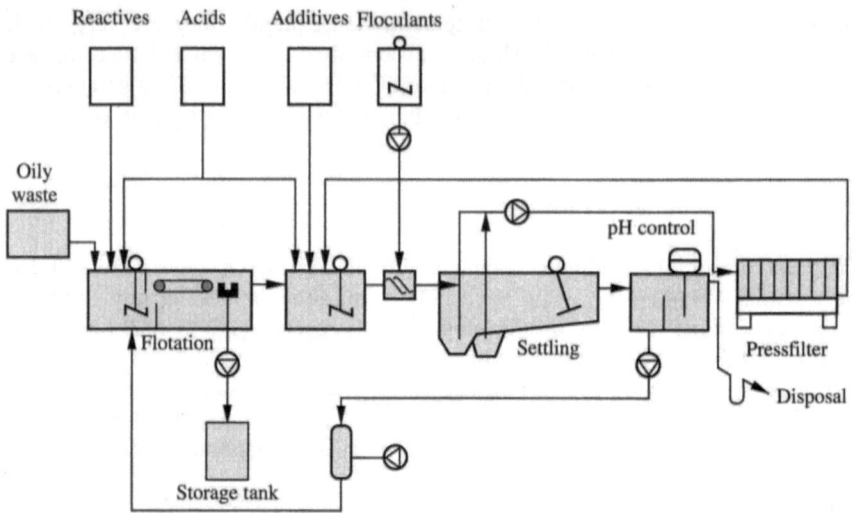

Fig. 2. Conventional treatment by flotation (from Eisenmann brochure)

4.7.3
Treatment Based on Membrane Technologies

Operating Principle

Procedures using membranes (ultrafiltration, reverse osmosis) can provide a partial remedy to some of the disadvantages presented by the conventional procedures.

The ultrafiltration of macro-emulsion type cutting fluids has been widely studied (Belkacem, 1993; Depeyre, 1990; Goldsmith, 1974; Lee, 1984; Lipp, 1988; Mahdi, 1990; Quemeneur, 1980).

It has been in common use for about ten years now, particularly in Germany, a county very conscious of environmental problems where, as early as 1979, there were over 250 ultrafiltration plants for the concentration of oily emulsions, with a capacity of 1 to 20 m³/day (Field, 1992).

In France, a recent study carried out by the national electricity company, EDF, (Valverde, 1994) showed that cutting fluid ultrafiltration units accounted for about 15% of the ultrafiltration plants in operation.

Ultrafiltration is a physical separation process that retains all the suspended matter (particles and colloids), bacteria, and oil drops, concentrates them and separates them from the aqueous phase. but all the soluble substances with low molecular weights (lower than the membrane cut-off threshold) can freely cross the ultrafiltration membranes.

Neutral or alkaline (if mixed with solutions coming from grease removal tanks) spent emulsions are first sent to a settling tank where the biggest suspended solids and the floating oil are removed. The next step of the treatment consists of coarse pre-filtration on band filters before the spent emulsion is pumped to an ultrafiltration unit.

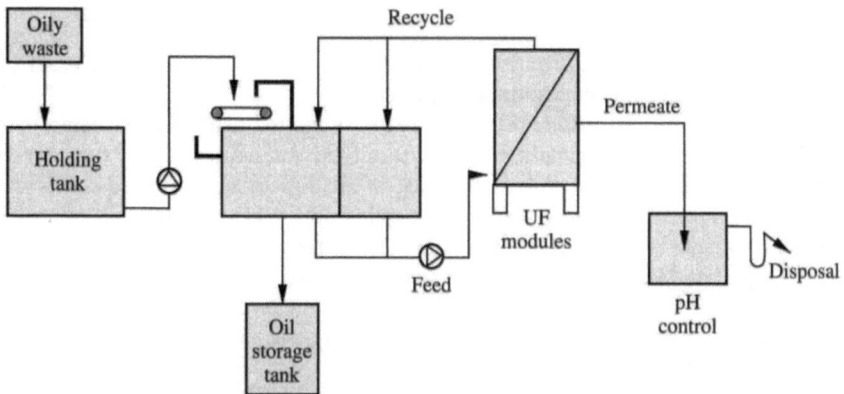

Fig. 3. Oil treatment by ultrafiltration (from Eisenmann brochure)

The membranes stop the oil drops and suspended matter while letting through the oil-free water containing soluble organic substances. When the concentrate has reached a concentration of 30 to 50%, it is sent to a used oil tank with the floating oil and may be neutralized then recycled or concentrated to 95% and destroyed by incineration. If necessary the pH of the permeate is readjusted and it is then discharged or used as process water if its COD, which depends on the raw effluent quality, permits this. Once a week, a cleaning solution is pumped through the ultrafiltration unit to unclog its.

The ultrafiltration method (Fig. 3) is competitive with the thermal or physical chemical methods described above (Fig. 2).

Interest and Limitations of Ultrafiltration Treatment

The method has many advantages:

- simplicity of the whole process, which comprises fewer individual operations than conventional processes;
- no chemical reagents needed (except possibly those used to balance the permeate);
- whatever the quality and flow rate of the effluent to be treated, the permeate has zero concentration of oil and suspended solids. This quality is not open to the risks connected with over-or under-doses of reagents;
- possibility of choosing between continuous and batch operation;
- process can be fully automated.

On the other hand, the productivity of membrane plants remains limited because of the well known problems of concentration polarization and membrane clogging and due to the fact that the oil concentration in the circulation loop increases with time.

Given the current price of membrane installations and the production flow rates considered for sizing them, certain engineering specialists assume that ultrafiltration is competitive in terms of investment for cutting fluid treatment if the flow to be treated is less than 5 m³/h. Beyond this, the conventional method using flotation is cheaper.

Treatment by Ultrafiltration and Reverse Osmosis

After conventional or ultrafiltration treatment, the effluent may in some cases be heavily loaded with soluble organic substances.

Some authors (Goldsmith, 1974) have already suggested using reverse osmosis to treat these aqueous effluents with a high TOD but few studies have so far been oriented towards this application, probably because of the high cost of reverse osmosis and the few incentives for depollution and water saving until recent times.

Reverse osmosis is a technique using dense membranes working under very high pressure (up to 80 bars) which are capable of rejecting dissolved salts and organic matter.

Examples of Industrial Applications of Membrane Processes

Example 1

This example concerns the treatment of the cooling and washing emulsions from a drink-can manufacturing unit (2700 l/h).

The unit set up by Eisenmann Company in Germany is used to treat an effluent containing synthetic oils in cooling and washing emulsions from mechanical engineering shops. It also serves to separate the emulsions obtained after passing through the main grease removing units. These emulsions contain sulphuric acid. The system chosen is based on the use of flat modules containing hydrophylic polymer membranes (TECH-SEP) and is particularly economical as 1600 l/h of oil-free filtrate can be recycled. On the basis of two-shift operation it was calculated that 6400 m³ of water would be saved. In fact, the production operates on three eight-hour shifts and a saving of 9600 m³ per year has been obtained. In comparison with a system using tubular modules (not shown), the installation would cut power consumption by 17.5 kW per hour of operation, representing an annual power saving of 105,000 kWh.

Example 2

After two years of feasibility trials on site, an industrial installation combining ultrafiltration and reverse osmosis was set up in 1976 in Pittsburgh, Pennsylvania, in the Alcoa mechanical engineering workshops (Sonksen, 1978). The installation treats a total of about 380 m³/day of oily effluent and produces an effluent that can be re-used. Since the investment had to be made quickly, the choice of membranes and operating conditions were not optimized. The ultrafiltration treatment system therefore remains relatively complex (see Fig. 4) and many problems were encountered when it was begun. The original feature of this installation is the use of reverse osmosis. The ultrafiltration permeate is pumped to a tank in which the reverse osmosis concentrate is recycled. About 20% of the reverse osmosis permeate is also recycled in this tank. The rest of the reverse osmosis permeate supplies an ion exchanger unit producing deionized process water, which has enabled exchanger regeneration to be reduced by at least 75%.

4.7.4
Limitations of the Process and Ways to Improve It

Although some membrane processes are already widely used for treating cutting fluids, ways of improving existing processes and innovations based on the use of other membrane processes can be envisaged.

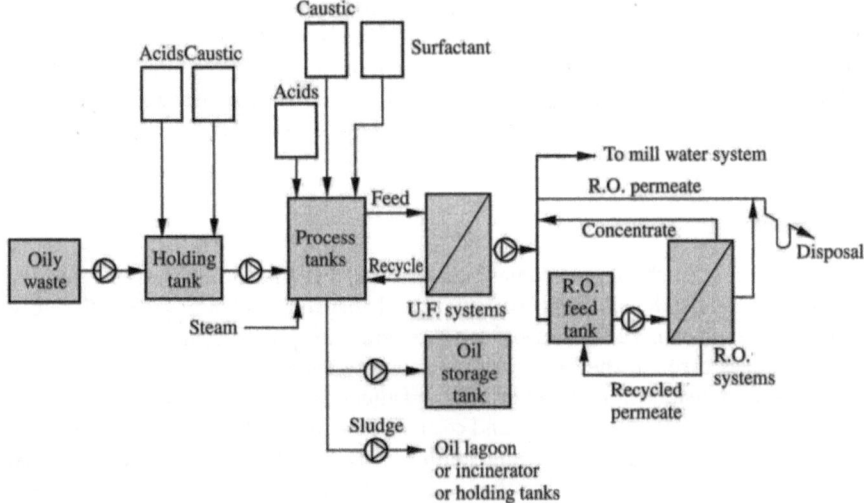

Fig. 4. Schematic of Davenport's membrane system for industrial oily wastetreatment

Three possible means of improving used cutting fluid treatment systems have been identified.

First Means

The first possible line of research consists of considering the cutting fluid treatment question by trying to anticipate the pollution problem and drawing up and proposing cutting fluid formulations that pollute as little as possible or are easy to treat. From a philosophical point of view and considering a global analysis of pollution fluxes, this is the most satisfactory method.

However, it can lead to changes in procedures for the users in mechanical engineering shops and changes in manufacturing methods, implying changes in industry costs. The challenge is then to meet the opposition to the development of these environmentally friendly products.

Second Means

Ultrafiltration is a technology widely applied to the treatment of macro-emulsion-type solutions but, for economic reasons, the method is reserved for units with low treatment capacities. Considering the advantages that ultrafiltration offers in terms of the quality of the effluents obtained and the ease with which they can be further treated, it appears important to eliminate the economic roadblocks to the process. These limits are imposed at present by the low permeate fluxes obtained, which lead to installations with too large a total surface area when the flow to be treated is greater than 5 m³/h. Another limitation of the process lies in the fact that the oil concentration operation necessary if the oils are to be recycled or destroyed by burning. This limit is imposed by the productivity of the membranes, which are too sensitive to concentration polarization effects that limit the mass transfer.

One way of improving the method would consist of putting forward processes or a set of processes which would reduce the concentration polarization and also increase the oil concentration in the treated effluents. These two objectives may seem contradictory at first glance, but they can be achieved, as will be demonstrated through the method entitled "Procedure for pushing back the economic limits of ultrafiltration treatment of macro-emulsions."

Third Means

Ultrafiltration is a technology that has been little used to date for the treatment of used micro-emulsions with very heavy surfactant and cosurfactant loadings. This is due to the fact that surfactants and cosurfactants are generally low molecular weight compounds that pass through the ultrafiltration membranes very easily. Ultrafiltration of these micro-emulsions thus allows the oily and aqueous phases to be separated, but the permeate obtained has a surfactant and cosurfactant content too great to comply with the discharge regulations concerning soluble TOD. Ultrafiltration thus does not satisfy the quality criteria imposed by the regulations. The second line of research explored in the field of cutting fluids was, therefore, to assess the limits of the application of ultrafiltration in the treatment of micro-emulsions and then to combine it with other membrane techniques showing greater selectivity with respect to surfactants and cosurfactants, such as reverse osmosis and nanofiltration, a technique that is currently gaining ground quickly thanks to the development of new membranes.

The third line of research consisted of gaining knowledge in the field of micro-emulsion treatment by different membrane techniques in an effort to propose and then optimize a procedure based on a combination of membrane processes suited to the treatment of micro-emulsions.

The following three lines of research correspond to three different but complementary ways of progressing in the treatment and understanding of cutting fluids.

4.7.5
Method 1: Development of Low-Pollution Cutting Fluids Known as "Green Products"

The aim of this prospective study was to develop new formulations of cutting fluid of the type called "zero pollution" after treatment, which should, in the long term, become a requirement in the European community.

The semi-synthetic cutting oils of the translucent micro-emulsion type are difficult to treat because they contain high levels of surfactants and co-surfactants since they have very little oil – around 5 to 15% – in the concentrates. If these micro-emulsion fluids are demulsified with the usual chemical techniques, various problems occur. The most difficult to overcome is the high residual water pollution due to the water-soluble alcohol-type cosurfactants used.

Similarly, for the cutting emulsions of the milky macro-emulsion type know as petroleum fluids, with a high mineral oil content (50 to 70% in the concentrate), alcohol cosurfactants are also included for auto-emulsification and stabilization of the emulsion. While it is easier to treat these macro-emulsions by chemical demulsification, the water-soluble residual pollution still exists due to alcohol cosurfactants.

The goals of the study were:

1. to develop new macro-emulsion formulations that create less pollution than those already on the market,
2. to develop new micro-emulsion formulations that are also less polluting and that are easier to treat,
3. to determine the most appropriate treatment for these new formulations and to compare their residual pollution to that of currently available cutting fluids.

Development of New Macro-emulsion Formulations

The mechanism of spontaneous emulsification (auto-emulsification) is dependent on the solvatation of the alcohols or cosurfactants in the water which, through diffusion, brings the oil towards the aqueous phase. This is how the spontaneous formation of cutting fluid emulsions occurs on mixing the oil with water.

This auto-emulsification mechanism occurring through solvatation of the cosurfactants lies at the origin of the majority of the pollution since, after chemical demulsification or ultrafiltration, the alcohols remain in the clear, purified aqueous phase and constitute the residual dissolved pollution. Research in limiting the use of highly water-soluble additives in cutting oil formulations was investigated (Yang, 1994). The first method sought to replace the emulsion stabilizer with a heavy, water-insoluble alcohol; auto-emulsification no longer being feasible, nonpolluting mechanical emulsification was used. This regression constitutes an unacceptable handicap from the commercial point of view. It is therefore important to develop new auto-emulsifying formulations to which the user has been accustomed.

The part of the investigation presented here attempted to discover a low-pollution formulation, if possible, as easy to use as products currently on the market; i.e. not requiring the addition of further compounds to the preparation water, thus not changing the user's habits. Therefore the method consisted of replacing the cosurfactants of the alcohol type with non-ionic surfactants which could offer the advantages of cosurfactants without presenting their disadvantages i.e. excessive water solubility. Owing to their non-ionic character, these surfactants can, like the alcohols, make the interface film of the oil microdroplets denser and thus contribute to the stability of the emulsions formed. Moreover, they have the advantage of presenting a very low CMC (Critical Micelle Concentration) and also reduce the CMC of the associated anionic surfactants. In addition, owing to the low interface tensions they give, they are often marketed under the name of dispersal agents.

After a systematic study (Yang, 1993) new satisfactory low-pollution formulations have been developed that could be used. For the sake of simplicity, only one is presented here. It was referred as NF7 and has the following composition:

- 81% basic oil (Nynas T 22)
- 15% anionic surfactants (Synacto 476, Exxon)
- 1% non-ionic surfactant (Simusol A, Seppic)
- 2% sodium oleate (soap)
- 1% water

It should be noted that this new formulation presents auto-emulsification properties similar to the equivalent commercial formulations, but the soluble residual pollution it presents after chemical demulsification with calcium chloride is much lower.

A commercial emulsion of type A (SARELF A, Elf France), with 4% volume of concentrate, after demulsification with calcium chloride, presents a soluble residual TOD of 3.6 g/l. In the same conditions, the new formulation presents a pollution of 0.3 g/l, i.e. 12 times less.

In addition, the NF7 emulsions are extremely stable. No destabilization was observed after three months of storage, the median diameter of the droplets measuring 0.5 micron.

Development of New Micro-emulsion Formulations

The principle mentioned above, i.e. the use of anionic surfactants as main surfactants and the replacement of the cosurfactants of the alcohol type, with non-ionic surfactants, was applied to the micro-emulsion concentrates.

The ratio cosurfactant/surfactant was fixed first. In the literature, most authors consider that the optimum weight ratio: cosurfactant/surfactant should equal 2 or above. The low value of 1 was chosen, since the aim was to replace the alcohol-type cosurfactants with non-ionic cosurfactants, which are relatively insoluble, but which possess poor anti-corrosion properties, and should be present in limited proportions in the concentrate.

The ratio basic oil/(cosurf + surf) was intentionally raised to a high level (equal to 0.75) compared to the products currently on the market – with values of 0.15 to 0.55 – for the following reasons:

• to lower the cost of the concentrates by limiting the quantities of emulsifying agents,
• to decrease the dissolved pollution,
• to limit foaming and sticking of the cutting fluids.

Numerous trails were carried out using anionic surfactants as main surfactants and non-ionic surfactants acting as cosurfactants and two basic oils (Nynas T22 – naphtene type and 100 Solvent paraffinic type from Elf Company). The formulation was considered interesting if:

• emulsification was easy,
• the emulsion was transparent,
• little foam was formed and, if formed, was unstable.

Then a certain number of basic formulations were selected which satisfy the three previously fixed criteria (Yang, 1993). Only two of these formulations are reported here:

Formulation 1: SA1

• 40.5% basic oil (100 Solvent – paraffinic type),
• 27% anionic surfactant (Synacto 476, Exxon),
• 27% non-ionic surfactant (Simusol A, Seppic),
• 5.5% bacteriocide (Sepicide HB, Seppic)

Formulation 2: SA4

• 41% basic oil (Nynas T22 – naphtene type),
• 27.5% anionic surfactant (Synacto 476, Exxon),
• 27.5% non-ionic surfactant (Simusol A, Seppic),
• 4% bacteriocide (Sepicide HB, Seppic)

Residual Pollution after Treatment of the New Formulations

In order to compare the residual pollution, the novel macro-emulsion (NF7) and the two micro-emulsions (SA1 and SA4) were treated using two different technologies; i. e. ultrafiltration and chemical demulsification with calcium chloride.

For separation by ultrafiltration, an Iris 3042 Tech-Sep membrane (Rhone-Poulenc) was used, which has a 50 kDa cut-off.

Table 1 gathers the results and compares the TOD of the two new formulations of micro-emulsions presented, SA1 and SA4, with respect to the commercial micro-emulsions, respectively C and D and the TOD of the macro-emulsion NF7 with respect to the commercial A macro-emulsion.

From these preliminary results, the following comments can be made:

- The new micro-emulsion concentrates SA1 and SA4 left much less pollution than the commercial micro-emulsions (refs C and D). The dissolved residual pollution was 2.5 to 9 fold lower.
- The lower residual pollution is obtained with the new NF7 macro-emulsion. The macro-emulsions (commercialized or not) are the least polluting fluids.

Conclusion

This study about new "green" cutting oils, has the merit of pointing out new directions to resolve the problem involved with the treatment of effluent form mechanical engineering workshops. In the long term, new, environmentally safe concentrate formulations should appear on the market, making the treatment of the effluents easier and enabling the users to treat own waste at lower cost.

4.7.6
Method 2: Procedure for Pushing Back the Economic Limits of Macro-emulsion Treatment by Ultrafiltration

The limits of mass transfer across the membranes have been clearly identified in the literature and correspond to the well known mechanisms encountered in most ultra-filtration applications; i. e. concentration polarization which may lead to the formation of a viscous gel on the membrane when the oil concentration reaches 30 to 50%, and flux limitation by adsorption of surfactants and cosurfactants on the surface or in the

Table 1. Comparison between the residual TOD of the new formulations NF7, SA4 and SA1 with respect to commercial formulations (A, C, D). Treated emulsions: 2% of concentrate in volume

Fluid type	Macro-emulsion		Micro-emulsion			
Designation	A	NF7	C	D	SA4	SA1
Chemical destabilization with salt						
$CaCl_2$, 2 H_2O (g/l)	1.5	2.0	2.0	2.0	4.0	4.0
residTOD (gO_2/l)	1.75	0.30	9.60	13.35	2.24	1.50
Ultrafiltration treatment						
residTOD (gO_2/l)	1.90	1.15	12.4	10.1	3.75	3.5

thickness of the membrane. These phenomena are aggravated by the oil concentration factor in the ultrafiltration loop.

The possibility of oil droplets penetrating certain pores of the membrane is also mentioned. This mechanism would lead to a marked fall in separation selectivity. This risk can be avoided if the pressure across the membrane is kept below the capillary pressure required for oil to penetrate the largest pores of the membrane.

The adsorption of organic compounds on or in the membrane is a slowly developing phenomenon that may cause a slow, gradual loss of permeability. The original permeability of the membrane can be restored by chemical regeneration procedures which are more or less complicated depending on the quality of the effluent to be treated and the type of membrane used. The frequency with which these procedures are necessary also varies with the composition and type of fluid treated. The problem of membrane regeneration is crucial as the procedures needed are often heavy, more or less effective, and produce an effluent that also has to be treated in most cases. This research theme has been previously studied (Belkacem, 1994a).

The principal questions to be considered for the sizing of an ultrafiltration installation are mainly connected with the concentration polarization and the incidence of the concentration factor in the ultrafiltration loop.

In an attempt to reduce the problems induced by polarization, this study sought to couple ultrafiltration with very partial chemical destabilization using a salt, the aim being to change the zeta potential of the oil drops, thus creating conditions favourable to an increase in drop size by coalescence.

Experimental Conditions

The trials were performed with an Amicon ultrafiltration cell working as a closed/batch separator and containing 350 cm^3 of solution to be treated. The cell was equipped with an IRIS 3042 (TECH SEP) polyacrylonitrile hydrophylic membrane with a cut-off threshold at 50 kDa. The useful surface area of the membrane was $3.7 \cdot 10^{-3}$ m^2. The mean transmembrane pressure was set at 1 bar with compressed air. The mixing speed was kept constant at 250 rpm. The permeate was continuously weighed on a precision balance, the output signal from which was sent to a microcomputer that calculated the instantaneous permeation flux.

At the end of each trial, the membrane was cleaned with a detergent solution, rinsed with distilled water and re-used for the next trial after a measurement of the permeability for ultrafiltrated water.

The solutions used for the study were prepared by diluting a cutting oil referenced SARELF A (marketed by ELF) which contained about 80% mineral oil and 20% surfactants, cosurfactants and other additives. The emulsions obtained after dilution contained 2 to 10% SARELF A oil by volume.

Experimental Results

Under the conditions stated above, during the ultrafiltration of an emulsion containing 4% oil, the permeation velocities fell very sharply in the first few minutes of operation and then stabilized at a low value of 13 l/h · m^2 at 20°C with a pressure of 1 bar transmembrane pressure.

When the trial was performed again in the same conditions, but following the addition of a salt such as calcium chloride to the emulsion at a concentration of 500 mg/l, the permeation velocities stabilized in 2 minutes, but this time at a value of 156 l/h · m² with the same transmembrane pressure. The simple fact of adding the calcium chloride had thus increased the membrane productivity by a factor of nearly 12.

For a better assessment of the effect of combining chemical destabilization and ultrafiltration, several tests were performed with various oil concentrations between 2 and 10 % and calcium chloride concentrations from 0 to 1500 mg/l, all other experimental conditions remaining the same.

Vp/Vpo was the ratio of the permeation velocities Vp obtained during ultrafiltration of the emulsion to the permeation velocity Vpo of clean water at the same pressure and mixing speed and at 20 °C. The analysis of the influence of the calcium chloride concentration on Vp/Vpo (Fig. 5) shows that, whatever the oil concentration, the addition of sufficient salt (the amount depending on the oil concentration) very significantly increases the permeation velocity.

For each oil concentration, the curve of Vp/Vpo against salt concentration shows three level portions observed for low, average, and high salt concentrations.

The first level is of no interest: it corresponds to salt concentrations so low that no improvement in mass transfer is obtained.

The third level, which appears for the highest salt concentrations and for which Vp/Vpo is equal to 1, expresses the disappearance of the polarization layer: the salt completely destabilizes the emulsion and the solution is no longer a diphasic mixture but a superposition of two layers: a layer of oil floating on a layer of water with no oil drops. The salt concentrations required for this total destabilization are of the same magnitude as those found in conventional chemical destabilization with stirring and settling but without membrane separation (Zhu, 1990).

The interesting operating zone is in the second level. For threshold salt concentrations, which we shall call the partial destabilization concentrations and which are much lower than the concentrations needed to completely break the emulsion, the permeation velocity can be multiplied by a factor of 12 to 16 relative to that obtained using the normal process without salt. In these conditions, the process is particularly attractive: it presents the advantage of physical separation processes, i.e. the gain of an oil-free permeate, while partly avoiding the limitations inherent in physical separation; i.e. concentration polarization. Its production flow rate also becomes competitive.

This beneficial effect arises as the coalescence of oil droplets chemically destabilize. The phenomena is furthered by the change of the zeta potential of the drops, the extraction of the anionic surfactants form the drops by the salt, and probably the reduction of the double layer repulsive forces due to the conducting role of the salt. In the case of coupling with ultrafiltration, these phenomena are amplified at the membrane and thus appear for lower salt concentrations because local shearing forces contribute to the coalescence. Also strong local concentration of oil leads to a higher probability of drops meeting each other. The drops coalesce on the membrane, which acts as a surface coalescer. When they reach a size, the limit value depending on the hydrodynamic conditions, they leave the membrane surface and contribute to the formation of a free oil layer floating on the aqueous phase. The local concentration at the membrane thus changes with time. The proof of these mechanisms is provided by the growing size of the drops and the formation of an oil layer whose thickness increases

Fig. 5. Evolution of Vp/Vpo versus CaCl₂ concentration for different oil-in-water compositions

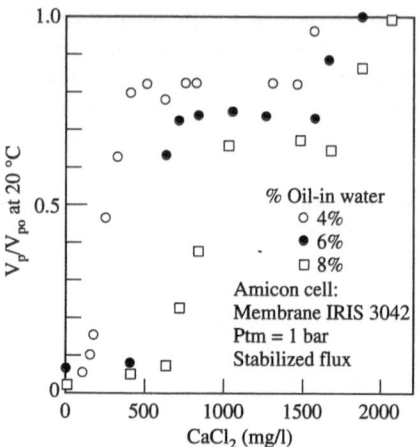

with time up to a limit value, together with the variation of the mean oil concentration (Belkacem, 1994).

The partial destabilization concentrations represent 29 to 62 % of the total breaking concentrations of the emulsion, depending on its oil content. The combination of chemical destabilization with ultrafiltration allows 28 to 71 % of the salt to be saved.

The higher the oil concentration in the emulsion and the more stabilizing surfactants and cosurfactants it contains, the higher the partial destabilization concentration/total breaking concentration ratio (cf. Table 2).

Beyond the partial destabilization concentrations the calcium content of the permeate is measurable and increases linearly with the mean calcium concentration in the concentrate. For salt contents lower than or close to the partial destabilization concentrations (500 mg/l for a 4 % emulsion) the permeate obtained by ultrafiltration shows a very low salt content. All the Ca⁺⁺ added must have been solubilized in the oil phase and used to extract the surfactants form the oil/water interface of the drops. Coupling with partial chemical destabilization does not, therefore, induce additional residual pollution of the aqueous effluent.

On the other hand, in all cases, the effluent treated by ultrafiltration has a high TOD (of the order of 3.5 g/l) which requires further treatment before discharge. This treatment may employ other membrane technology as discussed in the section below on micro-emulsions.

Table 2. Optimal CaCl₂ concentration for destabilisation in function of the oil-in-water concentration

Oil in water (%)	4	6	8	10	40
Cd = (CaCl₂) for destabilization (mg/l)	500	700	1000	1300	3200
Cb = (CaCl₂) for breakage (mg/l)	1700	1800	2200	2500	5200
Cd/Cb (−)	0.29	0.39	0.45	0.52	0.62
Cd/oil-in-water volumic ratio (gCaCl₂/l)	12.5	11.7	12.5	13.0	8.0
F/Fo (%)	80	72	65	61	47

Conclusion

This study has established that, in certain conditions, the productivity of ultrafiltration membranes applied to the treatment of cutting fluids may be greatly increased (multiplied by a factor of 12 to 16) by combining the ultrafiltration process with a partial chemical destabilization process. This does not add any more residual pollution to the treated affluents. This novel procedure has been patented (Belkacem, 1992). Most of the experimental results were obtained with a closed, batch operated ultrafiltration cell. The idea of combining partial destabilization and ultrafiltration is still being pursued by studying, in particular, the influence of parameters such as the membrane material and structure and the hydraulics of the ultrafiltration module on the performance levels to be expected from the treatment in terms of productivity and quality, in open, continuous ultrafiltration pilot and then in industrial plants. This new process could completely change the magnitude of the permeation velocity values presently used for sizing and evaluating industrial cutting fluid ultrafiltration installations, thus extending the field of application of ultrafiltration towards greater treatment capacities and entering into competition with flotation.

4.7.7
Membrane Based Processes for the Treatment of Micro-emulsions

This third line of research seeks to assess the potential of membrane processes for micro-emulsion cutting fluids, which are translucent, very finely dispersed emulsions with high soluble TOD loading. The strategy used to solve this problem consisted of dividing the treatment into two distinct steps. One concerned the concentration and separation of free oil and was based on phase separation, which could be done by ultrafiltration. The other was the retreatment of the aqueous phase obtained form the ultrafiltration (the permeate), with its high soluble TOD loading due to the surfactants and cosurfactants, in an attempt to bring it within the norms for industrial effluent discharge. The latter step could be performed using membrane techniques enabling the retention of compounds much smaller than those retained by ultrafiltration; i. e. reverse osmosis or nanofiltration.

Material and Methods

The experimental work required two types of installation: an ultrafiltration pilot plant and a reverse osmosis/nanofiltration pilot plant.

The ultrafiltration pilot plant
The ultrafiltration pilot plant shown in Fig. 6 had a UFP2 ultrafiltration cell (TECH-SEP) containing a flat IRIS 3042 hydrophylic membrane having a cut-off threshold of 50 kDa. the available filtering area was 0.01 m^2 and the membrane's initial permeability to ultrafiltered water was 270 l/h · m^2 bar at 20°C.

The untreated solution was stored in a 45-litre tank. The emulsion was pumped through the module by a variable speed volumetric pump. The circulating flow was measured with a magnetic flowmeter. The module output pressure could be controlled by means of the drain valve. Two taps up- and down-stream of the module allowed its

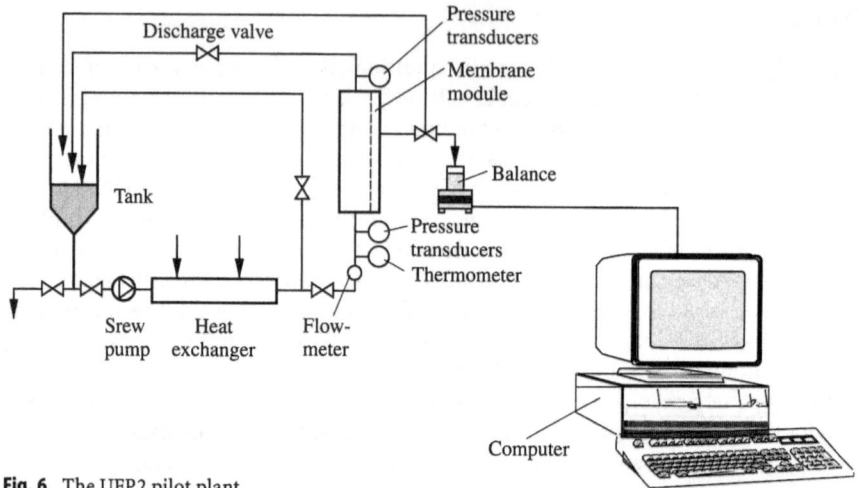

Fig. 6. The UFP2 pilot plant

input and output pressure to be measured. A heat exchanger at the pump outlet kept the micro-emulsion at a constant temperature of 20 °C. The mass of permeate obtained was measured continuously using a METTLER PM 6100 balance. A micro-computer made real-time calculations of the changes in permeate flow and the total cumulative volume produced versus time. The concentration of the micro-emulsion was kept constant in the ultrafiltration pilot plant by recycling the concentrate and permeate into the storage tank.

The emulsions were made by diluting straight cutting oil (noted SCO), supplied by ELF, with deionized water. This cutting oil contained about 40% mineral oil and anionic and non-ionic surfactants. It was biologically stable. The emulsions obtained contained 2 to 10% cutting oil in the form of fine droplets having a mean diameter of about 55 nm.

The reverse osmosis pilot plant
When the volume of permeate obtained after ultrafiltration was high enough (5 to 10 litres), the oil-free permeate was treated by the reverse osmosis/nanofiltration pilot plant. This installation (Fig. 7) comprised a stainless steel OSMONICS module (2) inserted in an anodised aluminum cell to which hydraulic pressure was applied. The available filtering surface was 0.0155 m².

For this study the membranes used were an OSMONICS MS 10 (polyamide material, cut-off 150 Da, initial permeability 2.2 l/h · m² · bar at 20 °C) and an OSMONICS SV 10 nanofiltration membrane (cellulose material, cut-off 600 Da, initial permeability 1.25 l/h · m² · bar at 20 °C).

The emulsion was stored in a 10-litre tank equipped with an internal, spiral thermostating circuit. A high pressure volumetric pump circulated the emulsion through the module. The filtration pressure was controlled with a drain valve and measured with a manometer. The measuring device (balance and software) of the ultrafiltration pilot plant was used to record the changes in permeate flow and total cumulative volume

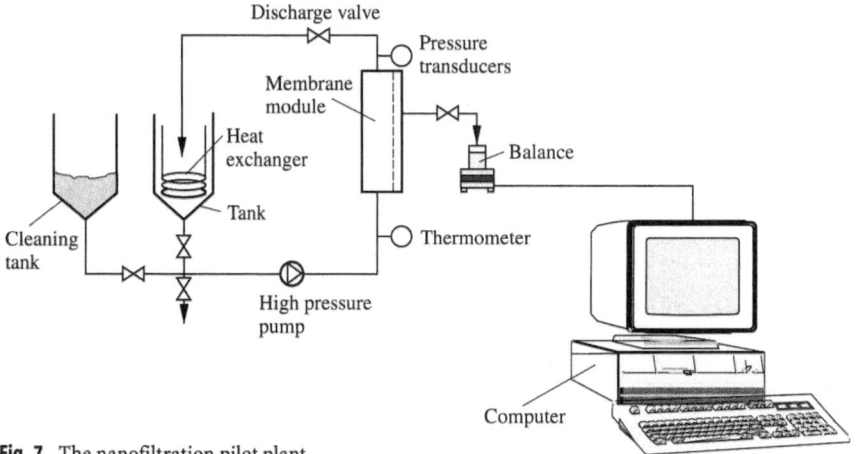

Fig. 7. The nanofiltration pilot plant

produced against time. For any given test, the emulsion concentration was kept constant by recycling the concentrate and permeate into the storage tank. The permeability of the membranes to water was checked before and after each test, after a water rinse.

The loading of the various concentrates and permeates in organic substances was determined by measuring the Total Oxygen Demand with an IONICS 1248 TOD meter.

Oil/Water Separation by Ultrafiltration

A rapid decrease in Vp/Vpo was observed in the first few seconds of filtration, followed by stabilization for the next 150 minutes. The higher the oil concentration, within the 2 to 10% range, the lower the flux obtained in the stable period. Under these operating conditions, the stabilized permeation velocity varied form 48 l/h · m² for an SCO concentration of 2% to 22 l/h · m² for an oil concentration of 10%. The curve of Vp/Vpo against concentration is given in Fig. 8.

It was observed that, whatever the oil concentration, the permeate obtained contained no free oil. The membrane had perfectly removed the fine oil droplets dispersed in the aqueous phase.

The low fluxes obtained could be explained principally by a reduction of the filtration by the polarization layer formed in the first few seconds. This was confirmed by the study of the influence of the mean circulation speed of the fluid to be treated in the ultrafiltration module.

Figure 9 plots the permeation velocity against time for an SCO concentration of 10% and a mean pressure of 1 bar across the membrane. The circulation velocity varied form 0.3 to 2.7 m/s.

For circulation velocities lower than 1.5 m/s, the permeation velocity varied in the usual way, i.e. a sharp fall during the first few seconds of filtration followed by stabilization. For a velocity of 1.5 m/s, a stabilized velocity of about 19 l/h · m² was obtained for the emulsion at 10%.

Fig. 8. Evolution of Vp/Vpo versus the SCO concentration in a new microemulsion

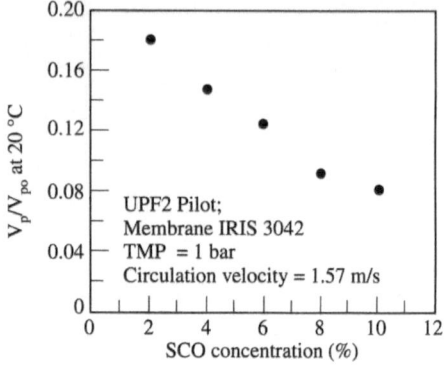

Fig. 9. Influence of the circulation velocity. Permeation velocity evolution versus time for a 10 % SCO emulsion-in-water

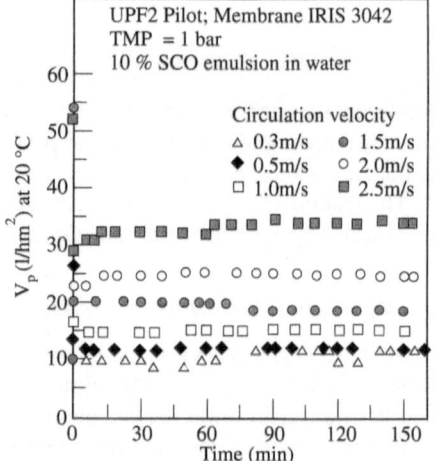

However, for circulation velocities greater than 1.5 m/s, the behaviour over time was most unexpected. The permeation velocity still fell sharply in the first few seconds but then rose by more than 10 % for a few minutes, continuing to rise gradually thereafter. This gradual rise was greatest for high circulation velocities and low oil concentrations. This behavior could be due to drops that had accumulated at the membrane surface being progressively dragged into the carrier fluid under the influence of drop coalescence and shearing forces or by a gradual change in the hydrophilic properties of the membrane during operation. The validity of these hypotheses is being investigated.

For circulation velocities between 0.3 and 2.7 m/s with a mean transmembrane pressure of 1 bar, increasing the circulation velocity results in a marked rise in the permeation velocity (see Fig. 10). For an emulsion concentration of 2 %, the curve of per-

Fig. 10. Comparison of the influence of the circulation velocity versus Vp for two new different SCO emulsion concentrations

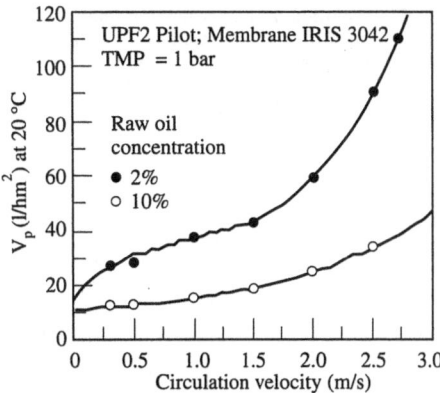

meation velocity against circulation velocity has still not leveled off at a circulation velocity of 2.7 m/s.

The well known ultrafiltration laws of the type $Vp = a V_c^b$ do not apply here. It is rather an exponential curve of Vp against Vc that is observed, which may be linked with the specificity of the behaviour of the permeation velocity over times for high circulation velocities.

As far as the quality of the permeate is concerned, there is no clear influence of oil concentration or circulation velocity on the purification efficiency, defined by:

Efficiency = $1 - (TOD_{permeate}/TOD_{concentrate})$

For circulation velocities between 03. and 2.7 m/s and oil concentrations between 2 and 10%, the purification efficiency remains around 82 to 85%. The permeates obtained do not contain any free oil but they do have an organic substance loading of about 6 or 7 gO_2/l for the 2% emulsion and 34 to 38 gO_2/l for the 10% emulsion. This very high soluble organic substance content is due to the passage through the membrane of low molecular weight surfactants and cosurfactants present in high concentrations in the emulsion. The ultrafiltration permeates do not satisfy the regulations concerning discharge standards and further treatment of the aqueous effluent is required.

Elimination of TOD by Reverse Osmosis and Nanofiltration

This second part of the study sought to assess the potential of reverse osmosis and nanofiltration to eliminate the TOD of the permeates resulting from the ultrafiltration of SCO emulsions.

A 10% SCO in water emulsion was ultrafiltered using the installation and membrane described above with a transmembrane pressure of 2 bars and a circulation velocity of 2 m/s at a temperature of 20 °C. Under these conditions, the permeate obtained had a residual TOD of 35 to 37 gO_2/l and a stabilized permeation velocity of about 38 l/h · m^2. The permeate was then treated by reverse osmosis and nanofiltration at pressures of 20 to 60 bars for the osmosis and 4 to 22 bars for the nanofiltration.

The flux stabilized in a few seconds for pressures of 20 to 30 bars and in 40 to 60 minutes for higher pressures (Fig. 11). The stabilized flux showed a linear variation

Fig. 11. Evolution of the permeation velocity versus time during reverse osmosis of a 10% SCO new UF permeate

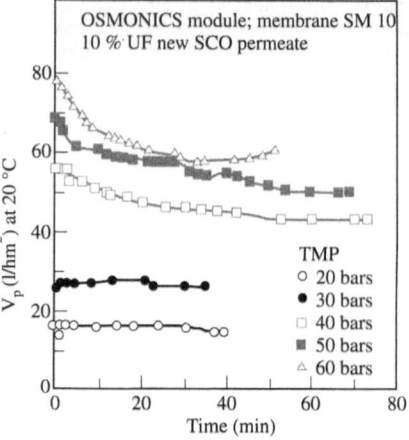

Fig. 12. Evolution of the permeate velocity versus the pressure applied in nanofiltration and reverse osmosis for a 10% SCO new UF permeate

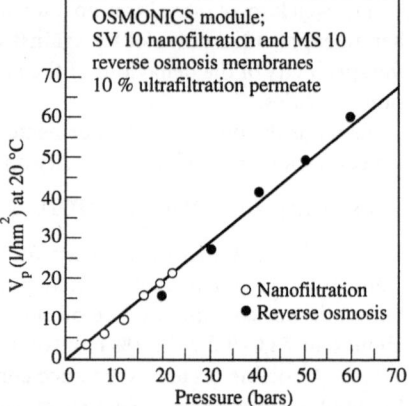

with pressure in the range of pressures studied (Fig. 12). The permeability obtained was about 2.2 l/h · m² · bar at 20 °C for the reverse osmosis and about 1.25 l/h · m² · bar at 20 °C for the nanofiltration.

The reverse osmosis permeate TOD (Table 4) was always higher than the nanofiltration permeate TOD (Table 3). Moreover the reverse osmosis permeate quality fell markedly when the filtration pressure rose (Table 4). The purification efficiency dropped form 94.1 to 63.0% and the TOD of the permeate increased about 6-fold when the pressure increased form 20 to 60 bars. For the most favourable case in quality terms, i.e. for a pressure of 20 bars, a combination of ultrafiltration and reverse osmosis reduced the TOD from 220 to 2.1 gO₂/l, an overall treatment efficiency of 99%. The power consumption for the whole treatment (UF + RO) was approximately 5.35 kWh/m³ of aqueous effluent treated. This figure needs to be validated and clarified on pilot installations with a greater production capacity, which will also allow the pump efficiency and pressure losses of industrial modules to be taken into account.

Table 3. Efficiency and nanofiltration permeate quality for a 10% SCO microemulsion

Pressure (bars)	4	8	12	16	20	22
TOD concentrate (gO₂/l)	37	37	37	37	37	37
TOD permeate (gO₂/l)	16	15.5	14.5	14.5	16.5	13.5
Treatment efficiency (%)	56.8	58.1	60.8	60.8	55.4	63.5

Table 4. Efficiency and reverse osmosis permeate quality for a 10% SCO microemulsion

Pressure (bars)	20	30	40	50	60
TOD concentrate (gO₂/l)	35.4	35.4	37.4	37.4	37.4
TOD permeate (gO₂/l)	2.1	2.1	7.9	8.1	13.1
Treatment efficiency (%)	94.1	94.0	76.5	77.0	63.0

Conclusion

Combining ultrafiltration and reverse osmosis or nanofiltration constitutes an efficient process for treating micro-emulsion cutting oils. The main role of the ultrafiltration is to eliminate and concentrate the free oil. It also acts as a pre-treatment that lowers the proportion of dissolved organic substances. The reverse osmosis or nanofiltration complete the treatment. With the ultrafiltration + reverse osmosis processes as a whole, TOD reductions of 94 to 99% have been obtained on a very fine emulsion (55 nm) concentrated at 10%. It has also been shown that treatment efficiency is greater for low reverse osmosis pressures. A technical and economical choice will be needed to find the best quality/operating cost/investment compromise.

Applying the combined treatment to used cutting fluid also yielded a purification efficiency of 99.3%, corresponding to a residual TOD of 1.01 gO₂/l for the treated aqueous effluent.

4.7.8
Conclusion

Pressure from the authorities to seriously consider the need for industrial oily effluent disposal, particularly for cutting oils, may led to the development and implementation of more efficient process hitherto considered too costly or bothersome or non-indispensable. The problem can be tackled in several complementary ways:

- formulation of environmentally friendly products that are non-polluting or easier to treat than the products on the market. This approach is conceivable even if it still requires investigation and optimization;
- improvement of ultrafiltration membrane effectiveness through progress and knowledge in membrane manufacturing techniques and ultrafiltration module design. This used in conjunction with mastery of the process and optimization of combination of various processes to take advantage of synergistic effects will allow the field of application of membrane techniques to be enlarged, all of which should ensure better quality of the treated effluent. In particular, the combination of partial chemical destabilization with ultrafiltration has been studied for treating macro-emulsion-type fluids. Such combinations of processes can increase the production capacities of ultrafiltration units and widen the potential market for these technologies in the treatment of cutting fluids. Our laboratory continues its studies to optimize the effects resulting from such combinations:
- use of combinations of different membrane techniques (ultrafiltration, microfiltration, nanofiltration and reverse osmosis) to treat fluids that are very finely dispersed micro-emulsions. The complementary nature of the separation processes makes it possible to obtain several different effluents, each with specific properties, which can be used to advantage or retreated independently in optimal conditions: an effluent composed of free oil that can be recycled or destroyed; an aqueous effluent containing a high concentration of organic compounds such as surfactants and can be discharged or used as process water. An example of a treatment sequence that could be achieved in the future is given in Fig. 13. We are continuing our studies so as to be able to put forward an optimal treatment plant. This approach shows the possibilities of integrating membrane processes in a clean process design which would take advantage to be taken of the products and ease the treatment of waste.

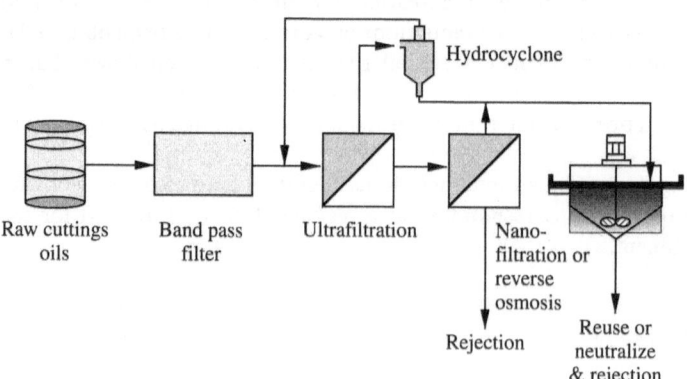

Fig. 13. Combination of different membrane technologies

4.7.9
References

Articles in Journal
Aurelle Y, et al. (1979) Traitement des huiles solubles par les polyélectrolytes. La technique de l'eau et de l'assainissement. 394: 27-35
Belkacem M, Matamoros H, Cabassud C, Aurelle Y (1994) New results in metal working treatment using membrane technology. J Membrane Sci 106 (1995) pp 195-205
Degrémont G (1972) Memento technique de l'eau. Degrémont
Field RW (1992) Oily-water clean-up: membrane processes and competitive technologies. 149-154
Goldsmith RL, Roberts DA, Burre DL (1974) Ultrafiltration of soluble oil water. Journal WPC 46: 2183-2192
Lee SB, Aurelle Y, Roques H (1984) Concentration polarization, membrane fouling and cleaning in ultrafiltration of soluble oil. J Membrane Sci, 19:23-28
Lipp P, Lee CH, Fane AG, Fell CJD (1988) A fundamental study of the ultrafiltration of oil-water emulsions. J Membrane Sci, 36:161-177
Quemeneur F, Schlumpf JP (1980) Traitement des huiles solubles par ultrafiltration. Entropie. 93: 22-29
Vandevenne L (1978) Procédés de traitement des émulsions usées d'huile soluble. La technique de l'eau et de l'assainissement; 382:27

Proceedings
Belkacem M, Matamoros H, Cabassud C, Aurelle Y, Cotteret J (1994) A new step forward in the field of metal working wastewater treatment using membrane technology. Engineering of Membrane Processes II, Tuscany, Italy, April 26-28
Belkacem M, Hadjiev D, Aurelle Y (1993) A model for calculating the steady state flux of organic ultrafiltration membranes for the case of cutting oil emulsions. ICOM 93. Heidelberg, Germany, August 30-Sep. 3
Depeyre D, Isambert A, Valter F, Mouihi M (1990) Fouling experimental studies in tangential ultrafiltration of oil/water emulsions. Vth World Filtration Congress
Mahdi S, Sköld R (1990) Membrane filtration for the recycling of waterbased synthetic metalworking fluids. Vth World Filtration Congress
Matamoros H, Cabassud C, Aurelle Y (1994) Traitement des huiles de coupe du type microémulsion par ultrafiltration et couplage ultrafiltration/osmose inverse. Interfiltra '94. Paris, France. Nov 15-17
Sonksen MK, MF Sittig MF, Maziarz FE (1978) Treatment of oily wastes by ultrafiltration reverse osmosis. A case history. Industrial Waste Conference. Lafayette
Valverde NC (1994) Synthèse de l'enquete Miss sur les techniques membranaires. Membrane et Environnement, Eurexpo, Lyon-France
Yang C, Canselier JP, Aurelle Y (1994) New, readily disposable, environmentally safe cutting oil formulations. International Seminar on surfactants/Detergents '94. Xian, PR China, Nov 10-13

Patents
Belkacem M, Hadjiev D, Julien E, Aurelle Y, Cotteret J (1992) Procédé d'ultrafiltration d'emulsions stabilisées. Eur Patent N° 92 07262

Thesis
Lee SB (1984) Contribution à l'étude de l'ultrafiltration des émulsions d'huile de coupe (Thesis). PhD INSA, Toulouse, France
Yang CZ (1993) Développement de nouvelles formulations de fluides de coupe peu polluants. Mise au point de techniques de traitement adaptées (Thesis). PhD INSA Toulouse, France
Zhu S (1990) Etude des traitements physico-chimiques d'épuration des émulsions d'huile de coupe. Influence de leur formulation (Thesis) PhD INSA Toulouse, France

4.7.10
Further Readings

Title: Filtration method efficiently desalts crude in commercial test
Corporate Source: USA
Source: Oil and Gas Journal v 91 n 20 May 17 1993. p 59-60
CODEN: OIGJAV ISSN: 0030-1388

Abstract: During 3 months of industrial testing of a filtration crude oil desalting method, a total of 120,500 metric tons (mt), or 1,475 mt/d (almost 11,00 b/d) of crude was processed. Rongxi Du, Kai Peng, and Li Wang, engineers at Wuhan Petrochemical Works, Wuhan, China, in an unpublished report indicate that they determined unit operating parameters and performed statistical analyses of desalting-efficiency data from the test run. The engineers also determined relationships between desalting efficiency and flow velocity, relative density, mixing pressure drop (MPD), filtration-tank pressure drop, and temperature. The desalting and dewatering level of single-stage filtration desalting was found to be equal to that of two-stage electrostatic desalting with remarkable benefits resulting from reduced power, water, and demulsifier requirements. This article describes the test, parameters and results.

Title: Treatment of oil in water emulsions by ceramic-supported polymeric membranes
Author: Castro, Robert P.; Cohen, Yoram; Monbouquette, Harold G.
Corporate Source: Univ of California, Los Angeles, CA, USA
Conference Title: Proceedings of the 1994 National Conference on Environmental Engineering
Conference Location: Boulder, CO, USA
Conference Sponsor: ASCE
Source: Critical Issues in Water and Wastewater Treatment National Conference on Environmental Engineering 1994, Publ by ASCE, New York, NY, USA. p 82–89
CODEN: NCEEDO ISSN: 0731-1516 ISBN: 0-7844-0031-8
Abstract: A novel membrane was developed by growing polymer chains from the surface of a porous ceramic support, resulting in a composite membrane which combines the mechanical properties of the inorganic membrane with the selective interactions of the polymer. The configuration of the grafted polymer brush layer is determined by solvent – polymer interactions with a hydrophilic polymer being stretched away form the surface by aqueous solutions and collapsed against the surface by organic solvents. This behavior of the grafted chains provides Ceramic-Supported Polymeric (CSP) membranes with unique properties for certain water treatment applications. One application envisioned for these CSP membranes, in which the selectivity is influenced by interactions between the solvent and the grafted polymer, is the cross-flow filtration of an oil-in-water emulsion. In this case, a hydrophilic grafted polyvinylpyrrolidone (PVP) brush layer expanded into the pore volume due to the affinity of the polymer for water. These extended grafted chains preferentially allow the passage of water over oil, producing a permeate stream with a lower total organic carbon content compared to an unmodified membrane. Another advantage of the CSP membrane, is in reducing permeate flux decline believed to be caused by the adsorption of oil onto the membrane surface. For the PVP-modified CSP membrane, the grafted polymer alters the membrane surface character from hydrophobic to hydrophilic, reducing the tendency or oil adsorption. This phenomenon was demonstrated by comparison of permeate flow rate behavior for both unmodified and graft polymerized (CSP) membranes. (Author abstract) 9 Refs.

Title: Cleaning of ultrafiltration membranes after treatment of oily waste water
Author: Lindau, J.; Jonsson, A-S.
Corporate Source: Lund Univ, Lund, Sweden

Source: Journal of Membrane Science v 87 n 1 – 2 Feb 23 1994.p 71–78
CODEN: JMESDO ISSN: 0376-7388

Abstract: The influences of different types of cleaning agents on a polysulphone ultra-filtration (UF) membrane which had been used to treat oily waste water were investigated. The cleaning experiments were performed with samples of polysulphone membranes removed from a commercial plant in which oily waste water is treated. Five different cleaning agents were used and their influences on the fluxes wee investigated. The influence on the flux when cleaning with the different gleaning agents in succession was also studied. Deposits on the membrane surface, before and after cleaning, were analyzed using different methods. Most of the analyses were carried out using scanning electron microscopy (SEM) in combination with energy dispersive X-ray (EDX) combined with a micro-analysis system permitting quantitative determination of elements. Some analyses were also performed using Fourier transform infrared spectroscopy (FTIR) and X-ray photoelectron spectroscopy (XPS). (Author abstract) 6 Refs.

4.8
Electrochemical Degradation of Organic Pollutants for Wastewater Treatment: Oxidation of Phenol on PbO$_2$ Anodes

ANDRÉ SAVALL · NOUREDINE BEL HADJ TAHAR

Abstract

The electrochemical oxidation of phenol for waste water treatment is studied on a PbO$_2$ layer deposited on three different substrates (graphite, titanium oxide and tantalum) in electrochemical cells without separator. Concentrations of phenol and reaction intermediates (hydroquinone, benzoquinone and aliphatic acids) were assayed by liquid chromatography. When using the titanium oxide or the tantalum substrate, no carboxylic acid accumulation has been observed after the aromatic ring rupture. Among the three electrodes studied, PbO$_2$ coated on a tantalum substrate gives the best performance for phenol degradation.

4.8.1
Introduction

The treatment of industrial waste water containing toxic and/or biorefractory organic substances is a priority due to the lack of satisfactory methods at present. In this case, electrochemical processes seem capable of offering an appropriate solution. The works of Stucki et al. (1991), Kötz et al. (1991) and Comninellis and Pulgarin (1991, 1993) have pointed out that the use of electrodes consisting of SnO$_2$ coated titanium allows a 90 % organic carbon (TOC) removal. On the contrary, under identical conditions, a Pt anode removes only 35 % of the TOC (Comniellis, Pulgarin, 1993). The SnO$_2$ having a poor conductivity, must be doped with antimony. It is prepared by spraying a SnCl$_4$ + SbCl$_3$ alcoholic solution on the titanium surface heated at 550 °C (Kötz et al., 1991). Although the technique of the preparation of these electrodes has been improved, their method remains insufficient for large scale use.

Among anode materials having a high oxygen evolution overpotential, it seems that PbO$_2$ can be used for organic waste water treatment by anodic oxidation. Its main advantage is its high electric conductivity. It can be easily prepared or regenerated via an electrochemical process, (Grigger et al., 1958; Narasimham and Udupa, 1976) and its electrocatalytic properties can be improved by doping (Gordon et al., 1994).

In this work, we studied the anodic oxidation of phenol in an acidic aqueous solution on three kinds of PbO$_2$ anodes. Two of the electrodes had been prepared via an electrolytic process in our laboratory.

4.8.2
Experimental Conditions

Electrode's Preparation

Electrochemical deposited PbO_2 with good electrocatalytic properties must be prepared. Thus, the deposit is carried out in a $Pb(NO_3)_2$ solution at a current density higher than 50 mA cm^{-2} (Grigger et al., 1958; Munichandraiah and Sathyanarayama, 1988). In this way, the formation of the β form is favored against the α one (Munichandraiah, 1992). The PbO_2 deposition mechanism consists of the following four steps (Chang and Johnson, 1989):

1) $$H_2O \rightarrow OH_{(ads)} + H^+ + 1e^-$$

2) $Pb^{2+} + OH_{(ads)} + OH^- \rightarrow Pb(OH)^{+}_{2(ads}$ (slow)

3) $Pb^{2+} + OH_{(ads)} + OH^- \rightarrow Pb(OH)^{2+}_{2(ads)} + 1e^-$ (slow)

4) $$Pb(OH)^{2+}_{2\ (ads)} \rightarrow PbO_{2(s)} + 2H^+$$

In this work, the PbO_2 electrodes have been electrolytically prepared at 65 °C using a current density of 50 mA cm^{-2} on a graphite substrate (diameter = 1 cm; S = 12,5 cm^2) from 150 cm^3 of a solution containing 350 g dm^{-3} of lead nitrate and 30 g dm^{-3} of copper nitrate (Narasimham and Udupa, 1976). Copper, which is less electropositive than lead, is reduced and prevents the formation of a dendritic lead deposit on the cathode. The counter electrode consisted of a lattic-work of iridized platinum (S = 75 cm^2) forming an homocentric cylinder around the anode. The solution's pH was set at 4–4.5 by addition of diluted nitric acid. During electrolysis, the acidity of the solution increases, because of the protons production (reactions 1 and 4). The produced acid has been continuously neutralized by progressive addition of lead carbonate.

The use of a small quantity of a wetting agent has proven very useful (Grigger et al., 1958; Narasimham and Udupa, 1976; Munichandraiah et al., 1987, 1992). It stabilizes the oxygen bubbles formed on the anode surface. These bubbles remain attached to the surface until they become large enough to be detached. The deposit obtained under these conditions is highly porous and has a specific area 15 to 40 times more important than its geometric surface (Munichandraiah, 1992). In this work, PbO_2 deposition has been carried out in the presence of sodium dodecyl sulfate. In order to obtain a solid deposit, with a satisfying adherence and good mechanical properties, the graphite substrate has been treated before electrolysis in a 10% (mass) NaOH solution 30 minutes and then neutralized in a 2 mol dm^{-3} nitric acid solution before being rinsed with distilled water. The electrochemical technique was also applied to prepare a PbO_2 layer on a tantalum substrate (plate of 4 cm · 1 cm, using an anodic current density of 20 mA cm^{-2}. A rod (diameter: 0.6 cm; S = 9 cm^2) of lead dioxide coated Ebonex (Atraverda, UK) was tested for comparison.

Electrolysis

Electrolysis has been carried out at controlled temperature in a pyrex glass cell using a 150 mA cm^{-2} anodic current density. The electrolyte was a phenol solution (V = 150 cm^3) with an initial concentration of 21 · 10^{-3} mol dm^{-3} and the pH was adjusted to

2 by addition of sulfuric acid. Two counter electrodes were used; a graphite rod (S = 12.5 cm^2) and an iridized platinum lattice-work (S = 75 cm^2). During electrolysis, temperature was kept constant at 70 °C to ensure the solubility of the formed products and avoid the covering of the anode surface with polymers formed by oxidation of phenol and its derivatives (Sharifian and Kirk, 1986).

Analysis

The advance of the degradation reaction was followed up by monitoring the phenol concentration as well as the concentration of its derivatives produced during this reaction. Samples of 0.5 cm^3 were taken regularly during electrolysis. They were twice diluted with a 0.1 mol dm^{-3} lactic acid solution (internal standard) and then rapidly analyzed by means of liquid chromatography (Hewlett Packard 1090) after filtration (0.45 µm). Products were separated through a column specific for organic acids (Hamilton PRP-X300, L = 25 cm and ϕ = 4.1 cm). To increase the lifetime of this column, a pre-column of the same characteristics (Hamilton PRP-X300, L = 2.3 cm) was used.

The wavelength used for the chromatograph's calibration and the analyses was 220 nm. For this choice, the absorbency of all species to be analyzed was taken under consideration. Analyses were carried out using a solvent which consisted of methanol in a $5 \cdot 10^{-2}$ mol dm^{-3} sulfuric acid solution, having a flow of 1.3 cm^3 min^{-1}. Methanol's proportion in this solvent varied linearly form 2 to 25% per volume during the first 10 minutes, then form 25 to 40% during the next 10 minutes and finally from 40 to 60% during the next 20 minutes. The fluent, very concentrated in sulfuric acid in the beginning of the analysis, achieved the separation of carboxylic acids during the first 10 minutes; thereafter, the increase of methanol's fraction in the solvent favored the separation of the aromatic products.

4.8.3
Results and Discussion

The mechanism of the phenol oxidation at SnO$_2$ electrodes was first studied by Kötz et al. (1991), Stucki et al. (1991) and then by Comninellis and Pulgarin (1991, 1993). Phenol was oxidized, first to hydroquinone and then to benzoquinone. In a second step, the opening of the aromatic ring produced aliphatic acids. The electro-oxidation of phenol at PbO$_2$ electrodes produced the same intermediates.

Electrolysis of Phenol on a Graphite Anode Covered with PbO2

Figure 1 shows the variation of the phenol concentration as well as of the concentration of the aromatic products of the oxidation during an electrolysis carried out with a PbO$_2$ coated graphite anode and a graphite cathode.

Phenol disappeared when a charge of 20 Ah dm^{-3} was applied. This result is comparable with the performance of platinum or tin dioxide anode (Comninellis and Pulgarin, 1993). At the same time, the hydroquinone and benzoquinone concentration increased, passing through a maximum at 20 and 10 Ah dm^{-3} respectively. Comparing these results to those of Comninellis and Pulgarin (1993) showed that on a PbO$_2$ anode, hydroquinone and benzoquinone concentrations were, respectively, 7 and 3 times greater than those obtained using a SnO_2 anode. In the same time, the degradation of these aromatic intermediaries was much longer on the PbO_2 anode than on the SnO_2 one (80 Ah dm^{-3} instead of 50 Ah dm^{-3} for the benzoquinone). On the other hand, no catechol was detected among the phenol oxidation products on the PbO_2 anode. It should be noted that, at the beginning of the electrolysis, benzoquinone formation rate was higher than that of hydroquinone (Fig. 1). When the phenol concentration was sufficiently decreased, the hydroquinone formation accelerated and its rate grew higher than that of the benzoquinone.

Maleic and fumaric acid appeared since the beginning of the electrolysis (Fig. 2). Their maximum concentrations wee reached after 20 and 10 Ah dm^{-3} respectively and were 7 and 8 times lower than those observed on SnO_2 (Comninellis and Pulgarin, 1993). Fumaric acid, though formed at the very beginning of the electrolysis, rapidly disappeared.

Chromatographic analyses show the formation of large quantities of formic and acetic acids (Fig. 3).

An evaluation of the number of carbon's atoms shows that the formation of these acids was not only due to the phenol degradation. Indeed, electrolysis of a sulfuring acid solution (10^{-2} mol dm^{-3}), using a graphite anode covered with PbO_2 and an iridio-platinum cathode showed the formation of the same quantities of oxalic and formic acids (Fig. 4) as found during the electrolysis of a phenol solution. So, formic acid resulted from the graphite oxidation.

The comparison of Figs. 3, 4 and 5 suggests that acetic acid formation resulted from the use of a graphite cathode.

The electrolysis of a phenol solution, under the same experimental conditions as those corresponding to Fig. 1 but with a Pt/Ir cathode, showed that hydroquinone is the main product formed (Fig. 6).

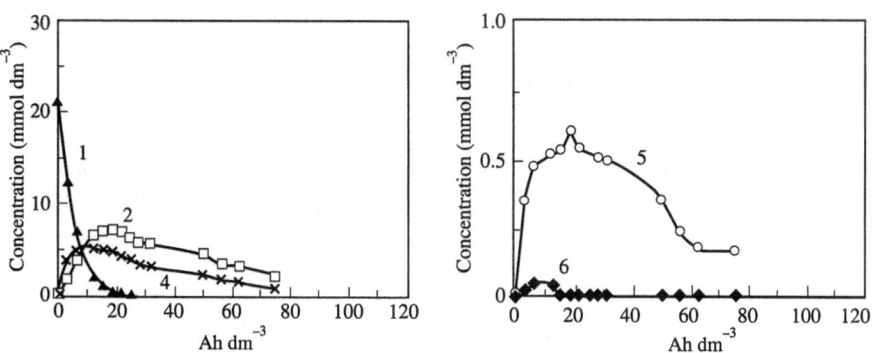

Fig. 1. Variation of the aromatic product concentration during electrochemical oxidation of phenol solution (V = 150 cm^3) on a PbO_2 (electrodeposited on graphite) anode (S = 12.5 cm^2). Anodic current density, j = 150 mA/cm^2; temperature, T = 70 °C; pH = 2. Initial phenol concentration: 21 mmol dm^{-3}. Graphite cathode. (1) phenol, (2) hydroquinone, (4) p-benzoquinone

Fig. 2. Variation of the aliphatic acids concentration during electrochemical oxidation of phenol solution on a PbO_2 (electrodeposited on graphite) anode. Graphite cathode. Same conditions as in Fig. 1. (5) maleic acid, (6) fumaric acid

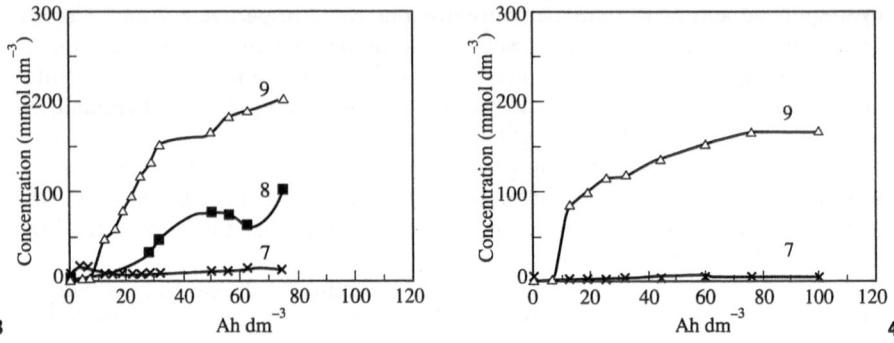

Fig. 3. Variation of the aliphatic acids concentration during electrochemical oxidation of phenol solution on a PbO$_2$ (electrodeposited on graphite) anode. Graphite cathode. Same conditions as in Fig. 1. (7) oxalic acid, (8) acetic acid, (9) formic acid

Fig. 4. Variation of the aliphatic acids concentration during electrochemical oxidation of sulfuric acid solution on a PbO$_2$ (electrodeposited on graphite) anode. Iridized platinum cathode. Same conditions as in Fig. 1. (7) oxalic acid, (9) formic acid

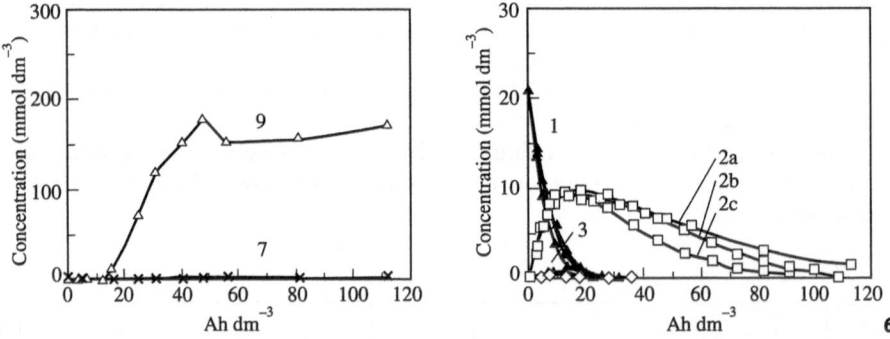

Fig. 5. Variation of the aliphatic acids concentration during electrochemical oxidation of phenol solution on a PbO$_2$ (electrodeposited on graphite) anode. Iridized platinum cathode. Same conditions as in Fig. 1. (7) oxalic acid, (9) formic acid

Fig. 6. Variation of the aromatic product concentration during electrochemical oxidation of phenol solution on different substrates covered with PbO$_2$: a) graphite; b) titanium oxide; c) tantalum. Iridized platinum cathode. (1) phenol, (2) hydroquinone, (3) catechol, (4) p-benzoquinone

In this case, it was not possible to detect benzoquinone by UV-adsorption at 220 nm because of its low concentration. It is assumed that the benzoquinone (BQ) formed in the course of the electrolysis was totally reduced into hydroquinone (HQ) on the Pt/Ir cathode:

$$BQ + 2H^+ + 2e^- \rightarrow HQ$$

Under the experimental conditions of Fig. 6, the complete disappearance of hydroquinone required more than 120 AH dm^{-3} (curve 2a).

Electrolysis of Phenol on Ebonex and Tantalum Anodes Covered with PbO_2

Figure 6 shows the variations of the concentration of phenol, hydroquinone and catechol in the course of electrolysis as functions of the applied charge in the case of the Ebonex electrode plated with lead dioxide; the cathode used in this case is made of Pt/Ir. Hydroquinone was the main product formed by oxidation under these conditions. Its concentration reached a maximum (10 mmol dm^{-3}) for a charge of about 15 Ah dm^{-3}. As previously observed, the Pt/Ir cathode bounded the 1,4-benzoquinone formation under its UV-detection limit. A very small amount of catechol was formed during the initial step corresponding to the phenol elimination (Fig. 6). The hydroquinone disappeared completely for a charge of about 110 AH dm^{-3}.

The use of a tantalum anode covered with PbO_2 (Fig. 6) produced similar results to those observed with the PbO_2 plated Ebonex, though the complete elimination of hydroquinone was attained more rapidly (at 90 AH dm^{-3}). Maleic and oxalic acids were weakly formed (C< 0.2 mmol dm^{-3}) on both Ebonex/PbO_2 and tantalum/PbO_2 anodes coupled with the Pt/Ir cathode.

The mean rates of concentration variation of hydroquinone and maleic acid were respectively –0.18 mmol dm^3 $(Ah)^{-1}$ and $4 \cdot 10^{-3}$ mmol dm^3 $(Ah)^{-1}$ for electric charges comprised between 20 and 60 Ah dm^{-3} (Fig. 6). Since a maleic acid molecule is formed form one aromatic molecule, it was assumed that the anodic oxidation on PbO_2 of this aliphatic acid into carbon dioxide occurred quickly and that the opening of the aromatic ring of 1,4-benzoquinone was the limiting step in the phenol degradation mechanism.

4.8.4
Conclusion

Lead dioxide was used as an electrochemical catalyst for the oxidation of aromatics in aqueous acidic solutions because of its high oxygen evolution overpotential. The anodix oxidation of phenol on PbO_2 electrodeposited on graphite produced hydroquinone and benzoquinone simultaneously when graphite was used as a cathode in a rector without separator. Phenol in aqueous solution, at the initial concentration of 21 mmol dm^{-3} disappeared completely after using an electric charge of 20 Ah dm^{-3}. The time necessary for a complete destruction of the aromatic intermediaries depended on the nature of the substrate and decreased as following: graphite >Ebonex>tantalum.

This work has shown that the electrochemical oxidation of phenol and of its derivatives, such as hydroquinone and especially benzoquinone which is highly toxic, needs higher electric charge on a lead dioxide anode than on a tin dioxide one. However, it is important to notice that electrochemical regeneration of the PbO_2 layer by oxidation of Pb^{2+} ions is easily attained. The performance of these electrodes could be improved by using either a separator or a graphite cathode.

4.8.5
References

Chang H, Johnson DC (1989) Electrocatalysis of anodic oxygen-transfer reactions. J Electrochem. Soc 136, 17–22

Comninellis Ch, Pulgarin C (1991) Anodic oxidation of phenol for waste water treatment. J Appl Electrochem 21, 703–708

Comninellis Ch, Pulgarin C (1993) Electrochemical oxidation of phenol for waste water treatment using SnO₂ anodes. J Appl Electrochem 23, 108–122

Gordon JS, Jr Young VG, Johnson DC (1994) Application of an electrochemical quartz crystal microbalance to a study of the anodic deposition of PbO₂ and Bi-PbO₂ films on gold electrodes. J Electrochem Soc 141, 652–660

Grigger JC, Miller HC, Loomis FD (1958) Lead dioxide anode for commercial use. J Electrochem Soc 105, 100–102

Kötz R, Stucki S, Carcer B (1991) Electrochemical waste water treatment using high overvoltage anodes. Part I: Physical and electrochemical properties of SnO₂ anodes. J Appl Electrochem 21, 14–20

Munichandraiah N (1992) Physicochemical properties of electrodeposited β-lead dioxide: effect of deposition current density. J Appl Electrochem 22, 825–829

Munichandraiah N, Sathyanarayama S (1987) Insoluble anode of porous lead dioxide for electrosynthesis: preparation and characterization. J Appl Electrochem 17, 22–32

Munichandraiah N, Sathyanarayama S (1988) Insoluble anode of α-lead dioxide coated on titanium for electrosynthesis of sodium perchlorate. J Appl Electrochem 18, 314–316

Narasimham KC, Udupa HVK (1976) Preparation and applications of graphite substrate lead dioxide (GSLD) anode. J Electrochem. Soc 123, 1294–1298

Sharifian H, Krik DW (1986) Electrochemical oxidation of phenol. J Electrochem Soc 133, 921–924

Stucki S, Kötz R, Carcer B, Suter W (1991) Electrochemical waste water treatment using high overvoltage anodes. Part II: Anode performance and applications. J Appl Electrochem 21, 99–104

4.8.6
Further Readings

Title: Electrochemical oxidation of phenol for wastewater treatment using SnO₂ anodes
Author: Comninellis, Ch.: Pulgarin, C.
Corporate Source: Swiss Federal Inst of Technologie, Lausanne, Switz
Source: Journal of Applied Electrochemistry v 23 n 2 Feb 1993. p 108–112
CODEN: JAELBJ ISSN: 0021-891X
Language: English
Abstract: The electrochemical oxidation of phenol for waste water treatment was studied on doped SnO₂ anodes. Analysis of reaction intermediates and a carbon balance has shown that the main reaction is oxidation of phenol to CO₂. This unexpected behaviour of the SnO₂ anode is explained by a change of the chemical structure of the electrode surface during anodic polarization. (Author abstract) 12 Refs.

Title: Electrochemical Oxidation of Refractory Organics
Author: Dixon, B.G.; Walsh, M.A.; Morris, R.S.
Source: Innovative Hazardous Waste Treatment Technology Series, Volume 2: Physical/Chemical Processes. Technomic Publishing Co., Inc. Lancaster, Pennsylvania. 1990. p 165–176, 5 fig, 2 tab, 9 ref.
National Science Foundation Grant No. ECE-842015 and EPA Contract No. 68-02-4428.
Abstract: As the nation's supply of clean water dwindles, the recycling and purification of all available water becomes of paramount importance. It has been projected that

industrial wastewater treatment alone will grow by 11 % per year at least through 1995. Clearly, new treatment processes will be required to supplement established technologies. A newly developed, capacitively driven electrochemical system, operating in the absence of added electrolyte, can efficiently degrade organic pollutants at low temperature, using ordinary AC current. Advantages include the removal and degradation of organic contaminants in extremely low concentrations at feast reaction rates through the direct activation of molecular oxygen, eliminating the need for chemical oxidant additions. The electrochemical cell is especially attractive at small waste-generation sites since it can be easily designed to fit the volume of solution to be handled and is easily retrofitted. A study has demonstrated that cyclic voltammetry can be effective in the evaluation of the electrochemical reactivity of pollutant compounds with a variety of catalysts. Of the catalysts evaluated, iron (III) tetraphenylporphorin, 2-amino-anthraquinone and the ruthenium/titanium mixed metal oxide spinel showed the most significant levels of activity with the widest range of pollutants, and is worthy of further investigation. The study also found that it was possible to induce the disappearance of toluene or phenol form a dilute solution in the absence of added electrolyte by AC current. Ongoing investigations involve examinations of new catalyst-support combinations and careful study of the effects of the rate of alteration of the cell current upon pollutant degradation rates and products. When fully developed, this technology may be useful at petroleum-spill sites, though much work remains to be done before a practical electrochemical cell can be marketed. (See also W92-05065) (Doyle-PTT)

The Electrochemical Destruction of Waste Reprocessing Solvent and Ion Exchange Resins

D.F. Steele, AEA Technology Dounreay, Caithness, Scotland, UK, J. P. Wilks and W.Batey; Univ of Calif/et al 1992

Incineration Cof, Albuquerque, NM, May 11-15, 92, p167(8) conf paper

Abstract: A process developed by AEA Technology in the UK for the treatment of organic wastes is described. The process involves the generation of highly oxidizing species, usually bivalent silver, in an electrochemical cell which attack and destroy the organic. The by-products formed are carbondioxide and water. The electrochemical cell is detailed, as is the process chemistry. The process has been shown to be effective in destroying aliphatic and aromatic hydrocarbons, phenols, and PCBs. Results are presented from studies on the oxidation of solvents and ion-exchange resins. Electrochemical oxidation operates at or below atmospheric pressure and at relatively low temperatures, and little or no volatilization of the low molecular weight species is realized, so the environmental safety of the process appears high. (6 diagrams, 7 references, 2 tables)

4.9
Treatment of Aqueous Organic Wastes by Molecular Oxygen at High Temperature and Pressure: Wet Air Oxidation Process

JEAN-NOËL FOUSSARD · MEHREZ CHAKCHOUK · GÉRALDINE DEIBER · HUBERT DEBELLEFONTAINE

Abstract

Aqueous wastes containing organic pollutants can be efficiently treated through wet air oxidation (WAO), i.e. oxidation by molecular oxygen in the liquid phase, under high temperature and pressure. The process is well designed for wastes fairly concentrated ($10 < COD < 100$ kg O_2/m^3), refractory to biological treatment, toxic or containing mineral salts in addition to the organic pollutants. Wastes from other treatment processes, such as biological sludges or concentrates from ultrafiltration, can also be treated by this method.

During the course of the treatment, high molecular weight organic pollutants are turned to free fatty acids like acetic acid, refractory to oxidize, but biodegradable. Temperature, usually ranging from 200 up to 325 °C, is the key parameter according to all kinetic studies. To offset the water vapor pressure, the working pressure value is 50 to 150 bar. During oxidation of heteroatomic organic pollutants, organic compounds containing chlorine, sulfur and phosphorus lead to almost total mineralization as chloride, sulfate and phosphate. Treatment of compounds that contain organic nitrogen mainly produces ammonium.

The current development of the process is limited due to the associated capital cost and the strictness of the operation under high pressure. By introducing continuous small amounts of hydrogen peroxide and ferrous salts, it is possible to promote the initiating step of the oxidation chain reaction. In the same way, heterogeneous catalysts (Mn/Ce oxides) enhance the oxidation kinetics. These two techniques are currently under study at the laboratory and are hopeful methods for decreasing the working constraints and for turning the high pressure process into a medium pressure one.

4.9.1
Introduction

Because of the recent trends towards the recycling of industrial wastewater and of the increasing severity of waste disposal regulations, it has become necessary to efficiently remove the polluting load of the aqueous industrial wastes; even though various methods have been developed to minimize waste generation. Four types of oxidative processes are known to treat wastewater polluted with organic matters:

- Chemical oxidation at ambient temperature using strong oxidizers (Doré, 1989, Masschelin, 1991 and Roques, 1990) like chlorine or ozone, well adapted to the very low concentrations.
- Biological treatment, neither suited for toxic waters nor for chemical oxygen demands (COD) above about 10 kg/m^3. This process also produces large amounts of sludge.
- Incineration, appropriate for effluents having a COD greater than about 100 kg/m^3, but plagued by energy cost, corrosion problems and the ecological perception of a process which can disseminate dust in the atmosphere.
- Wet oxidation in the liquid phase, under high temperature and pressure. This enclosed process, with limited interaction with the environment, is well adapted for COD loads ranging from 10 up to 100 kg/m^3. Toxic wastes can be treated and energy recovery is feasible via this method.

The premise of the wet air oxidation (WAO) process (Zimmermann and Diddams, 1960) is to enhance contact between molecular oxygen and the organic matter to oxidize. Temperature conditions convert the organic matter to carbon dioxide and water. The liquid phase is maintained by high pressure, which also increases the concentration of dissolved oxygen (Himmelblau, 1960) and then increases the oxidation rate. Typical conditions are 200 to 325 °C for the temperature, 50 to 150 bar for the pressure and 1 hour for the residence time. The process can treat any kind of wastes, produced by various branches of the industrial activity or sludges produced by the conventional treatment processes (physico-chemical, biological, etc.). WAO is one of the few processes that do not turn the pollution from one form to another, but makes it essentially disappear. This process was run under conditions that yield nearly an entire oxidation of the organic matter and that allow the generation of mechanical power (Chou and Verhoff, 1981) due to energy released in the reaction. This process was also used to treat biological sludges (Seiler, 1987) and night soil, but under milder conditions that yield a limited COD reduction despite an alteration of the molecules' structures (biodegradability is frankly improved).

This paper reviews laboratory results from which it is concluded that the WAO process is very efficient, but only under severe temperature conditions. To lower this constraint, various types of catalysts were tested; copper, iron, cobalt and manganese salts were used as homogeneous catalysts. More promising results were obtained by the continuous injection of hydrogen peroxide with ferrous salts in a batch reactor (Debellefontaine et al., 1992). The addition of about 10 ppm of ferrous iron with less than 15 % (of the amount for a stoichiometric oxidation) of hydrogen peroxide can turn the high pressure WAO process into a medium pressure one. With heterogeneous catalysis, namely Mn/Ce oxide, it was possible to achieve a total oxidation of acetic acid at only 200 °C and to transform ammonium nitrogen into molecular nitrogen at 260 °C.

4.9.2
Literature Survey

The Conventional WAO Process

Fundamental studies have produced extensive knowledge regarding the dependency of the oxidation rate on the various operating parameters and the mechanism of oxidation (Foussard et al., 1980, Day et al., 1973). At about 250 °C, nearly all the compounds can be completely transformed but acetic and propionic acids. Under such conditions, the final product is not only carbon dioxide, but various carboxylic acids, mainly acetic, according to Pujol et al. (1980) and Imamura et al. (1988). Acetic acid requires the most extreme conditions (320 °C) to oxidize to carbon dioxide. The partial order with respect to the organic compound is usually 1 to 1.4. It is close to 0.4 with respect to dissolved oxygen (Foussard et al., 1989 and Willms et al., 1987) but Taylor and Weygant (1974) and Joglekar et al. (1991) reported values close to 1 which could be typical of an oxygen transfer limitation or of an auto-catalytic effect of phenolic compounds. The oxidation proceeds according to a chain reaction mechanism, presented on Fig.1, which points out the particular role of acetic acid.

Catalyzed Oxidation in the Liquid Phase

Free radicals in the mechanism search for radicals reactions promoters, mainly as catalysts. For heterogeneously catalyzed oxidation of phenol using copper oxide, the mechanism appears to clearly involve free radicals with an initiation step on the catalytic surface. Inoue et al. (1986) have mentioned that a cobalt oxide based catalyst was successfully used to transform, at 250 °C, acetic acid to carbon dioxide and ammonia to molecular nitrogen. With the same catalyst, he reported that wet oxidation, at the same temperature, of aniline, pyridine and acetonitrile led quantitatively to molecular nitrogen. For homogeneous catalysis, ions like Fe^{2+}, Fe^{3+} and Cu^{2+}, associated with peroxides, appear attractive (Chowdury and Ross, 1975). The temperature can be lowered down to 200 °C and the rate of oxidation dependency on the oxygen partial pressure is less marked. In every case, acetic acid is also identified as a final product besides carbon dioxide, therefore it can be concluded that the mechanism is similar to that for non catalyzed oxidation.

Reactors

Only tubular continuous reactors have been developed industrially. Zimpro has developed a bubble column reactor approximately 15 m high. Ten or so units have been set up for industrial wastes treatment. Ciba Geigy, for its plant of Monthey (Switzerland) has designed a series of two reactors. They are 30 m high, their diameter is 1 m and they are coated with titanium to prevent corrosion. For the plant of Stignaes (Denmark), Krüger has built a reactor which is 3800 m long. The diameter is 0.11 m. The mean residence time is about 1 hour and oxygen injection is distributed in various parts of the tube in order to control the temperature rise. For the treatment of domestic sludges, vertical reactors have been developed (Kaufman, 1986), as in Appeldorm (The Nederlands). A shaft, approximately 1500 m deep is fitted with concentric tubes. Mechanical

Fig. 1. Simplified sketch diagram for wet air oxidation

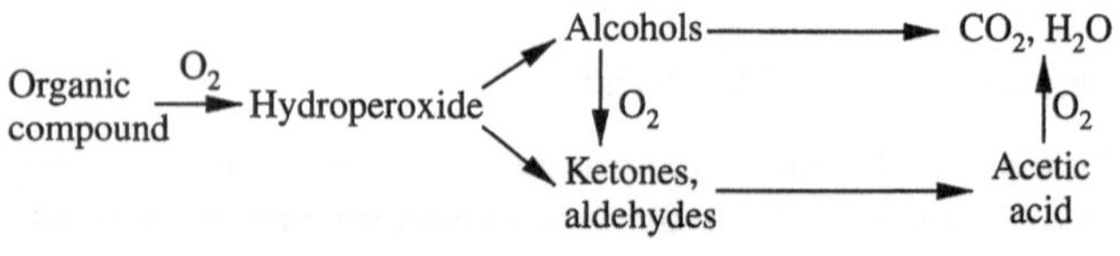

compression of the gas is avoided, as it dissolves only by the hydrostatic pressure while flowing down with the incoming waste. The reactor is also a countercurrent heat exchanger since the oxidized waste directly heats the incoming flow. The energetic self-sufficiency of the system is obtained.

Supercritical Wet Oxidation

Supercritical conditions are obtained when the water temperature and pressure are above critical conditions (Tc = 374 °C ; Pc = 220 bar). In that region, the specific weight of water is smaller than 320 kg/m^3 and the viscosity is smaller than 7.10^{-5} Pas. The solubility of various organic materials is drastically increased and gases, like oxygen and carbon dioxide, are infinitely soluble. In such an homogeneous medium, the physical step for oxygen transfer is suppressed. But, tremendous corrosion problems are reported, especially with chlorine containing wastes. These problems are especially hard to solve, even with special materials and, the associated capital cost is especially high. Eckert et al. (1990) have reported the oxidation of p-chlorophenol under supercritical conditions with residence times ranging from 5 to 60 s in a tubular reactor. The change in the oxidation rate is only moderate (50%) when changing the conditions from subcritical (340 °C, 140 bar) to supercritical (400 °C, 240 bar). The main product is carbon dioxide, but carbon monoxide, ethane, methane and ethylene are also produced in small amounts.

Oxygen Transfer During WAO

The overall wet air oxidation mechanism includes two steps. The first, a physical step, is the transfer of oxygen from the gas phase to the liquid. The second is the chemical reaction between organic matter and dissolved oxygen. When designing a wet air oxidation reactor, one usually considers that oxygen diffuses rapidly within the gas phase. The only significant transfer resistance is located at the gas-liquid interface (film model). The 3 limiting cases hereafter can be observed:

- Oxygen reacts within the film because of a rapid chemical reaction. In that way, the oxygen transfer rate is enhanced.
- Oxygen reacts within the bulk liquid where its concentration is close to 0. The overall rate is equal to the physical step of oxygen transfer.
- Oxygen concentration within the bulk liquid is equal to interface (or equilibrium) concentration. The overall rate is the chemical step rate, usually low.

The actual conditions within an industrial reactor will depend on its hydrodynamics. At the laboratory, only a high mixing efficiency (corresponding to the third case) will allow the determination of unbiased kinetic rates.

Solubility of Oxygen

The dependency of the Henry's law constant H on the temperature (Himmelblau, 1960), is reported on Fig. 2. It is clear that, above 250 °C, the solubility is greater than at room temperature. While solubility usually decreases as the temperature increases, in WAO reactors dissolved oxygen concentrations as high as 3 kg/m^3 are easily obtained.

Oxygen Transfer at High Temperature

Due to the scarcity of data about oxygen transfer at high temperature, an experimental study was conducted using the sodium sulphite method. The Hatta's number value ensured an enhancement factor close to 1. For an agitated vessel, the relationship (1) allows the determination of oxygen transfer capability $K_l \cdot a$ at any temperature T ranging from 20 to 240 °C. If the oxygen transfer capability is known at a reference temperature T_{ref}, it is given through:

$$(K_l \cdot a)_T = (K_l \cdot a)\ T_{ref} \cdot \exp\left[\gamma\ (1/_{ref} - 1/T)\right] \tag{1}$$

One can see on Fig. 3 that the capabilities estimated throughout the relationship (1) agree with those measured. The γ value is 2740 K for transfer through unruffled liquid surface. It is 1530 K when gas bubbles are dispersed within the bulk liquid by an impeller. This model leads to results similar to those previously attained at room temperature; but these cannot be extrapolated above 50 °C.

Oxygen Balance for a Batch Reactor

The actual concentration of oxygen within a completely mixed batch reactor (autoclave) is governed by the following set of differential equations (relationships (2) to (4)) dealing, respectively, with the oxygen balance for the liquid phase, the chemical rate of oxidation and the oxygen balance for the gas phase:

$$dC_l/dt + K_r \cdot (COD)^\mu \cdot (C_l)^\nu = K_l \cdot a \cdot (C_l^* - C_l) \tag{2}$$

$$Rc = -d\,(COD)/dt = K_r \cdot (COD)^\mu \cdot (C_l)^\nu \tag{3}$$

$$(V_g \cdot H/V_l \cdot R \cdot T) \cdot dC_l^*/dt + K_l \cdot a \cdot (C_l^* - C_l) = K_l \cdot a \cdot C_l \tag{4}$$

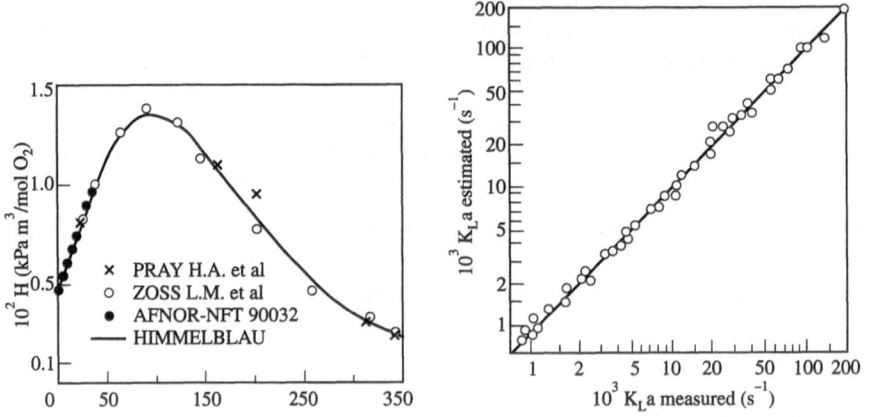

2   **3**

Fig. 2. Dependency of the Henry's law constant for oxygen solubility on the temperature

Fig. 3. Comparison between estimated and measured oxygen transfer capabilities for various conditions of mixing. Temperature ranges from 20 °C to 240 °C

This can be solved only by a numerical method (Range-Kutta-Merson) after specifying the initial conditions. The evolution of the chemical rate of oxidation R_c, the transfer rate of oxygen R_t, the equilibrium concentration C_l^* and the actual concentration of dissolved oxygen C_l are presented on Fig. 4 for phenol wet air oxidation. This typical result, obtained with the organic that is most easily oxidized, provides evidence that the chemical rate R_c is observed under the usual experimental conditions, except for the short period (20 seconds), during which the steady state for oxygen transfer is established.

4.9.3
Wet Air Oxidation Studies

Conventional Wet Air Oxidation

Several kinds of wastes have been studied in a batch reactor to determine the dependence of the reaction rate on the temperature, the oxygen pressure and the organic compound concentration, including:

- Biological sludges and paper-mill black liquors, which are of industrial concern.
- Retentate from ultrafiltration of dye-works wastewaters.
- Various acidic wastes from the pharmaceutical and the phyto-sanitary industries.
- Oxalic, acetic and formic acids which are significant intermediaries for oxidation of industrial wastes, and model molecules.

Using a numerical method for parameters adjustment, all the data obtained for a specific waste was demonstrated using a power function model. Then, the relationships (3) and (5) can describe the dependency on the various parameters:

$$K_r = K_{r0} \cdot \exp\left(- E/RT\right) \tag{5}$$

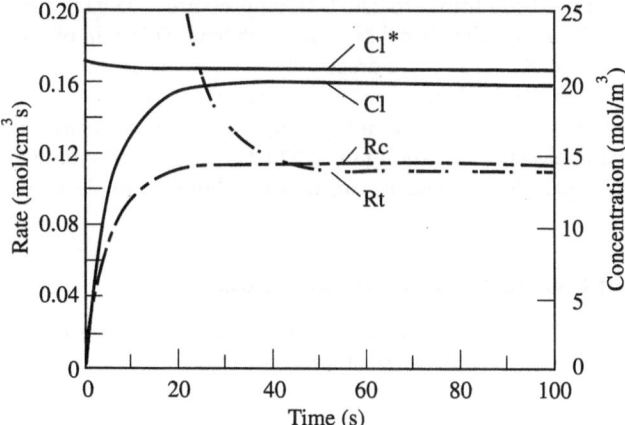

Fig. 4. WAO of a phenol solu-tion into a batch reactor. Typical time evolution of the rates and oxygen concentrations. Conditions: $T = 180\,°C$; $(COD)0 = 4.8\ kg/m^3$; $(P_{O2})^0 = 21\ bar$

Table 1. Kinetic parameters for conventional WAO

Compound	K_{r0} $(s^{-1} \cdot (kg/m^3)^{1-\mu-\nu})$	E (kJ)	μ	ν
Black Liquor	$1.24 \cdot 10^{10}$	135	1	0.38
Formic Acid	$4.70 \cdot 10^{10}$	143	1.33	0.46
Acetic Acid	$2.00 \cdot 10^{11}$	170	1	0.37
Oxalic Acid	–	155–134	1–2	0.48–0.31

Table 1 summarizes the various parameters and, under a synthetic form, Fig. 5 compares various organic compounds according to their ability to oxidize. This chart was established using results from the literature in addition of those determined at the laboratory (underlined). Acetic acid appears as the most refractory chemical. Meanwhile, by increasing the temperature up to 310 °C, all the compounds, including acetic acid, can be oxidized over 90 % within 1 hour.

4.9.4
Catalyzed Wet Air Oxidation

Homogeneous Catalysis Using Dissolved Salts

Because of its industrial significance (Bayer process for alumina recovery) and because of its medium situation on the chart Fig. 5, oxalic acid was selected as a test compound for catalyzed oxidation. The non-catalyzed experiments pointed out that for high concentrations of oxalate, the partial order μ was 2 and changed down to 1 for lower concentrations (see Table 1), a very similar result was indeed observed under catalytic conditions. Table 2 summarizes the effect of various catalysts. One can easily see that the catalysts have a marked influence for the entire oxidation period and that ferrous salts are very efficient. Using the reference values of Table 1, it is clear that the addition of ferrous salts raised conditions to 230 °C (total pressure = 35 bar), a result expected only at 285 °C without catalyst (total pressure = 75 bar). This gain of more than 50 °C for temperature and 40 bar pressure drastically changes the operating constraints and on the capital cost. In addition, very similar conclusions were deduced from runs at various temperatures. As indicated on figure 6, increasing the catalyst concentration can markedly increase the oxidation rate at 230 °C. But an overloading of the catalyst will result in separation problems. So, it was advisable to develop substitution methods.

Heterogeneous Catalysis by Composite Metallic Oxides

Imamura et al. (1987) have reported the catalytic efficiency of such oxides. We have tested various types for the treatment of refractory compounds, acetic acid and ammonia.

Acetic Acid Oxidation

The various oxides CuO, MnO_2, CeO_2 and Fe/Cu are inefficient, but the manganese and cerium composite oxides (here after Mn/Ce) possess a significant catalytic activity. It

Fig. 5. Oxidability of various organic compounds using WAO under standard conditions.
$T = 260\,°C$; $P_{O2} = 20$ bar; Residence time = 1 hour.

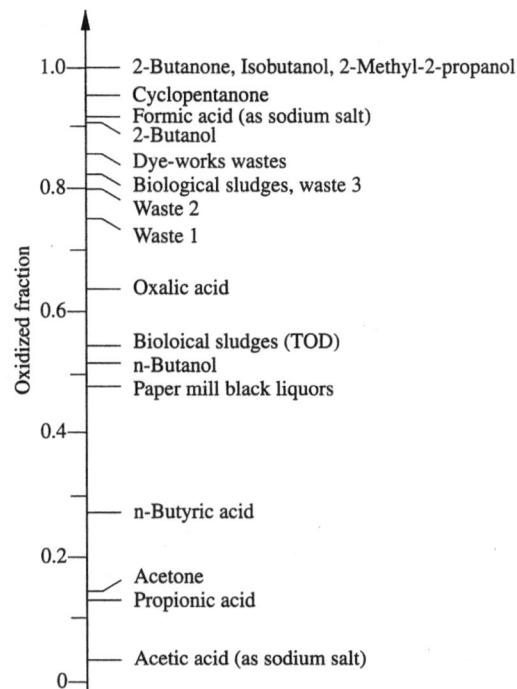

Table 2. Effect of various metallic salts as catalysts on potassium oxalate WAO. Conditions: reaction time = 2.5 hour; $T = 230\,°C$; $P_{O2} = 7.5$ bar

Catalyst concentration (mol/m³)	Final yield %	$10^3 \cdot K_{r1}$	$10^3 \cdot K_{r2}$	Mean magnifiying ratio
Without	18	1.0	8.7	1
$Cu^{2+}/3.5$	31	2.5	14.6	2.1
$Co^{2+}/1$	37	2.9	19.6	2.6
$Fe^{3+}/2$	47	3.6	30.7	3.6
$Mn^{2+}/3$	88	13.1	126.9	13.8
$Fe^{2+}/4$	98	33.5	269.0	32.2

is seen in Table 3 that coprecipitation with ammonia leads to the best result. When using soda, undesirable sodium ions can be trapped into the precipitate, affecting its structure and then its catalytic properties.

Ammonia Oxidation

Considering the promising results obtained during acetic oxidation, the Mn/Ce composite oxide was tested during ammonia WAO. The results of three tests are listed in

Fig. 6. Potassium oxalate WAO. Effect of the catalyst concentration.

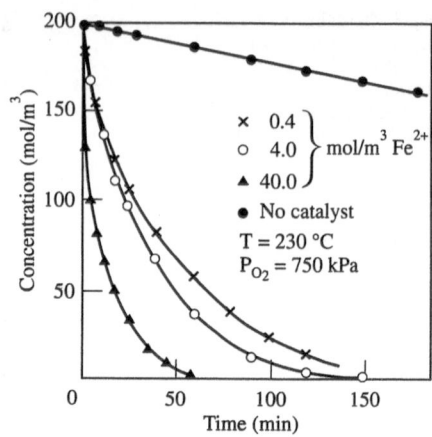

Table 3. WAO of acetic acid solutions with metallic oxides. Conditions: T = 200 °C; Acetic acid = 4 kg/m³; pH = 3; Catalyst = 3.6 kg/m³

Catalyst	no catalysis	CuO or MnO₂ or CeO₂ or Fe/Cu	Mn/Ce (1/1) coprecipitate with soda	Mn/Ce (1/1) coprecipitate with ammonia
% removal	0	0	73.4	91

Table 4. WAO of ammonia solutions with Mn/Ce composite oxide. Conditions: N–NH₂ = 2 kg/m³; pH = 10.5; Mn/CE (7/3) = 20 mM

T (°C)	t reaction (mm)	% removal of N–NH₃	N–NO₂⁻ (%)	N–NO₃⁻ (%)	N–N₂ (%)
200	60	22	10	6	7
260	60	66	2.5	1	65
260	240	91	0	0	91

Table 4. At 200 °C, a limited (22 %) removal is obtained, but increasing the temperature up to 260 °C allows an almost complete (91 %) removal after 4 hours, with neither nitrite nor nitrate production. Figure 7 details the results summarized on the second line of Table 4. The reaction order with respect to ammonia appears to be close to 1. In a first step, significant amounts of nitrite are observed, but they are eliminated as the run progresses. Only very small amounts of nitrates are detected. Molecular nitrogen is the main reaction product (by chromatographic analysis of the gas phase of the reactor). At the same time, small amounts of protoxide N_2O are detected, but, despite special care, no NO_x could be detected. It can be concluded that the composite oxide Mn/Ce is suitable for the removal of ammonia and leads to molecular nitrogen, a form absolutely innocuous for the environment.

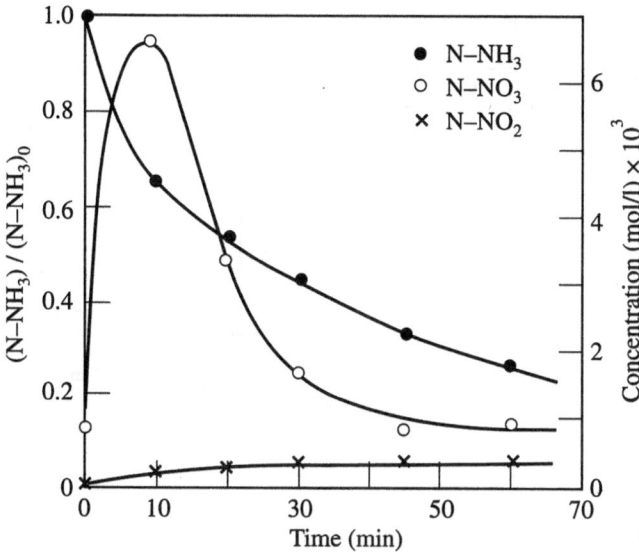

Fig. 7. Ammonia WAO using Mn/Ce composite oxide. Time dependency concentration. Conditions: $(N-N^{H3}) = 2$ kg/m³; $^{Mv/Ce}(7/3) = 20$mM; $T = 260$ °C; pH = 10.5

Hydrogen Peroxide Promoted Wet Air Oxidation

This technique is adapted from wet air oxidation and still uses molecular oxygen as the oxidizing agent. In the meantime, a low dosage of hydrogen peroxide is added to promote the radical reactions. The test compound (phenol) was selected due to its frequent occurrence in the wastewater of refineries, steel works and chemical industries. This method was tested using a completely mixed batch reactor (stirred autoclave). The cold reactor was loaded with a phenol and ferrous sulphate (0.01 kg/m³) solution at a convenient pH value (3.5). After heating at the rated temperature, the run was begun by injecting instantaneously a large amount of oxygen (4 to 8 times the amount necessary). At the same time, a dosing pump was begun and fed hydrogen peroxide continuously within the reactor all along the run (90 minutes). The total amount injected (Q_i) was usually 10% of the amount necessary for a stoichiometric oxidation (Q_e). The promoting effect of hydrogen peroxide on molecular oxygen oxidation at 160°C is clearly evidenced on Fig. 8. The initiating period is shortened and the COD removal efficiency is increased. TOC analysis supported similar conclusions (final removal increased from 76% up to 90%). Hydrogen peroxide obviously promotes molecular oxygen oxidation as the oxidation efficiency actually observed (curve 3) is greater than expected by adding the efficiencies of molecular oxygen and hydrogen peroxide if separated (curve 2). After 90 minutes WAO promoted with hydrogen peroxide produced better oxidation efficiencies at 160°C than conventional WAO at 220°C, turning a high pressure process into a medium pressure one. This method was successfully used (Chakchouk et al., 1994) to improve the biodegradability of olive mill wastewaters, regarded as very difficult to treat.

Fig. 8. Effect of hydrogen peroxide during WAO of phenol at 160 °C.
$Q_i = 0.1 \ Q_e \cdot$ (Å: no peroxide; É: with peroxide; Ç: calculated curve, addition of the theoretical effect of peroxide to curve Å).

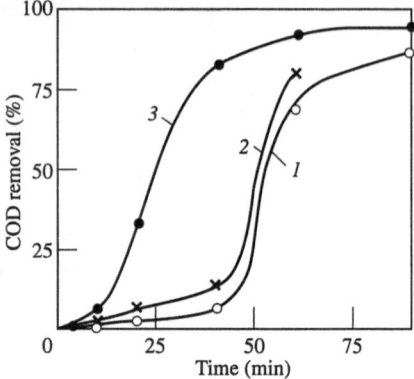

4.9.5
Conclusions

In recent years, our laboratory has acquired significant knowledge about the conventional WAO process. It has very good efficiency, as the COD reduction can easily reach 95% for most of the wastes. This process is well adapted to organic aqueous wastes that are fairly concentrated, nonbiodegradable, toxic or containing an associated mineral load. During the course of the treatment, the high molecular organic compounds are oxidized first to free fatty acids and then to carbon dioxide. These free fatty acids (mainly acetic) appear to be refractory to the WAO treatment, but their biodegradability is well established. According to numerous kinetic studies, temperature is the key parameter. The rate dependency of the oxygen partial pressure is usually limited. The use of pure oxygen versus air is an economic disadvantage. In a continuous process, the incoming effluent is heated by the outcoming treated flow. Because of the exothermicity of the process, a set of vapor and gases turbines can allow power generation from the outcoming gaseous flow.

However, temperature and ensuing pressure constraints deserve the industrial development of the process, because of the costs associated and the severity of the technology required. We are convinced that this urgency must drive the development to detoxify liquid wastes which are prohibited from land disposal because of environmental impacts. Attention must focus on modified oxidative techniques that allow lower constraints and then facilitate the industrial development. Then, catalysis may be necessary. During heterogeneous catalysis, composite oxides Mn/Ce lead to very satisfactory results for the WAO of refractory molecules, like acetic acid and ammonia, which is transformed to molecular nitrogen. Meanwhile, no extensive studies have been reported about industrial wastes treatment by this method. During homogeneous catalysis, it was established that H_2O_2/Fe^{2+} can act as an oxidation promoter. The high pressure WAO conventional process can be turned to a medium pressure process, thus more easily affordable to industries. The optimal temperature for this new process is from 180 up to 205 °C. Peroxide should be introduced in the amounts of 5 to 15%, on a stoichiometry basis.

List of Symbols

C_l = Actual concentration of dissolved oxygen
C_l^* = P_{O2}/H; Equilibrium concentration of oxygen at the interface
COD = Organic concentration, expressed as the chemical oxygen demand
E = Activation energy for the chemical reaction
H = Henry's law constant for oxygen solubility
$K_l \cdot a$ = Overall oxygen transfer coefficient
K_r = Rate constant for chemical oxidation
K_{r0} = Pre-exponential factor for the rate of chemical reaction
P_{O2} = Oxygen partial pressure
Q_i = Amount of peroxide actually injected
Q_e = Amount of peroxide needed for a stoichiometric oxidation
R = Ideal gas constant
R_c = $K_r \cdot (COD)^\mu \cdot (C_l)^\nu$; Rate of chemical oxidation
R_t = $K_l \cdot a \cdot (C_l^* - C_l)$; Oxygen transfer rate
T = Temperature
t = Time
V_g = Gas phase volume
V_l = Liquid phase volume

Greek Letters

γ = Activation energy parameter for the $K_l \cdot a$ dependency on the temperature
μ = Partial order with respect to the organic compound for the chemical rate
ν = Partial order with respect to oxygen for the chemical rate

Subscripts

ref = Reference condition for the temperature
0 = Initial condition

4.9.6
References

Chakchouk M, Hamdi M, Foussard JN, Debellefontaine H (1994) Complete treatment of olive mill wastewaters by a wet air oxidation process coupled with a biological step. Env Technology 15, 323–332

Chou CL, Verhoff FH (1981) Process for power generation from wet air oxidation with application to coal gaseification wastewater. Ind Eng Chem Process Des Dev 20, 12–19

Chowdury AK, Ross LW (1975) Catalytic wet oxidation of strong wastewaters. AIChE Symp Ser 71(151), 46–58

Day DC, Hudgins RR, Silveston PL (1973) Oxidation of propionic acids solutions. Can J Chem Eng 51, 733–740

Debellefontaine H, Striolo P, Chakchouk M, Foussard JN, Besombes-Vailhé J (1992) Nouveaux procédés d'oxydation chimique pour l'élimination des rejets aqueux phénolés. Revue des Sciences de l'Eau 5, 555–572

Doré M (1989) Chimie des oxydants et traitement des eaux. Tec & Doc editors, Paris, ISBN 85206-562-2, 488–493

Eckert CA, Leman GW, Yang HH (1990) Homogeneous catalyst for wet oxidation: design and economic feasability of a mobile detoxification unit. Hazardous Materials Controls 3, 20–33

Foussard JN, Debellefontaine H, Besombes-Vailhé J (1989) Efficient elimination of organic liquid wastes: wet air oxidation. J Environ Engng ASCE 115(2), 367–385

Foussard JN, Talayrach B, Besombes-Vailhé J (1980) Oxydation en phase aqueuse de l'acètate de sodium par l'oxygène molèculaire à tempèrature et á pression èlevèes. Bull Soc Chim Fr 11–12, 427–433

Himmelblau D (1960) Solubilities of inert gases in water, 0 °C to near the critical point of water. J Chem Eng Data 5(1), 10–15

Imamura S, Fukuda I, Ishida S (1988) Wet oxidation catalysed by ruthenium supported on cerium (IV) oxides. Ind Eng Chem Res 27, 721–723

Imamura S, Nishimura H, Ishida S (1987) Preparation of Mn/Ce composite oxide catalysts for the wet oxidation of acetic acid and their catalytic activities. Sekiyu Crakkaishi 30, 199–202

Inoue H, Komiyama H, Tanaka H, Tamura N, Matsuo M, Morishita T, Miyazawa J (1986) Treatment of pollutants in wastewater by catalytic wet oxidation. World Congress III of Chemical Engineering, Tokyo, 13 b-153, 596–599

Joglekar HS, Samant SD, Joshi J B (1991) Kinetics of wet air oxidation of phenol and substituted phenols. Water Research 25(2), 135–144

Kaufmann LA, Peterscheck H (1980) Modeling Vertech's mile long multiphase reaction vessel. Chem Eng Science 41, 685–692

Masschelein WJ (1991) Ozone et ozonation des eaux. Tec & Doc editors, Paris, ISBN 2-85206-689-0, 317–380

Pujol C, Talayrach B, Besombes-Vailhé J (1980) Modélisation de la cinètique d'oxydation en phase liquide de la liqueur noire de papeterie par l'oxygène de l'air. Water Research 6, 967–979

Roques H (1990) Fondements thÇoriques du traitement chimique des eaux. Tec & Doc editors, Paris, ISBN 2-85206-614-9, 715–807

Seiler GS (1987) Twenty years of sludge management by wet oxidation. Sludge Management Series 17, 100–105

Taylor JE, Weygandt JC (1974) A kinetic study of high pressure aqueous oxidation of organic compounds using elemental oxygen. Can J Chem 52, 1925–1933

Willms RS, Balinsky AM, Reibel DD, Wetzel DM, Harrisson DP (1987) Aqueous phase oxidation. The intrinsic kinetics of single organic compounds. Ind Eng Chem Res 26, 148–154

Zimmermann FJ, Diddams DG (1960) The Zimmermann process and its application in the pulp and paper industry. TAPPI 43(8), 710–715

4.9.7
Further Readings

Title: Nitrotoluenesulfonic acid: UV, IR and NMR properties and rate studies of wet air oxidation
Author: Phull, Kotu K; Hao Oliver J
Corporate Source: Univ of Maryland, College Park, MD, USA
Source: Industrial & Engineering Chemistry Research v 32 n 8 Aug 1993. p 1772–1779
Publication Year: 1993
CODEN: IECRED ISSN: 0888–5885

Abstract: Environmental fate of aromatic sulfonates is relatively unknown. This study reports the findings of wet air oxidation (WAO) of several sulfonic compounds, specifically 5-nitro-o-toluenesulfonic acid (NTSA). WAO is effective in the oxidation of NTSA, a compound similar in structure to dinitrotoluenesulfonates which are major components of TNT red water. The WAO reaction is first-order with respect to NTSA; the order with respect to oxygen pressure is 0.6. The activation energy is approximately 21 kcal/mol. The reaction rate is significantly enhanced with the addition of 5 mg/L Cu(II) catalyst. The effect of initial pH on rate of NTSA oxidation is complex. Higher reaction rates are observed both at low and high pH's, with a better overall rate being achieved at low pH. Some properties of NTSA (e.g., UV, IR, and NMR) were obtained to provide additional information. (Author abstract) refs.

Title: Factors affecting wet air oxidation of TNT red water: rate studies
Author: Hao Oliver J; Phull, Kotu K; Chen Jin M; Davis Allen P; Maloney Stephen W
Corporate Source: Univ of Maryland, College Park, MD, USA
Source: Journal of Hazardous Materials v 34 n 1 Apr 1993. p 51–68
Publication Year: 1993
CODEN: JHMAD9 ISSN: 0304-3894
Abstract: Preliminary experiments have demonstrated wet air oxidation (WAO) to be feasible for TNT red water treatment. This paper presents the results of rate studies for the evaluation of temperature, partial oxygen pressure P_{O2}, initial red water concentration, salt concentration, and catalyst/initiator addition on WAO performance. Results show the WAO efficiency to be a function primarily of temperature, and to a lesser extent, the initial P_{O2}. A significant initial (usually less than 5 minutes) rapid reduction in total organic carbon (TOC) or chemical oxygen demand (COD) is observed in all experiments. The extent of reduction varies with the experimental conditions: the harsher the condition, the higher the initial reduction. At lower temperatures, the subsequent WAO of red water proceeds as a first-order reaction with respect to TOC or COD. Under harsher temperature conditions, the reaction follows two distinct first-order phases. High salt concentrations (Na_2SO_4 and $NaNO_3$) slightly enhanced the overall oxidation. Addition of Cu(II) as a catalyst results in rate enhancement. Several issues regarding application of WAO are discussed. (Author abstract) 21 Refs.

Title: Anaerobic treatment of sewage sludge treated by catalytic wet oxidation process in upflow anaerobic sludge blanket reactors.
Author: Song JJ; Takeda N; Hiraoka M
Corporate Source: Kyoto Univ, Kyoto, Jpn
Conference Title: Proceedings of the 16th Biennial Conference of the International Association on Water Pollution Research and Control – Water Quality International '92
Conference Location: Washington, DC, USA Conference Date: 1992 May 24–30
Source: Water Science and Technology v 26 n 3 "ƒ" 4 1992. p 867–875
CODEN: WSTED4 ISSN: 0273-1223
Treatment, sewage sludge was treated by catalytic wet oxidation process (CWOP). The CWOP is designed to treat sewage sludge with the aid of a newly developed catalyst. It is a treatment process, by which concentrated COD components and suspended solids in various kinds of sludge can be simultaneously oxidized and treated with great efficiency in a single step without dilution. After CWOP, this supernatant was treated by upflow anaerobic sludge blanket (UASB) for recovery of methane. This experiment was to evaluate the acclimation of UASB process to the supernatant of sewage sludge treated by CWOP. For the granule growth and accumulation, supernatant was introduced stepwise (25%, 33%, 50%, 100%). The supernatant was 7,200 mg center dot COD/L which contained 53% carbonic acids. The reactor was operated at a volumetric loading rate from 1.8 (kg center dot COD/m³ center dot day) to 14.4 (kg center dot COD/m³ center dot day) and overall HRT of less than 24 hrs throughout the experiment. In the conventional anaerobic biochemical process of methane gas recovery was limited to

50% of COD recovery. As a result of this study, the COD recovery to about 93% was obtained by the combination of CWOP and UASB process. The following conditions were enough for pretreatment: temperature, 270 degree C; pressure, 86 kg/m^2; stoichiometric air ratio, 1.1; reaction time, 24 min. These results indicate that the CWOP-UASB process may be attractive as an alternative sewage sludge treatment. (Author abstract) Refs.

5 Hazardous Waste Management

5.1
Hazardous Wastes Treatments

GÉRARD ANTONINI

Abstract

A short state of the art review is presented concerning the hazardous waste treatments, including physico-chemical treatments by non thermal chemical oxidation and chemical reduction together with thermal treatments by incineration and pyrolysis or by wet air oxidation. A case study concerning a mobile incineration unit under oxygen enriched condition is presented.

5.1.1
Introduction

A hazardous waste may be defined as a waste, generally a solid waste, that poses substantial threat to human health or to the environment. These wastes are generally ignitable, corrosive, reactive or toxic and they have needed the development of a "Cradle to Grave" chain, in an attempt to track hazardous waste form its generation joint (the "cradle") to its ultimate disposal point (the "grave"). This imposes regulatory requirements to the generator and to the transporter of hazardous waste and also for the treatment, storage and final disposal of waste.

The priority in managing hazardous waste include:

- reduction of the amount of waste generated on site. This can be obtained by changing process conditions, or by changing the process itself in order to minimize the production of toxic materials or by- product,
- waste concentration: evaporation or dewatering minimize the volume of wastes,
- waste separation, recycling and reuse, energy valorization.

These operations tend to minimize disposal costs. Also separation of solvent wastes can provide the feed stock for some other industries:

- detoxification and neutralization of wastes. This consists in chemical or biological reactions which permit the waste to transform in a less hazardous form. Acids and bases can be converted to non-hazardous materials by simply neutralizing them with an appropriate base or acid
- treatment/destruction of hazardous wastes.

There are two basic processing technologies:

- thermal and chemical/physical/biological,
- stabilization/solidification of final residues,
- land disposal.

5.1.2
Physico-chemical waste treatments

Physical treatment processes are essentially devoted to phase separation of the hazardous wastes. They are essentially based on:

- adsorption (activated carbon and resins),
- centrifugation, filtration,
- dialysis, electrodialysis,
- filtration, flocculation (precipitation and sedimentation), ultra filtration,
- flotation.

The treatments are generally considered as pre-treatments as no destruction is obtained. Chemical treatments are likely to transform hazardous waste in a less or non-hazardous form.

Apart from the neutralization treatments, they are essentially based on:

Chemical oxidation (non-thermal)

Liquids are the primary waste forms treatable by chemical oxidation, but slurries, tars and sludges may also be treated in this manner. This chemical oxidation requires the adjustment of the pH, followed by the addition of the oxidizing agent and mixing with the waste. This agent can be in the form of a gas (chlorine gas) or a liquid (hydrogen peroxide), or a solid, if strongly dispersed.

Consider the cyanide destruction reaction using chlorine gas.

$$2\,NaCN + 5\,Cl_2 + 12\,NaOH \rightarrow N_2 + 2\,Na_2CO_3 + 10\,NaCl + 6\,H_2O$$

or by sodium hypo chlorite

$$2\,NaCN + 5\,NaOCl + H_2O \rightarrow N_2 + 2\,NaHCO_3 + 5\,NaCl$$

Once reacted, the final oxidized solution can be precipitated or filtered.

Also, complete oxidation of carbon-carbon bonds can be obtained by ozone which appears as an extremely reactive gas. Cleavage of aromatic cycles can be obtained, as in the conversion of phenol in oxalic acid.

Radiation induced chemical oxidation should also be mentioned. For instance, UV light induces oxidation of organic contaminants by the generation of hydroxyl radicals through UV photolysis of various oxydants such as H_2O_2, O_3... (W.E. Schwinkenberg, L.L. Nenninger, 1994).

Also ultrasound affects organic oxidation primarily through cavitation which, via high localized transient temperatures and pressures, enhancing the rate of chemical decomposition reaction.

Finally electrolytic/electrochemical oxidation can be utilized in an electrochemical cell in which oxidizing species are generated at the anode, migrate into the bulk electrolyte to oxidize and destroy the organic compounds (W.E. Schwinkenberg, L.L. Nenninger, 1994).

Chemical reduction

This type of treatment is of interest because metals often can be reduced to their elemental form for recycling or converted to less toxic oxidation states.

For instance, hexavalent chromium (Cr(VI)), a very toxic material can be reduced to trivalent chromium Cr(III) and precipitated. A variety of reducing agents can be utilized including sulfur dioxide (SO_2), sodium bisulfite ($NaHSO_3$) or ferrous sulfate ($FeSO_4$). For instance with SO_2, the reaction steps are:

$$SO_2 + H_2O \rightarrow H_2SO_3$$
$$2\,H_2\,C_r\,O_4 + 3\,H_2SO_3 \rightarrow Cr_2(SO_4)_3 + 5\,H_2O$$

Because soluble chromium III are themselves toxic they must be precipitated in $Cr(OH)_3$, using calcium hydroxide.

$$Cr_2(SO_4)_3 + 3\,Ca(OH)_2 \rightarrow 2\,Cr(OH)_3 + 3\,CaSO_4 \text{ which can be easily separated.}$$

5.1.3
Thermal treatments of hazardous waste

Thermal processes offer many advantages as hazardous waste treatment options. They aim at the volume reduction and/or toxicity of organic wastes by exposing them to high temperatures in controlled environments designed to encourage material breakdown into simpler and less toxic forms. Thermal treatment can take many forms which are reviewed here.

Incineration (Yen-Hsiun Kian, A.M. Metry, 1982)

Incineration is a controlled process that uses combustion to convert a waste to a less bulky material. Traditional excess air incineration is an exothermic process, operating at temperatures of 800 to 1200 °C. The temperature is controlled by adjusting the feed rate and the amount of excess combustion air. The flue-gas volume is of the order of 7000–8000 Nm³/t, containing carbon dioxide, water and flying ashes with toxic gas that have to be removed like CO, SO_x, NO_x, halogens. This imposes secondary treatments such as after burning, filtration, scrubbing to lower concentrations to acceptable levels prior to atmospheric release. The solid and liquids effluents generated by the secondary treatment will require treatment prior to ultimate disposal.

The basic types of incinerators are:

- multiple hearth which consist of a vertical cylindrical shell lined with refractory material, divided in smaller horizontal zones. When fed from the top, the waste is pushed downwards by a rotary shaft. The mean residence time within the incinerator can very form 0,25–1,5 h. They are well adapted for solid and sludges treatments (Fig. 1).
- rotary kiln; they consist of a refractory lined shell mounted with its axis at a slight slope from the horizontal. They are well adapted for the incineration of a wide variety of liquid and solid wastes. An afterburner using gaseous or liquid fuels or

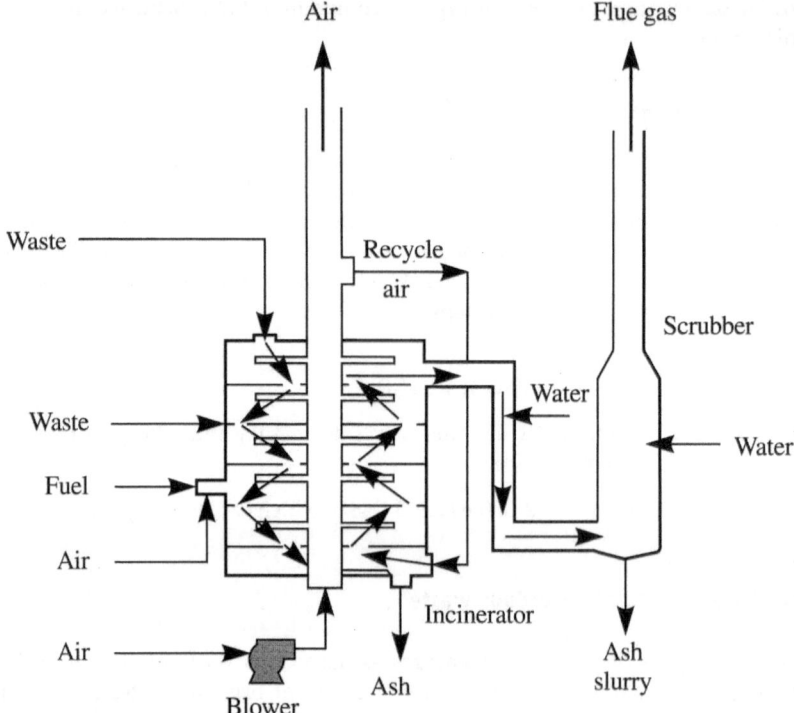

Fig. 1. Multiple hearth incinerator

wastes to generate a high temperature oxidizing environment is almost always required to complete the gas-phase combustion reactions. It is connected to the discharge end of the kiln. The mean residence time for gases varies from 1–10 s while the MRT for solids can reach hours.

- fluidized bed (Fig. 2): they consist of a refractory lined cylindrical vessel containing a bed of inert granular material on a perfored metal plate with air passing through it to "fluidize" the bed. Liquids or granular wastes are injected into or above the bed for combustion. Because of the turbulence of the bed, heat is rapidly transferred to the waste and thermal destruction can be obtained within a few seconds. Two types of fluidized bed systems are available: the stationary bed and the circulating system which the whole bed circulates between two vessels.

The temperatures within the bed ranges from 500 to 900 °C with relatively low NO$_x$ generation.

Pyrolysis (A. G. Buekens, J. G. Schoeters, 1986)

Pyrolysis is defined as the chemical decomposition or change brought about by heating in the absence of air, or oxygen. This decomposition is enothermic and generally obtained for temperature ranging form 400 to 800 °C.

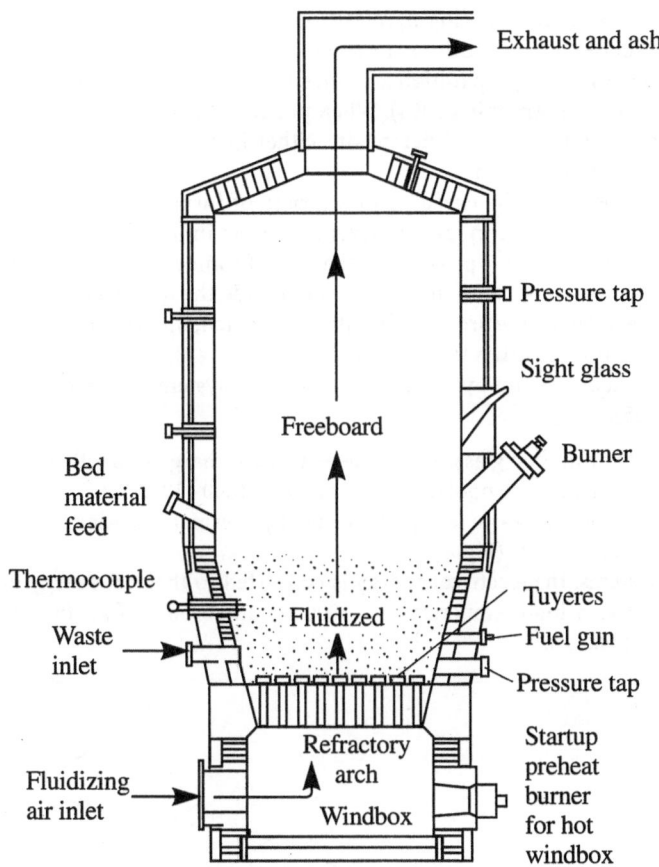

Fig. 2. A cross section of a fluidized bed incinerator

In a pyrolyzer the waste material is heated indirectly in an oxygen starved condition, to volatilize the organic (combustible gases, water vapor, etc....) separately from a non volatile char and ash. In a second step, volatile components are burned under proper conditions to assure the complete destruction of all hazardous components. The recovery of the heat generated can be utilized to maintain the pyrolytic reactions (G. Gaulard, G. Antonini, G. Martin, 1993). The advantages of this type of process is its ability to be very closely controlled (2 steps operations) to minimize the flue gas production (smaller equipments). Its low temperatures of destruction avoids metal and salts volatilization. The absence of air feeding minimizes the flying ashes to be recovered.

The thermal treatment during pyrolysis renders most of the solid-waste residues from the pyrolyser non-hazardous. In some cases, it is still possible to extract the remaining heavy metal by separation tectonics, based for instance, on a selective agglomeration process (G. Antonini, 1994). When purified, the char may be re-utilized for direct energy production in a different site, either in a dry combustible form or formulated as a slurry (liquid fuel).

Pyrolysis process can be executed in a variety of furnace designs, such as rotary kiln, with positive pressure in the pyrolyzer to prevent infiltration of air.

If a small amount of air is provided at the base of some multizone systems, like vertical shaft, internal combustion hot gases can provide the heat necessary to pyrolysis in an upper part. This is the process of gasification which, when hot air injection is utilized, allows the bottom ashes to melt (Fig. 3).

Other high temperature pyrolytic decomposition systems are in development (H. Freeman, 1985).

- Molten salts or molten glass, where the wastes are charged directly into a combustion chamber above the pool of molten medium (800 – 1200 °C),
- Electric reactor: the waste is rapidly heated by thermal radiation to temperatures between 2200 – 3000 °C (Fig. 4).
- Plasma systems: the electrical energy is converted to thermal energy in a plasma state gas at very high temperature (> 5000 °C). In absence of air this creates flash pyrolysis of wastes.

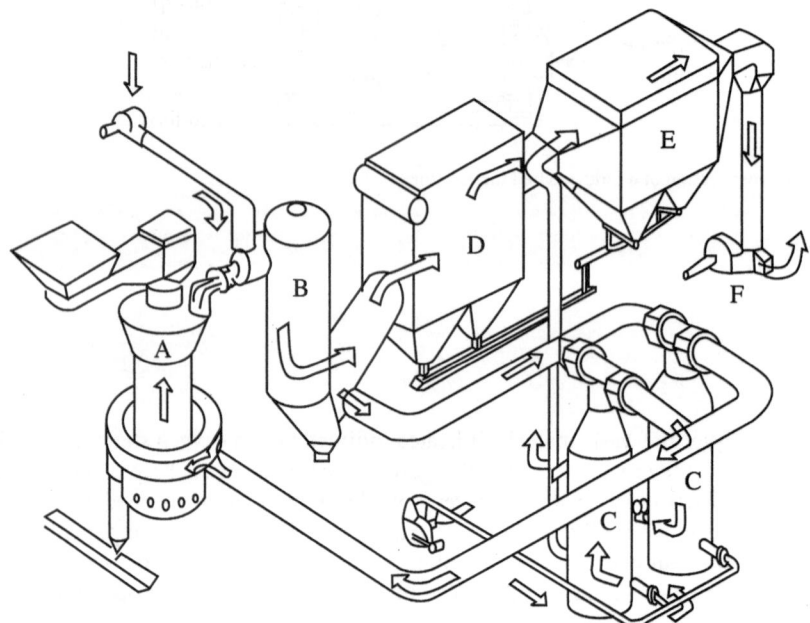

Fig. 3. Andco-Torrax process: (A) gasifier; (B) combustion chamber; (C) cowper; (D) steam boiler; (E) electrostatic precipitator; (F) suction fan

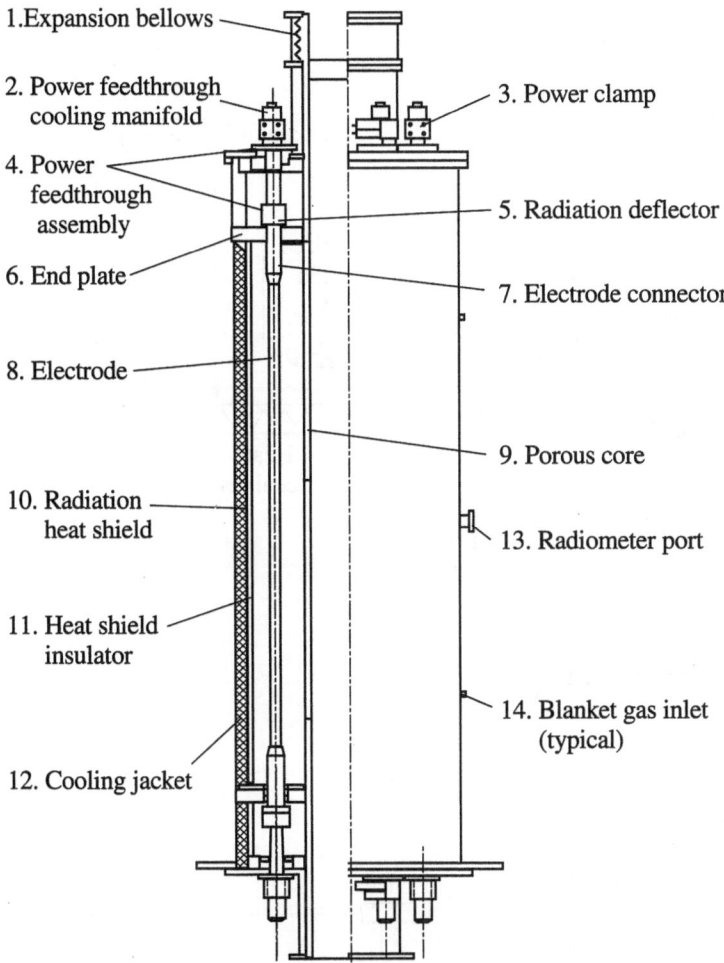

1. Expansion bellows

2. Power feedthrough
   cooling manifold

3. Power clamp

4. Power
   feedthrough
   assembly

5. Radiation deflector

6. End plate

7. Electrode connector

8. Electrode

9. Porous core

10. Radiation
    heat shield

13. Radiometer port

11. Heat shield
    insulator

12. Cooling jacket

14. Blanket gas inlet
    (typical)

Fig. 4. Vertical cross section of a typical fluid-wall electric reactor

Wet air oxidation (wao), supercritical water (C. C. Lee, 1983)

Wet air oxidation is a process for oxidizing organic contaminants in water at elevated temperatures and pressions. Water serves to promote, oxidation reactions at a relatively low temperature (175–340 °C) without evaporation, in a thermally self-sustaining condition. air or oxygen is bubbled through the liquid phase in the pressurized hot water containing the pollutants (Fig. 5).

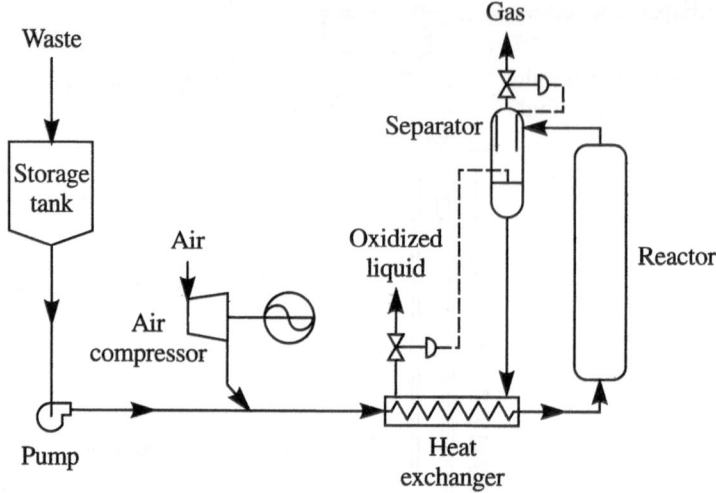

Fig. 5. Flowsheet of wet air oxidation process

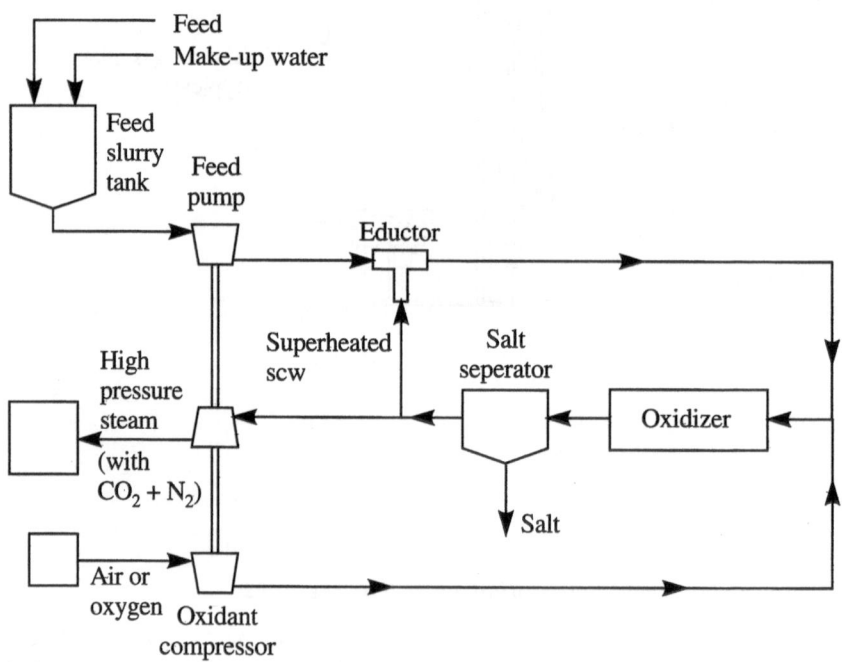

Fig. 6. Schematic diagram of the Modar supercritical water process

This process is designed primarily of dilute aqueous wastes, too dilute to be incinerated economically and yet too toxic to be treated biologically.

When the temperature exceeds 379 °C (218 atm), the water is in a supercritical state and oxygen is then completely miscible with water. Organic are then oxidized rapidly, while inorganics are practically insoluble in supercritical water, and can be easily removed from the water stream (Fig. 6).

5.1.4
A case study: mobile incineration unit (G. Antonini et al., 1994)

In some cases, the low amount of hazardous wastes cannot justify an on-site treatment operation. Also, these wastes cannot be transported for regulations reasons. It is then viable to develop the concept of a mobile unit for treatment. This was the case for leaching effluents form the maintenance of a PWR steam generator. These effluents consist essentially of aqueous wastes with dissolved organic and metallic compounds, with a total weight content of about 100 g/l leading to a radio-activity of 8000 Bq/l (^{60}Co, ^{137}Cs). A total amount of 2000 m^3/year has to be treated. A new mobile incineration unit has been developed for this purpose by Société des Techniques en Milieu Ionisant/Université de Technologie de Compiègne (STMI/UTC).

This process is based on the waste preparation in an aqueous solution containing a surfactant, and if necessary, a carbonizing agent, to modify it into a low pressurized foam and then to inject it into a cyclone furnace run on oxygen, for thermal destruction (Fig. 7).

The preparation stage permits (loaded foam) an homogenization of the wastes prior its destruction. Sorbing agents (de-SOx) can be added within the formulation for in suit desulfurization. Also, a small amount of light fuel oil is added within the triphasic mixture to adjust the calorific value of the mixture.

The utilization of pure oxygen permits the:

- minimization of the flue gas volume, which, in turn permits utilisation of small scale treatment units
- minimization of the calorific value necessary for "flame holding", (800 kcal/kg instead of 3000 kcal) thus lessening the amount of needed co-combustible
- minimization of NO$_x$ generation.

The utilization of cyclone furnace permits high turbulence levels, a direct gas/solid separation, enhancement of the solid residence time as compared to gas.

A mobile incineration unit (900 kg/h) has been designed and erected which permits high performances in thermal destruction of the wastes, although of small scale.

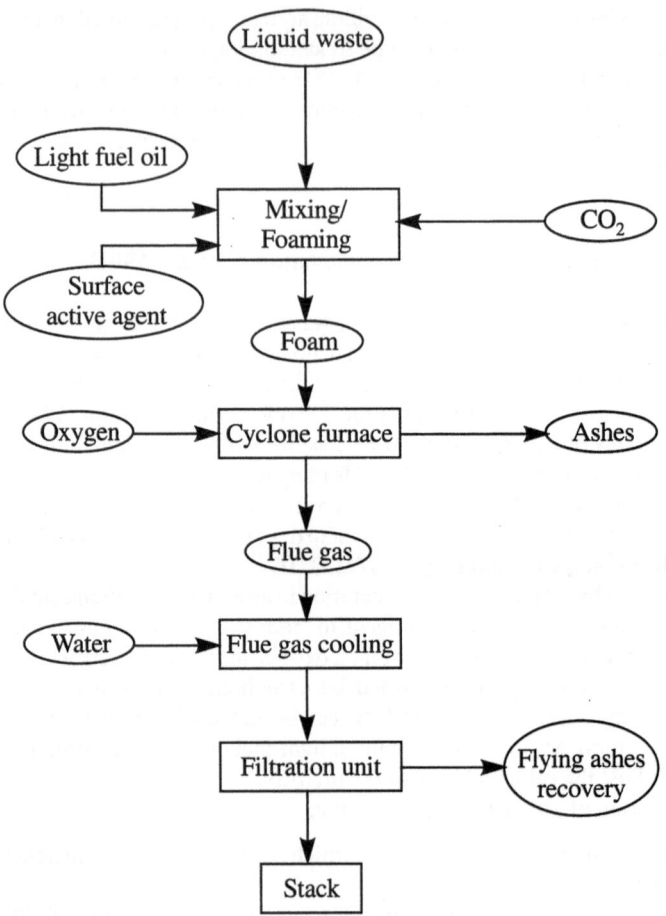

Fig. 7. General synoptic

5.1.5
References

Antonini G (1994) Pré-purification des cokes de pyrolyse en vue de leur valorisation énergétique. Colloque Int. La Thermolyse, une Technologie de Recyclage et de Depollution, 24–25 Mars, ISSEP Liège, Beligque

Antonini G et al. (1994) Mobile incineration wit for liquid waste thermal treatment. Int Incineration Conference Proceeding, p 441 May 9–13, Houston, USA

Buekens AG, Schoeters JG (1986) European Experience in the Pyrolysis and Gasification of Solid Wastes. Conservation and Recycling 9, n · 3, p 83

Freeman H (1985) Innovative thermal processes for the destruction of hazardous wastes. In: Incinerating Hazartsoud Wastes. Ed Harry M Freeman, Technomic Publisher

Gaulard G, Antonini G, Martin G (1993) Traitement des Déchets par Thermolyse. 4e Congrès de Génie des Procédés, 21–23 septembre, Grenoble

Lee, CC (1983) A comparison of innovative technology for thermal destruction of hazardous waste. 1st Annual Hazardous Materials Management Conference. Proceeding Philadelphia, July 12 – 14, p 278

Schwinkenberg WE, Nenninger LL (1994) Evaluation of Alternatives to Thermal Treatment for DOE Mixed Waste. Int Incineration Conference. Proceedings, p 13, May 9-13, Houston, USA

Yen-Hsiung Kiang, Metry AM (1982) Hazardous Waste Processing Technology. Ann Arbor Science Publisher

5.1.6
Further Readings

Hazardous Waste Treatment Technologies

Byung J. Kim, US Army Construction Engineering Research Laboratories, Champaign, IL. Shaoying Qi and Richard S. Shanley: Water Environ Res, Jun 94, v 66, n 4, p 440 (16) journal article

Abstract: A large number of studies were conducted in 1993 on hazardous-waste treatment technologies. Overall data are reviewed in the areas of biological treatment, including biofiltration, bioremediation, bioventing, and biodegradation; chemical and physical treatment, including advanced oxidation, membrane technologies, vapor sparging/extraction, and flushing/washing/extraction by solvents; and thermal treatment, such as combustion/incineration, thermal desorption, plasma/nitrification, and wet oxidation/wet-air oxidation/supercritical-water oxidation. Relevant data are presented form the studies on each of the technologies cited. (256 references)

Hazardous waste treatment and environmental remediation research

Performing Organization: Los Alamos National Lab., NM.
Report No: LA-SUB-93-264
Sponsoring organization: Science Applications International Corp., San Diego, CA (United States); Department of Energy, Washington, D.C.
NTIS No: DE93040571/HDM
Contract No: W-7405-ENG-36
Notes: Sponsored by Department of Energy, Washington, D.C.
Date: 29 Sep 89 Pages: 51p NTIS Price Code: PC A04/MF A01
Language: English Country: United States

Abstract: Los Alamos National Laboratory (LANL) is currently evaluating hazardous waste treatment and environmental remediation technologies in existence and under development to determine applicability to remediation needs of the DOE facilities under the Albuquerque Operations Office and to determine areas of research need. To assist LANL in this effort, Science Applications International Corporation (SAIC) conducted an assessment of technologies and monitoring methods that have been demonstrated or are under development. The focus of this assessment is to: (1) identify existing technologies for hazardous waste treatment and environmental remediation of old waste sites; (2) identify technologies under development and the status of the technology; (3) assess new technologies that need development to provide adequate hazardous waste treatment and remedial action technologies for DOD and DOE sites; and (4) identify hazardous waste and remediation problems for environmental research and development. There are currently numerous research and development activities underway nationwide relating to environmental contaminants and the remediation of waste sites. To perform this effort, SAIC evaluated current technologies and monitoring methods development programs in EPA, DO, and DOE, as these are the primary agencies through which developmental methods are being demonstrated. This report presents this evaluation and provides recommendations as to pertinent research needs or activities to address waste site contamination problems. The review and assessment have been conducted at a programmatic level; site-specific and contaminant-specific evaluations are being performed by LANL staff as a separate, related activity.

Emerging Technologies for Hazardous Waste Management: an Overview

D. William Tedder, Georgia Inst of Technology, Atlanta and Frederick G. Pohland, Univ of Pittsburgh, PA
ACS Symp 518 Emerging Technol in Hazard Waste Manag III, Atlanta, GA, Oct 1-3, 91, pl(15) conf paper
Abstract: The status and applicability of methods emerging for the management and treatment of hazardous liquid and solid wastes are surveyed. Topics covered include physical and chemical wastewater treatment, biological treatment, soil remediation and treatment, volatile compounds treatment, and selected mixed-waste treatment applications. Separation and destruction options can be helpful in decontaminating waste streams. Waste stream recycle and prevention may be more attractive as release limits become more stringent. Biological treatment schemes feature considerable potential because of low costs and benign characteristics. (1 graph, 49 references)

Hazardous waste treatment trends in the U.S.

Author: Skinner, John
Corporate Source: U.S. Environmental Protection Agency, Washington, D.C., USA
Source: Waste Management & Research v 9 n 1 Feb 1991 p 55-63
Language: English
Document Type: JA; (Journal Article) Treatment Code: G; (General Review); M; (Management Aspects)
Abstract: Developments in hazardous waste treatment and waste minimization technologies in the U.S. are described. They have been spurred largely by the Resource Conservation and Recovery Act (RCRA) and Superfund. RCRA's land ban program and the enormous quantities of hazardous waste requiring remediation under Superfund are necessitating a rapid expansion in U.S. hazardous waste treatment capacity and technology innovation. To meet these needs, incineration capacities have been built and thermal treatment capacities have been diversified. Bioremediation, soil washing, solvent extraction and in-situ stripping has also been advanced. Improved standards on the construction, design, and operation of hazardous waste landfills have also been put forward under the RCRA. The purpose of U.S. water policy is to reduce or eliminate the generation of hazardous waste at their source, wherever feasible.

5.2
Advanced Method for the Treatment of Organic Aqueous Wastes: Wet Peroxide Oxidation – WPO®, Laboratory Studies and Industrial Development

Hubert Debellefontaine · Magali Falcon · Katia Fajerwerg ·
Philippe Reilhac · Philippe Striolo · Jean-Noël Foussard

Abstract

A wide range of organic effluents produced by the chemical industry cannot be efficiently treated by biological methods, often because of their toxicity. A new liquid phase oxidation process, which is derived from the Fenton's reaction, using hydrogen peroxide at a high temperature (120 °C) and iron salts as catalyst, has been developed to help solve this problem. This process, named "Wet Peroxide Oxidation" (WPO®), is well suited for effluents containing between 0.5 and 10 g/l of carbon (TOC). A limited capital cost associated with a reasonable running cost are its main characteristics, as well as flexibility. It can be run alone before releasing-off the treated wastes or associated with other methods as either a pre-treatment or a post-treatment step.

Several pure test compounds, wastes and by-products from various industries such as paper, petroleum, chemical and pharmaceutical, have been treated by the WPO® process. In most cases, an almost total oxidation was obtained. However, the Fenton's reaction efficiency is limited by the accumulation of volatile fatty acids such as oxalic, malonic, succinic and acetic acids (hereafter OMSA). In order to improve the efficiency of the original process, various transition metal ions have been tested as catalysts. The experimental results indicated that the system using simultaneously homogeneous Fe, Cu and Mn is a promising one. This new catalytic system allows efficient elimination at a temperature not greater than 100 °C to avoid pressurization of the treatment reactor.

A pilot plant unit has been developed, which was mainly used for polluted water tables treatment during industrial sites reclamation and industrial wastewater treatment. The efficiency of the method was established on a large scale basis. At the same time, economical data were assessed.

5.2.1
Introduction

The treatment of organic refractory wastewaters released by the chemical industry is a field of investigation towards which numerous studies have been directed. To this end, several inadequate methods currently exist. The biological treatment processes are not suitable for concentrated wastewaters or for those with a high toxic level. Incineration is effective, but penalized by a high operating cost. Therefore, it is an unrealistic treatment for wastewater in which the COD value is less than 50 or 100 g/l. Moreover, incineration is penalized by its unfavorable ecological perception due to the production of dust in the atmosphere (open process).

Oxidation processes in the liquid phase seem to provide a suitable solution to the problems arising with aqueous organic pollution, as they are enclosed processes. The WPO® process (Wet Peroxide Oxidation) developed at the laboratory allows the treatment of wastewater whose COD value is less than 20 g/l (Debellefontaine et al., 1993). The reaction leads, in an acidic medium, in the presence of the Fenton's reagent (Fe(II)/H₂O₂) and at a high temperature (100 °C to 150 °C), to the elimination of the initial organic compounds. The treatment reactor is slightly pressurized to avoid vaporization of the solution. The TOC and COD reduction (Haddoud, 1989) in this case is much more significant than at room temperature. The process was used for phenol-like compounds and other organic pollutants oxidation. During the reaction, limited amounts of species refractory to the oxidation accumulate. Since these are mainly volatile fatty acids (oxalic, acetic, etc.), it was necessary to improve the efficiency of this conventional process. Efficiency could be obtained by using new catalysts; the association of Fe, Cu and Mn salts lead to an efficient removal of a synthetic mixture of oxalic, malonic, succinic and acetic acids (OMSA) and of actual industrial pollutants (Falcon et al., 1993). The tests were carried out under milder conditions (100 °C) than was the case for the usual WPO® process with iron salts, then avoiding any pressurization of the system. Cu(II), Mn(II) or Fe(II) alone have a slight catalytic effect, while the association is very effective during the oxidation of the mixture of carboxylic acids (OMSA). This apparent synergistic effect existing among them can be explained by the main specific roles of Cu(II) and Mn(II) ions. They respectively enhance the oxidation rate of acetic and oxalic acids.

Wet Peroxide Oxidation Catalyzed by Metal Salts; Literature Review

Hydrogen peroxide is a cheap and relatively strong chemical oxidant. These essential properties make it an attractive oxidizing agent for the treatment of industrial effluents. At the same time, its decomposition will lead only to harmless final products (H₂O, O₂). Yet hydrogen peroxide reacts slowly with some organic substrates. To take advantage of its strength, it may be required to activate the peroxide with transition metal ions at a high temperature. Activation means improving the reactivity towards a substrate through the formation of a more strong oxidizing species like hydroxyl (OH·) and hydroperoxyl (HO₂·) radicals.

5.2.2
Catalysis by Iron Salts

Effect of Temperature – Fenton's Reagent Limits

The commonly known Fenton's reagent uses iron salts in acidic conditions to promote the formation of OH· radicals, which are the main active species. It was noticed that the iron (II) and (III) ions show the most pronounced catalytic activity among all the transition metals ions (Murphy et al., 1989 ; Song-Yang Wei et al., 1990 ; Hocking et al., 1985). Murphy and coworkers compare the catalytic activities of different transition metal ions during the oxidation of formaldehyde at room temperature. The oxidation of phenol involves the formation of various by-products, as in such carboxylic acids as succinic, malonic, muconic, acetic and oxalic acids (Striolo, 1991). According to Al Hayek and Doré (1985), their catalytic oxidation is very difficult at room temperature whereas a more satisfactory removal of acetic acid can be obtained under a higher temperature (55% at 130 °C) (Haddoud, 1989).

Mechanism of the Oxidation Process

Mechanism of the oxidation process was first described for the Fenton's reagent at ambient temperature (Barb et al., 1951; Doré, 1989). At the approximate pH level of 3, when combined with a convenient catalyst like iron salts, hydrogen peroxide decomposes following a radical mechanism and leads to the formation of strong oxidizing radicals OH$^{\cdot}$. These radicals react on the organic compounds through addition or elimination reactions. A step by step oxidation finally leads to carbon dioxide or to refractory carboxylic acids.

Hydroxyl radical formation

$$Fe^{2+} + H_2O_2 \rightarrow Fe^{3+} + OH^{\cdot} + OH^{-}$$
$$OH^{\cdot} + H_2O_2 \rightarrow H_2O + HO_2^{\cdot}$$

Addition or elimination of OH$^{\cdot}$ on organic compounds

$$R + OH^{\cdot} \rightarrow {}^{\cdot}R\,(OH) \rightarrow \text{Hydroxylation products}$$
$$RH + OH^{\cdot} \rightarrow {}^{\cdot}R + H_2O \rightarrow \text{Oxidation products}$$

A broad diagram presented on Fig. 1 explains the competition between two reactions: The main reaction (number 1) where OH$^{\cdot}$ reacts with organic compounds to form oxidation products like carboxylic acids and, finally, carbon dioxide and the side-reaction (number 2) where the OH$^{\cdot}$ radicals react all together to form inert decomposition products like oxygen and water. A good oxidation efficiency is obtained when the organic compounds are able to trap all the radicals as soon as they are produced. This is generally observed when the concentration of the radicals OH$^{\cdot}$ in the reactive medium is lower than that of the organic species. This will result in a step by step addition of the peroxide during oxidation of a batch loaded pollutant. In a continuous process, this will result in the distribution of the peroxide feed all along the reactor.

Catalysis by Various Transition Metal Salts

Several transition metal ions can react with hydrogen peroxide as initiators of Fenton's type reactions. Among them, Co(II) can hydroxylate aromatic compounds with H_2O_2 (Strukul, 1992). Cr(II) ions can react with hydrogen peroxide by a Fenton's type reaction to produce hydroxyl radicals, but chromic salts are known as highly toxic and cannot be used for wastewaters treatment purposes (Song-Yang Wei et al., 1990). The Cu(II) ions show a weaker catalytic activity than Fe(III) ions during phenol oxidation with hydrogen peroxide. In the case of copper ions, the active oxidizing specie could be something other than HO$^{\cdot}$ (Johnson et al., 1988). The formation of Cu(III) species by a reaction of H_2O_2 with Cu(I) was suggested. On the other hand, Ni(II) and Mn(II) used alone have no catalytic effect during phenol oxidation because they do not significantly catalyse the decomposition of the hydrogen peroxide (Song-Yang Wei et al.,

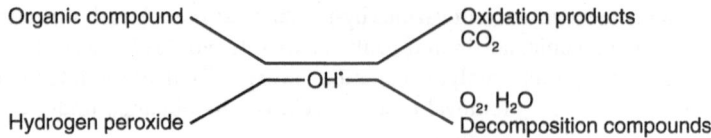

Fig. 1. Competition between two reactions during hydrogen peroxide oxidation of organics

1990). In numerous studies, H_2O_2 oxidation, in combination with two of the metals mentioned above, has been developed to improve the oxidation of various organic compounds.

Catalysis by the Combination of Metals

The association of transition metal ions such as Fe/Mn and Fe/Cu are efficient catalysts for the homogeneous catalytic auto-oxidation of aqueous S(IV). An apparent synergism exists among them (Reda, 1988). An industrial process, which allows hydroxylation of aromatic compounds for the synthesis of hydroquinone and catechol, uses H_2O_2 in combination with iron and cobalt salts (Strukul, 1992). The catalytic oxidation of phenol is slower with the Fenton's reagent than with a solid catalyst iron – copper/alumina (Al Hayek and Doré, 1985). The solubility of the supported metals suggests a catalytic synergism in the homogeneous phase. Wastewaters are chemically oxidized using the Fenton's reagent in the presence of a small amount of a transition metal ion (Cu(II), Mn(II), Co(II), ...) in addition to Fe(II) as the major catalyst (Mitsubishi, 1978). This results in a four times reduction of the amount of the major catalyst Fe(II) or in a two times reduction of the reaction time. These combinations of transition metal ions seem to be interesting ways for chemical wastewater treatment with hydrogen peroxide. The Fenton's reaction has a limited interest because the oxidation of organic acids proceeds very slowly and various homogeneous catalysts have been studied in order to oxidize refractory acids into CO_2 and H_2O (Falcon et al., 1993). This laboratory work demonstrated the efficiency of a new catalytic system of three metal salts: Fe(II), Cu(II), Mn(II). This association leads to a very important synergistic effect.

5.2.3
Main Characteristics of the Oxidation Process

The "Wet Peroxide Oxidation" process (WPO®) uses a liquid oxidizing agent (hydrogen peroxide). This process is adapted from the classical Fenton's reagent, but it is conducted under temperature so that an important TOC removal efficiency can be obtained. The tests are conducted in a completely mixed semi-continuous mode as shown on Fig. 2. The organics to oxidize are batch loaded within the reactor at the fixed operating conditions (catalyst concentration, pH, temperature, etc.). With the aim of minimizing its side-decomposition, the hydrogen peroxide is continuously fed into the reactor with a flow rate in direct ratio with the experimental volume remaining within the reactor after drawing-off samples. The reaction pH is monitored through the injection of sulphuric acid or soda. The treated solution is brought to basic pH; the metal-

Fig. 2. Sketch diagram of the process

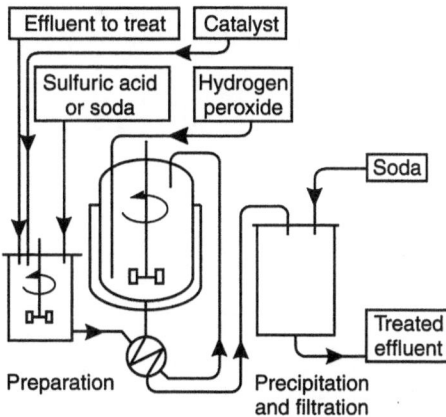

lic hydroxides are thus precipitated and separated from the solution. This technique allows us to follow the evolution of the species into the solution during oxidation (organic substrate, peroxide, catalyst) through dosages carried out on the samples taken over the time.

Optimal Conditions for the Process

Due to particular conditions, this process allows immediate oxidation of initial organic compounds and an efficient TOC removal. Under such conditions, the parasistic decomposition of the peroxide, and then the running cost, is minimized. The main operating parameters are listed in Table 1 where the actual amount of peroxide is compared to that necessary for a stoichiometric conversion of organics to carbon dioxide and water. An acidic pH is necessary and only small quantities of iron are used.

Phenol Treatment

Phenolic wastes are one of the most prevalent forms of chemical pollutants in industry. A HPLC analysis allows identification of hydroquinon, pyrocatechol and formic and oxalic acids as the main intermediaries during phenol oxidation. At 70°C the phenol is quickly oxidized but, during the reaction, carboxylic acids (mainly formic and oxalic) accumulate. Because of the high concentration of these acids, the TOC removal is not complete at the end of the reaction, as indicated on Fig. 3. On the contrary, Fig. 4 shows

Table 1. Optimal conditions for the WPO® process

	Working range
Temperature (°C)	100 – 150
H_2O_2 stoichiometric ratio	1.1 – 1.5
Reaction time (mn)	15 – 60
pH	2.5 – 4
Catalyst (mg/l)	10 – 100

Fig. 3. Phenol oxidation at 70 °C. Carboxylic
acids accumulation

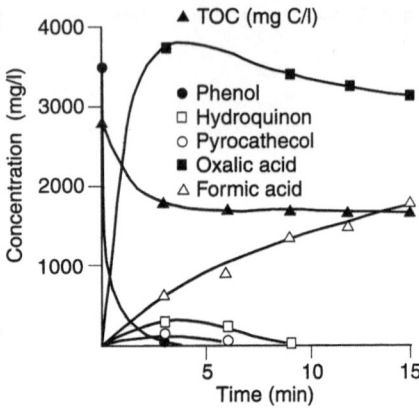

Fig. 4. Phenol oxidation at 140 °C. Complete
TOC removal

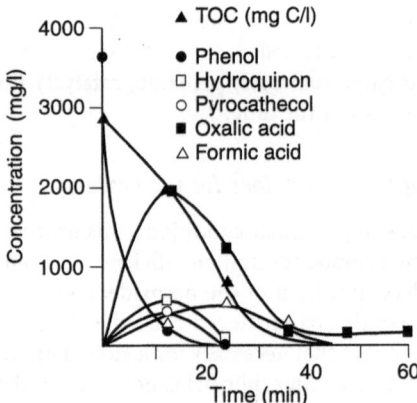

that, at 140 °C, the phenol oxidation rate is the same but that the TOC removal is more
important because carboxylic acids are oxidized. Increasing the temperature over
150 °C will result in an important parasitic decomposition of the peroxide (see Fig. 1)
and then in a loss of efficiency.

Treatment of Various Organics

Various organic compounds have been tested for treatment by the WPO® process.
These are classical pollutants: industrial pollutants, solvents, pesticides, etc. The standard conditions of treatment were as follows:

pH = 3.5 t_r = 60 mn H_2O_2 stoichiometric ratio = 1
T = 120 °C   [Fe^{2+}] = 30 mg/l

On Figs. 5 and 6 these organics have been arranged according to the result of the treatment. A total elimination of the initial pollutant is obtained, whereas the TOC removal
does not exceed 60 to 90 %. When the initial product is a carboxylic acid, its oxidation

Fig. 5 Removal of the initial organic compounds during oxidation.
pH = 3.5; t_r = 60 mn; H_2O_2 stoichiometric ratio = 1;
T = 120 °C; ($Fe^{2}+$) = 30 mg/l

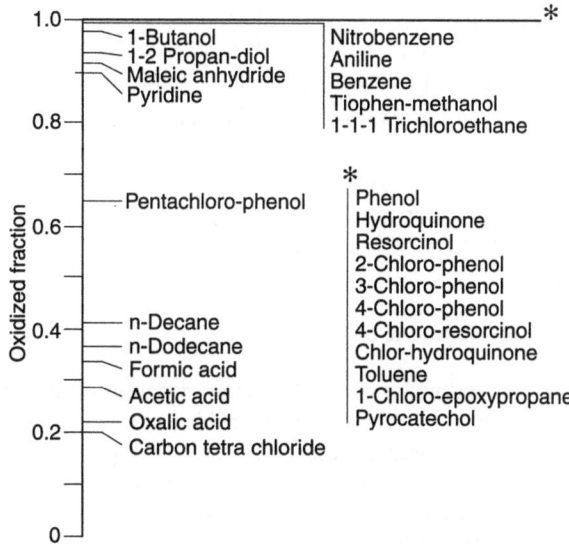

Fig. 6. TOC removal during oxidation
pH = 3.5; t_r = 60 mn; H_2O_2 stoichiometric ratio = 1; T = 120°C; ($Fe^{2}+$) = 30 mg/l

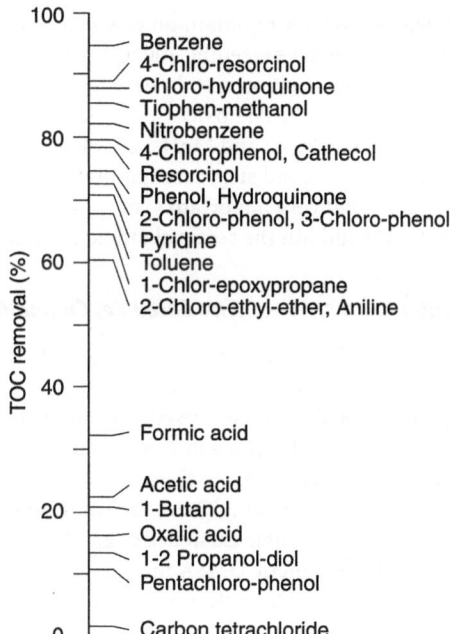

is not complete (20 to 40%) and the TOC removal observed corresponds to 15 and 30%. These results confirm that these products are refractory to oxidize. Three categories clearly appear among the compounds tested:

- Easily oxidizable compounds: aromatics (TOC removal 65 to 100%),
- Carboxylic acids (acetic, formic and oxalic acids) and insoluble hydrocarbons (TOC removal: 15 to 30%),
- Organic compounds without C–H bonds like pentachlorophenol and carbon tetrachloride with unsatisfactory TOC removal.

5.2.4
New Development of the Process

The conventional process allows the oxidation of the initial organic compounds and the formation of the volatile fatty acids (oxalic, malonic, succinic, acetic, ...) refractory to oxidation. It was necessary, therefore, to improve the efficiency of this classical process. This could be obtained by using new catalysts. Their catalytic activity was tested during the treatment by H_2O_2 of a synthetic mixture of oxalic, malonic, succinic and acetic acids (OMSA).

Preliminary Studies

It was important to obtain an efficient removal of the refractory OMSA mixture at a temperature not exceeding 100 °C in order to run the process at atmospheric pressure. So, various transition metals were tested alone as catalysts according to the conditions listed in Table 2. The oxidation efficiency is never satisfactory. Nevertheless, results suggest that Cu(II) and Mn(II) are able to increase the efficiency (16 and 22% respectively). In a second step, associations of 2 or 3 metals were tested. The Fig. 7 evidence suggests that very important synergistic effects can be obtained. When using a mixture of Fe, Cu and Mn the removal efficiency can be increased up to 91%.

Catalytic Activity of Associated Fe, Cu and Mn Ions

The second step was the optimization of the proportions of each of the 3 metals. For this purpose, an optimal design method specially developed for mixtures analysis (Mathieu et al., 1992) was used. This method allows researchers to determine a predicting model for the efficiency as a function of the proportions of each of the catalytic components. At the same time, this optimal design limits the number of experimental tests as far as possible. Then, the conditions for the best oxidation yield are determined and the possible existence of catalytic synergism is evidenced.

For the synthetic OMSA mixture, the best proportions of the metals were approximately 23/50/27% (Fe/Cu/Mn) and a removal efficiency of 91% could be obtained (Falcon et al., 1994).

Metal–metal synergism

When the metals are used separately the oxidation efficiencies are low. In the best case, i.e. in the presence of manganese, a 22% oxidation yield is obtained. Consequently, the

Table 2. Oxidation catalyzed by various metallic salts. $TOC_{[OMSA]} = 5$ g/l; pH = 3.5; H_2O_2 stoichiometric ratio = 1.5; T = 98 °C; P = 1 atm; t_r = 60 mn; $[MSO_4]$ = 5.4 mM

Metal	Fe	Cu	Mn	Co	Ni
% TOC removal	2	16	22	3	1

Fig. 7. Catalytic activity of metal ions Fe(II), Cu(II) and Mn(II) alone or associated.
$TOC_{[OMSA]} = 5$g/l; H_2O_2 stoichiometric ratio = 1.5; pH = 3.5; T = 98 °C ; P = 1 atm ; t_r = 60 mn; $[MSO_4]$ = 5.37 mM

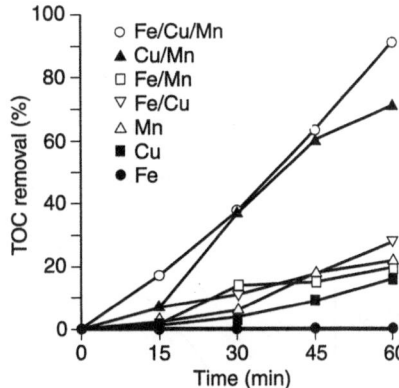

association of these three salts can constitute a catalytic system giving better performances than those observed for the salts used separately, or two by two, even if the structure of the active catalytic species is not known.

Metal – substrate synergism

Previous tests carried out at the laboratory supported the conclusion that the Fe/Cu/Mn catalytic system totally decomposes the hydrogen peroxide in the presence of an organic substrate (Falcon et al., 1993). It is therefore possible that the substrate plays the role of a ligand which forms an efficient organo-metallic specie in order to decompose H_2O_2. Previous results suggest important affinities between the organic substrate and the metal. Thus, the catalytic system (Cu(II)/H_2O_2) shows a high activity for the removal of acetic acid (see Fig. 8a) whereas Mn(II) ions are very effective on oxalic acid oxidation (Fig. 8b). These specific behaviours probably involve the formation of complexes which are active catalytic species. The reactivity of the usual Fenton's reagent is high for succinic (Fig. 8c) and malonic (Fig. 8d) acids, limited for acetic acid and almost zero for oxalic acid. This is confirmed by HPLC analysis: the degradation of succinic and malonic acids with the Fenton's reagent is not complete because of the refractory oxalic acid as shown in Fig. 8c and 8d. The main role of Mn(II) ions is to improve this specific oxalic acid oxidation. The reaction of (Mn(II)/H_2O_2) with oxalic acid is characterized by a transient pink color. Using a spectrophotocolorimetric method, it was possible to establish that this color is related to the MnO_4^- specie (Falcon et al., 1994). The reaction between oxalic acid and permanganate is used for the titration of oxalic acid (Benson, 1972). Permanganate and oxalate ions react in acidic solu-

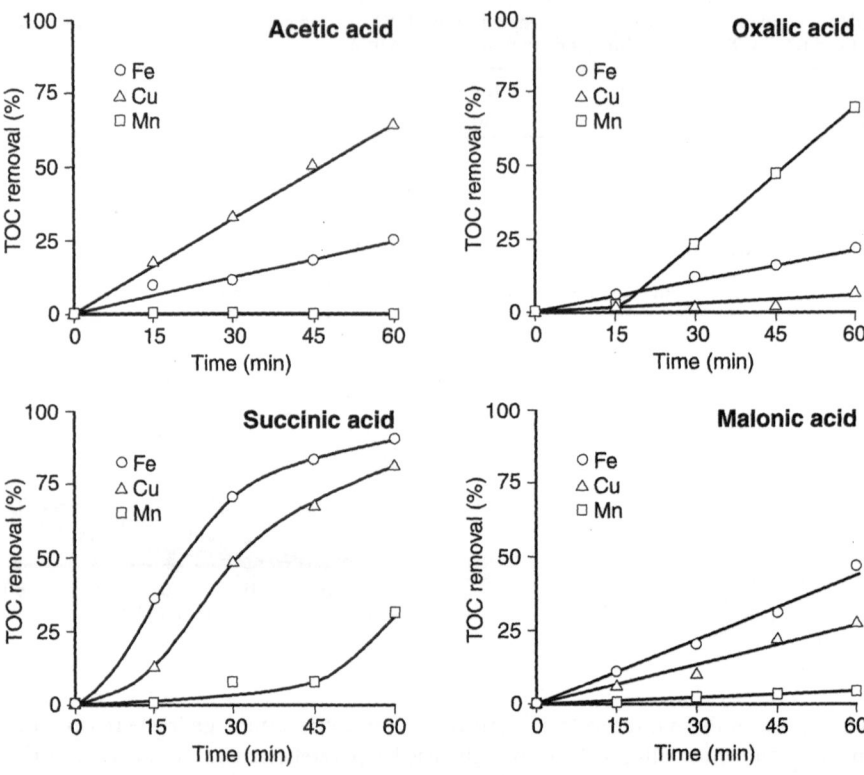

Fig. 8a–d. Carboxilic acids oxidation by the systems M(II)/H$_2$O$_2$. [Carboxylic acid] 0 = 0.1 mol/l; T = 98 °C; 3.5 ≤ pH ≤ 4.5; H$_2$O$_2$ stoichiometric ratio = 1.5; [M(II)] 0 = 5.37 mmol/l. **a** Acetic acid, **b** Oxalic acid, **c** Succinic acid, **d** Malonic acid

tion as according to the reaction (1). During the reaction with the OMSA mixture, each metal acts on a specific carboxylic acid, thus allowing a quite complete TOC removal.

$$2\,MnO_4^- + 5\,H_2C_2O_4 + 6\,H^+ \rightarrow 2\,Mn^{2+} + 10\,CO_2 + 8\,H_2O \tag{1}$$

5.2.5
Industrial Development of the Process

The oxidation of several industrial effluents was studied by the various catalysts presented here. The results obtained at the laboratory with these catalytic systems are reported on Table 3. The difference between the efficiency of the new catalysts and the conventional one is shown on this Table 3. The Fe/Cu/Mn catalytic system reveals higher performances than those observed when using Fe salt alone.

On the other hand, a full scale pilot was designed and constructed, using mainly standard equipment, pictured on Fig. 9. It has a nominal capacity of 5 cubic meters per hour. This unit is transportable (road trailer) and self-powered (electric generator, boiler). It is composed of a 2-stages mixed reactor (glass coated), of various tanks (acidification, buffer, ...) and of heat exchangers. The rated temperature is 125 °C and the

Table 3. Treatment of various industrial wastes. T = 100–130 °C; [MSO₄] = 50–100 ppm; H₂O₂ stoichiometric ratio = 1–1.7

Effluent	Pre-treatment (g/1)	COD init	COD Oxidation yields (%)	
			Catalyst Fe (T = 130 °C)	Catalyst Fe/Cu/Mu (T = 100 °C)
Pesticides and solvents	Removal of materials in suspension	19.2	70	98.5
Polluted groundwaters		0.5	53	80
Long chains alcohols production unit			49.4	97.7
Printing ink waste		12.8	30	81

Fig. 9. Trailer mounted industrial unit for WPO® treatment of industrial organic liquid wastes

rated pressure is 4 bar. This pilot plant was successfully operated in Spain during the year 1992 for treating polluted waters (COD = 10 kg/m³) within an EEC program for industrial sites reclamation. It was also operated for tests using various industrial wastes. A typical result obtained during a 1 week test in an organic synthesis factory is pictured on Fig. 10. The COD standard requested for directing the treated wastes of the various units to the biological treatment facility (extended aeration) of the factory was 500 ppm. The raw waste of a special unit, typically containing a 3000 ppm COD, is particularly refractory to the biological treatment. Figure 10 shows that the standard was easily met with the exception of a lack of peroxide, during a pre-treatment by the WPO® process. The total capital cost was about $500,000 US dollars and for this first

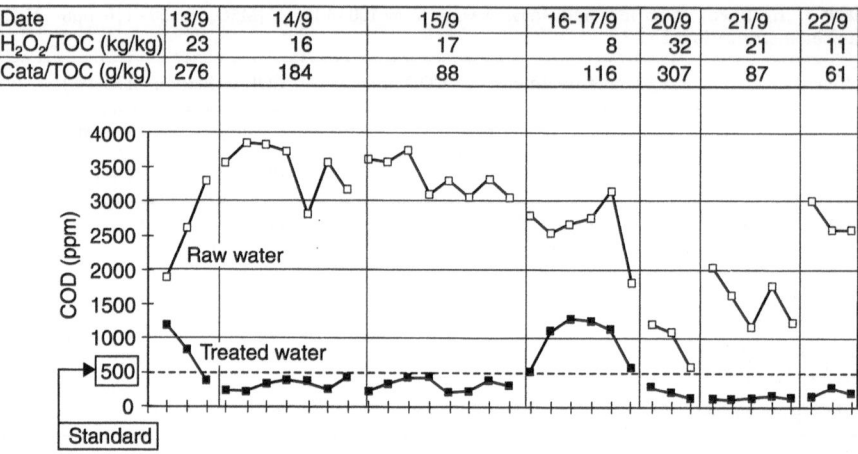

Date	13/9	14/9	15/9	16-17/9	20/9	21/9	22/9
H$_2$O$_2$/TOC (kg/kg)	23	16	17	8	32	21	11
Cata/TOC (g/kg)	276	184	88	116	307	87	61

Fig. 10. On-site treatment of a waste from the chemical industry

demonstration, the total cost, including investment, running and peroxide supply was as expressed in Relationship (2). Hydrogen peroxide (concentration 50%) was estimated to be commercially available at the price of $650 US dollars/ton. This process favorably compares with the wet air oxidation process (WAO), using molecular oxygen for oxidation in a liquid phase at high temperature and pressure. The process compares especially well with incineration, because wastes can be treated without gaseous release of dust.

$$\text{Total price (US \$/cubic meter treated)} = 23 + 7 * \text{COD (kg/m}^3) \tag{2}$$

5.2.6
Conclusion

The WPO® process provides an original and efficient method for the treatment of wastewaters; available methods are too expensive or have a limited efficiency. Essentially, the process uses hydrogen peroxide catalysed by ferrous iron salts in an acidic medium and at high temperature. In comparison with the classical treatment by the Fenton's reagent (Fe(II)/H$_2$O$_2$) at ambient temperature, this process allows concurrent oxidation of the initial organic compounds, an important COD and TOC reduction and the formation of volatile fatty acids (oxalic, malonic, succinic, acetic, etc.) refractory to oxidation. The latter are totally degraded by hydrogen peroxide in the presence of salts of the metals Fe/Cu/Mn. An optimal design method applied to mixtures was used in order to determine the optimal proportions of the three components. Therefore, this method reveals a considerable synergism independent of the proportions. Indeed, the association of these three salts results in an oxidation yield of about 90%. This catalytic system shows higher performances than those observed when using the separated salts. Phenomena such as the metal-substrate affinity arise during the reaction. In the reaction mixture, each salt has a specific role in relation to the acids involved. The iron probably ensures the attack of "long" chains and leads to the accumulation of oxalic and acetic acids. It is hypothesized that copper and the manganese complex themselves with acetic and oxalic acid respectively, and thereby promote their degradation. The Fe/Cu/Mn$-H_2O_2$ process is more efficient than the conventional WPO® process for the treatment of most of the industrial pollutants. More, it can be run at a temperature below 100 °C and then without pressurization of the reactor.

5.2.7
References

Al Hayek N, Doré M (1985) Oxydation des composés organiques par le réactif de Fenton: possibilités et limites. Env Techn. Letters 6, 37–50

Barb WG, Baxendale JH, George P, Hargrave KR (1951) Reactions of ferrous and ferric ions with hydrogen peroxide. Trans Faraday Soc 47, 462–591

Benson D (1972) Introduction aux mécanismes des réactions inorganiques en solution. Masson et Cie Editeur, 153–154

Debellefontaine H, Peyrille B, Tissot D, Foussard JN (1993) Traitement des eaux à fort niveau de pollution par le procédé WPO®. Informations Chimie, n° 352, 86–89

Doré M (1989) Chimie des oxydants et traitement des eaux. Technique et Documentation Editeur, Paris, ISBN 2-85206-562-2, chap VII

Falcon M, Peyrille B, Reilhac P, Foussard JN, Debellefontaine H (1993) Oxydation en voie humide de la pollution organique aqueuse par le peroxide d'hydrogène. Procédé Wet Peroxide Oxidation (WPO®). Etude de nouveaux catalyseurs. Rev Sc Eau, 6, 411–426

Falcon M, Fajerwerg K, Puech-Costes E, Foussard JN, Debellefontaine H (1994) Oxydation catalytique par le peroxyde d'hydrogène d'une fraction organique réfractaire au procédé WPO. First International Research Symposium on water treatment by-products. 29–30 Septembre, Poitiers, France

Haddoud F (1989) Utilisation du peroxyde d'hydrogène pour l'oxydation de déchets industriels à une température comprise entre 100 et 150 °C: étude de la dégradation des acides carboxyliques légers. Thèse de docteur ingénieur, n° 183, INSA Toulouse, France

Hocking MB, Intihar DJ (1985) Oxidation of phenol by aqueous hydrogen peroxide catalyzed by ferric ion-catechol complexes. J Chem Tech Biotechnol 35A, 365–381

Johnson GRA, Nazhat NB, Saadalla-Nazhat RA (1988) Reaction of the aqua copper (I) ion with hydrogen peroxide: Evidence for Cu^{III} (cupryl) intermediate. J Chem Soc, Faraday Trans 1, 84, 501–510

Mathieu D, Phan-Tan-Luu R (1992) New Efficient Methodology for Research using Optimal Design. (Software NEMROD), LPRAI, Centre St Gérôme, Université d'Aix Marseille

Mitsubishi Heavy Ind KK (1978) Wastewater treatment by oxidation using Fenton's reagent in presence of ferrous ions and transition metal ions. Japan Patent. – A-53 099 6S7

Murphy AP, Boegli WJ, Price MK, Moody CD (1989) A Fenton's like reaction to neutralize formaldehyde waste solutions. Env Sci Technol 23(2), 166–169

Reda M (1988) Homogeneous catalytic oxidation of aqueous sulfur (IV) by transition metal ions, part III: effect of iron (III), copper (II) and manganese (II) and synergistic catalysis. Wat Sci Tech 20 (10), 45–47

Song-Yang Wei, Yung-Yun Wang and Chi-Chao Wand (1990) Factors influencing the catalytic effect of transition metal ions on the oxidation of phenol with hydrogen peroxide. J of the Chin I Ch E 21(4), 263–268

Striolo P (1991) Oxydation d'effluents organiques aqueux par le peroxyde d'hydrogène à haute température (procédé WPO®). Thèse de docteur ingénieur, n° 178, INSA Toulouse, France. Chap III

Strukul G (1992) Catalytic oxidations with hydrogen peroxide as oxidant. Kluwer Academic Publishers, Dordrecht, Boston, London, ISBN 0-7923-1771-8, chap IV, 114–115

5.2.8
Further Readings

Wet Oxidation of Organic Ion-Exchange Resins with Hydrogen Peroxide – Radwaste Process Development
RA Baxter, CEGB Berkeley Nuclear Labs, UK, RG Brown, DJ Hebditch, AM McCabe and MG Segal; Br Nucl Energy Society/et al Radioact Waste Manag
2 Int Conf, Brighton, UK, May 2–5, 89, v1, p61(5) conf paper

For the treatment of CEGB radioactive wastes in the UK, the metal-catalyzed, hydrogen peroxide wet oxidation method has been chosen due to its use of low pressures and temperatures, good partitioning of the radioactivity within the liquid phase, and simplicity of operation suitable for a power station site. This oxidation method is described, and pilot-scale oxidation tests are detailed. In the tests, eight runs were performed in which 10 kg of Lewatt DN resin were oxidized at 100EC under reflux for approximately 5 hr. A preliminary flowsheet is described. A volume reduction factor of five is expected as compared to direct encapsulation. (2 diagrams, 4 references)

Title: Demonstration test and evaluation of Ultraviolet/Ultraviolet Catalyzed Peroxide Oxidation for Groundwater Remediation at Oak Ridge K-25 Site

Final report (March 16, 1993 – March 16, 1994); Progress rept
Performing Organization: Schafer (WJ) Associates, Inc., Chelmsford, MA.
Report No: DOE/OR/22000-1
NTIS No: DE94013173/HDM
Sponsoring Organization: Vulcan Peroxidation Systems, Inc., Tucson, AZ
(United States).; Department of Energy, Washington, DC.
Date: Mar 94 Pages: 247p NTIS Price Code: PC A11/MF A03

Abstract: We demonstrated, tested and evaluated a new ultraviolet (UV) lamp integrated with an existing commercial technology employing UV catalyzed peroxide oxidation to destroy organics in groundwater at an Oak Ridge K-25 site. The existing commercial technology is the perox-pure(trademark) process of Peroxidation Systems Incorporated (PSI) that employs standard UV lamp technology to catalyze H_2O_2 into OH radicals, which attack many organic molecules. In comparison to classical technologies for remediation of groundwater contaminated with organics, the perox-pure (trademark) process not only is cost effective but also reduces contaminants to harmless by products instead of transferring the contaminants from one medium to another. Although the perox-pure(trademark) process is cost effective against many organics, it is not effective for some organic contaminants of interest to DOE such as TCA, which has the highest concentration of the organics at the K-25 test site. Contaminants such as TCA are treated more readily by direct photolysis using short wavelength UV light. WJSA has been developing a unique UV lamp which is very efficient in the short UV wavelength region. Consequently, combining this UV lamp with the perox-pure(trademark) process results in a means for treating essentially all organic contaminants. In the program reported here, the new UV lamp lifetime was improved and the lamp integrated into a PSI demonstration trailer. Even though this UV lamp operated at less than optimum power and UV efficiency, the destruction rate for the highest concentration organic (TCA) was more than double that of the commercial unit. An optimized UV lamp may double again the destruction rate; i.e., a factor of four greater than the commercial system. The demonstration at K-25 included tests with (1) the commercial PSI system, (2) the new UV lamp-based system and (3) the commercial PSI and new UV lamp systems in series.

5.3
Heavy Metals Recovery by Electrolyzing Technique: The 3.P.E. Technology

GERMAIN LACOSTE

Abstract

The problem of treatment of dilute solutions has led to the use of granular electrodes, which are found to be highly efficient in that they provide a large interfacial area to the treated solution. These electrodes work in cells with a pulsed flow a liquid which permit movement of the solid phase, preventing the risk of sealing. A computer system based upon the NEC V 25 microprocessor has been developed to control the evolution of the metal deposit. It requires certain analog inputs connected to sensors for the acquisition of data on electrical parameters (total current I and the voltage E between the cathode and electrode reference with a large entry impedance). With this acquisition, the computer system may control the hydrodynamic parameters (steady state velocity of percolation v_0 and pulsation of amplitude a and of frequency f) and the electrical parameters. The control actions associated to this control system are: the frequency of the pulsation motor, the constant voltage U and the flow rate of the solution to be treated.

5.3.1
Introduction

The recovery of metal present in chemical solutions is very important from both environmental and economic viewpoints. The classical electrochemical technique used poses a problem since only a weak density of current can be passed through the solution to be treated. The volumic electrodes (already called the flow-through parous electrodes) avoid this problem as they present a large surface area due to their volumic character. This method could be applied in different domains, such as organic electrosynthesis, oxidation, recovery of valuable metals form effluents or dilute solutions and the electrochemical depollution (Wragg, 1977; Olive et al., 1980).

The electrodes are made up of a good conductor with a porous structure through which the solution to be treated flows. The porous structure may be of different types, such as a bed of granules or conducting grains or a packing conductor (metal grids, rolled wire-netting or sheet of metal).

In comparison with classic plane electrodes, the principal advantages of the volumic electrodes are the following:

- a large surface area of the interface matrix solid-solution due to the volumic aspect at the specific area,

- a high mass transfer coefficient due to the fact that the solution percolates through the porous structure.

However, such a system has a few drawbacks:

- the plugging of fixed beds during their long term utilization for the recovery of metals (autonomy of the reactor),
- the necessity of providing the potential of the electrode in such a manner that only the desired reaction takes place (selectivity of the rector).

5.3.2
Development on the Porous Electrodes

The technology has evolved to solve the principal difficulties. In order to avoid the disadvantages, such as the drop of potential in the liquid phase and the problem of sealing, a new type of rector shown in Fig. 1 and called the Pulsed Porous Percolated Electrode (3. P. E.) has been put into use in our laboratory (Duverneuil et al., 1987; Ratel et al., 1988). It could be considered in compromise between the fixed bed electrode (good productivity) and the fluidized bed electrode (continuous operation without risk of sealing).

The new reactor eliminates the problem of plugging of the bed by moving the matrix conductor during a fraction of time of the functioning of the rector. This is done by using pulsed columns, which are largely used in different processes in chemical engineering. In this method, a periodic movement comparable to a sinusoidal one, is applied to the fluid.

Depending on the values of the pulsation parameters (frequency and amplitude), it is possible to leave the particles suspended during a small fraction of the pulsation period, thus provoking their dislocation, while during the other fraction of the period they look like the fixed bed. In this manner, one could expect to obtain the advantages of both the fixed and fluidized bed.

5.3.3
Description of the Installation

In the 3. P. E. technique, two elements may be schematically distinguished, namely:

- The complete control mechanism, which could be subdivided into two parts, mechanical and electrical. The mechanical part is also divided in two systems: first, a motor, powered by a triphase current, communicates the rotation to a connecting rod-crank system (1) and second the liquid peculation system (2) which consists of a centrifugal pump and an electrically operated valve. The electrical part (3) permits the input of a direct current to the electrochemical cell.
- The electrochemical cell which is fed by a liquid flow having a pulsated movement provided by the connecting rod-crank system.

The cell is a cylinder made of plexiglas which contains: an homogenization element, whose principal role is to ensure an uniform distribution of the electrolyte on all the section of the reactor; a porous bed which constitutes the cathode (the element of the reactor to be studied) and a titan platinum anode placed above the bed (in the case of

Fig. 1. Porous Pulsed Percolated cell

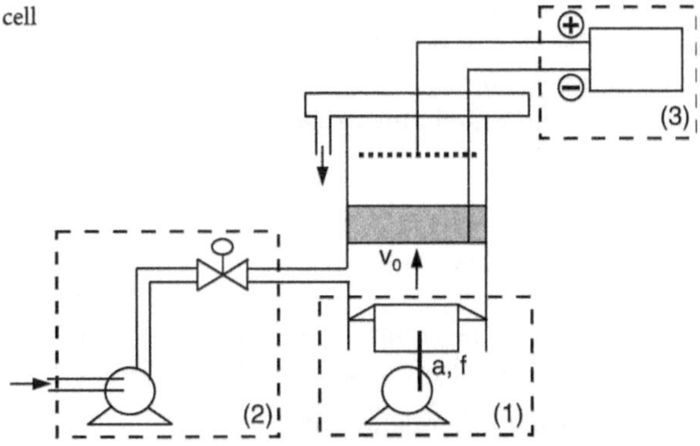

axial type). The bed of particles having an axial configuration is placed on a grid. On this grid is located the input current electrode.

The porous bed is made up of conducting particles. A probe of negligible external diameter (in comparison with the tube diameter) is linked to a reference electrode made of mercurous sulfate and saturated potassium sulfate (S.S.E.) by a capillary tube. This enables the voltage at the top of the bed, which constitutes the volumic electrode, to be fixed.

The pulsation movements of the liquid in the original method were performed by a circular solid piston. The teflon piston is connected to a connecting rod-crank arrangement to translate the circular rotation by a motor, supplied by a D.C. current, to a pulsation movement. The connecting rod-crank system transmits the pulsation to the liquid through an intermediate made up of an acid resistant (elastomer) membrane.

A stabilized voltage is applied between the anode and the cathode. A voltmeter measures the voltage drop between cathode and reference. The total instantaneous intensity of the current in the anodic circuit is measured by means of a known resistance placed in the circuit.

5.3.4
Control of the 3.P.E. Cell

The application of electrolysis reactor necessitates control of the distribution of electric potential at the cathode, in such a way that this distribution is maintained in the zone where the desired reaction takes place. This potential distribution is a function of hydrodynamic parameters (a, $\omega = 2\pi f$ and v_0) and the metal ion concentration. This is the reason why the effective potential difference in the cell, which determines the efficiency and specificity of the reaction, must be continuously monitored.

The metal is deposited on the granules which constitute the cathode, owing to the reduction of the ions present in the solution. The density of the bed and the diameter of the particles increase as a function of the metal quantity deposited; they influence, therefore, the percolation velocity necessary for the movement of particles. The pulsation parameters (particularly the frequency) and the rate of steady flow of the fluid should evolve in parallel so as to have periodically proper fluidization of the cathode.

A compromise has to be met between the matrix destabilization due to the pulsed flow (favoring the mass transfer and the redistribution of the granules), and the preservation of the granules in a fixed state in order to conserve a good conductivity. In the present section, the preliminary studies conducted in the automatic control of the 3. P. E. by the regulation of the potential and hydrodynamic parameters will be developed.

5.3.5
Flow Hydrodynamics

Two non-dimensional groups (Saint James et al., 1961; Krasuk, 1983) are generally used to represent the flow hydrodynamics of 3. P. E. They are:

- the Strouhal number defined by:

$$Sr = 2\pi \, a \, f/v_0 \tag{1}$$

where a and f are respectively the amplitude and the frequency of the pulsing movement superimposed on the steady flow, which is characterized by the average velocity v_0. This number denotes the relative intensity of the pulsation to the steady flow.

- the Womersley number given by:

$$Wo = (\omega \, \varrho_p/\mu)^{0.5} d_p \tag{2}$$

where d_p is the average diameter of the particles of granular cathode, and ϱ_p and μ are respectively the density of the particles and the viscosity of the liquid. This dimensionless number, also called pulsed Reynolds, relates the hydrodynamic state of the flow around the particle. A study of these parameters shows the particularity of a 3. P. E. reactor because the complete porous bed feels the immediate effect of the pulsation imposed by the liquid phase. Following the characteristics of the pulsation parameters, it can be supposed that in a pulsed liquid flow through a percolated bed, the bed may or may not stay fixed in view of the characteristics of the flow. In the case of a steady upward flow, depending on the influence of the maximum velocity created by the pulsation (a ω), the flow may change its direction during the period.

Depending on the height of the bed and its hydrodynamic, the electrochemical reactor (3. P. E.) functions like a Continuous Agitated Reactor (C. A. R.), if the Strouhal number Sr is equal or greater than 4; if Sr is smaller than 4, a piston type flow is observed. When the reactor functions like a C. A. R., the concentration at the cathode stays constant and low in a manner that the current line penetration becomes large (the "active" volume of the electrode is then greater).

The use of a 3. P. E. in a working range is shown in Fig. 2. The limits of this range are defined in the following manner:

a) the technical requirements impose a maximum value of 0.3 m/s for the product a ω curve (1). This means that the amplitude of the piston can not exceed 2.5 cm and that it is not advantageous to impose a frequency of pulsation greater than 2 Hz.

b) in practice, high mass transfer rates are obtained when the Strouhal number is high (10 < Sr < 20 or more precisely Sr = 15). The curves (2) and (3) represent the limits of the flat respectively for Sr = 10 and Sr = 20,

c) if a maximum possible velocity v_0 is imposed on the solution flow (for the needs of production), the time for the fluidization of the granular matrix (for the reasons of

Fig. 2. Working range of a porous pulsed percolated electrode

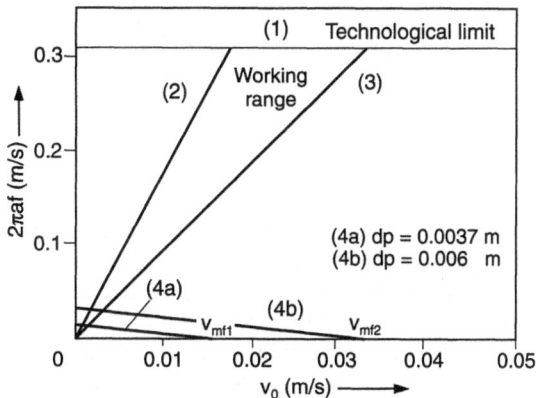

conduction) should be limited, by fixing a maximum value of the instantaneous velocity just greater than the minimum velocity of the fluidization of particles, given by (Riba, 1978):

$$v_{mf} = 1.54 \cdot 10^{-2} \frac{\mu}{dp\, \varrho_1} \left(\frac{d_p^3 \varrho_1^2 g}{\mu^2} \right)^{0.66} \left(\frac{\varrho_s - \varrho_1}{\varrho_1} \right)^{0.7} \tag{3}$$

where μ and ϱ_1 are respectively the viscosity and the density of liquid, and ϱ_s is the density of particles. This functioning limit is represented in Fig. 2 by the curves (4a) and (4b) as a function of the particles' diameters.

Thus three relations define the limits of the working range of the electrode:

$\alpha\omega < 0,3$	(technical limit)	
$\alpha\omega = 15\, v_0$	(mass transfer)	(4)
$\alpha\omega > v_{mf} - v_0$	(fluidization aspect)	

It can be observed from this figure that the working range reduces when the minimum velocity of fluidization increases (particularly due to increase of the diameter and the density of the particles). Additional parameters should also be considered such as the initial working conditions, the total mass of the granular matrix, the particle diameters of the composants. The known of initial conditions of the matrix, d_{p0} and M_0, and of the reactor hydrodynamic characteristics, flow rate, section of the cell, a and f lead to determine the final mass Δm of the deposit that the reactor is able to contain.

5.3.6
Intensity through the Electrode

The intensity through the electrode is representative of the mass transfer occurring between the liquid and the solid phase, because a deposition of one mole of metal is accompanied by a liberation of n mole of electrons.

Observation of the signal intensity as a function of the frequency has permitted the characterization of the drop in intensity due to the destabilization of the bed, which determines the minimum frequency necessary to suspend the particles. The various signals observed for the intensity differ from one another in depth and width of the falls. An increase in the frequency diminishes the width and increases the depth.

5.3.7
Selectivity of the Reactor

One of the advantages of the electrochemical reactor is the possibility of favoring the principal reaction of deposition, by the application of a known potential difference to the solid matrix without spoiling the electrolytic solvent (faradic efficiency equal to one). This value of potential difference characterizing the selectivity of the reactor is called the potential working zone ΔE, and has been measured for different electrochemical systems; ΔE is calculated by the relation:

$$\Delta E = E_0 - E^* \tag{5}$$

where E^* is the greater cathodic electrode potential and E_0 is the smaller cathodic electrode potential (close to the equilibrium potential).

During polarization, the instantaneous potential E is regulated so that the minimum value lies near E^* (in absolute values, this means the maximum value E_{max}).

5.3.8
Control of Potential

Two methods which can be used to control the potential, are:

• the direct method in which, by the use of potentiometer-regulator, the difference between the potential drop solution/reference and the record value is measured and is then used in the regulation of the potential;
• the indirect method, where the measurements and regulation are made on other parameters such as intensity or concentration (Fritze).

For all the indirect methods, the automation of the reactor and the control of the internal potential drop require a model connecting the parameter to be controlled and the functioning of the reactor.

The direct method ensures good selectivity, since it controls the potential where the active element is deposited. This type of regulation will be used to control the polarization in the volumic electrodes.

The measurement of the potential drop is compared to a data base. As a function of the difference between them, a regulator commands a regulating system, which acts on the power amplifier which, in turn, feeds that anode-cathode circuit. The power supplier receives a low current voltage and gives a voltage of high power that produces a high current intensity.

5.3.9
Experimental Setup and Working Mode

The terminal is a computer system using a NEC V25 microprocessor that is developed so as to control and supervise the behavior of 3. P. E.-type reactor. It requires a certain number of analogic inputs connected to appropriate probes and a certain number of outputs emitting electrical signals, to reach the control objective.

The electrical connections for the input of measurements and output of commands for the system response are shown in Fig. 3. These three output electrical commands are:

- for the regulation of flow. An electric valve is used, where flow is proportional to the opening of the valve. It is supplied by a 0–10 V voltage,
- for the pulsation. The electric motor running the pulsation systems supplied by a frequency variation device in such a way that the frequency evolves from 0 to 100% of the nominal value of the frequency of the motor for a 0 to 10 V variation of the input voltage,
- for the polarization. The voltage U (potential drop between anode and cathode) is provided by the regulated power supplier and controlled by an external signal from the terminal.

5.3.10
Presentation of the Behavior of the Regulator

Taking into account the data, the terminal should perform the following functions:

- the possibility of driving the reactor automatically or manually by using the key and of memorizing time at the beginning and at the end of the automatic mode,
- supply of the polarization voltage,
- measurement of intensity so as to regulate the pulsation parameters,
- calculation of the mass parameters,
- stockpiling the parameters linked to the mass of the deposit at times recorded in the computer system memory.

The control of the potential is carried out by a direct method. The measurement of the intensity is performed according to the hydrodynamics (pulsation regime or steady flow) and deals with the determination of the drop intensity ΔI.

The variations of the instantaneous intensity are, by themselves, an indication of the state of the contact in the solid matrix. If the latter rests fix, the intensity is practically

Fig. 3. Hardware of the terminal

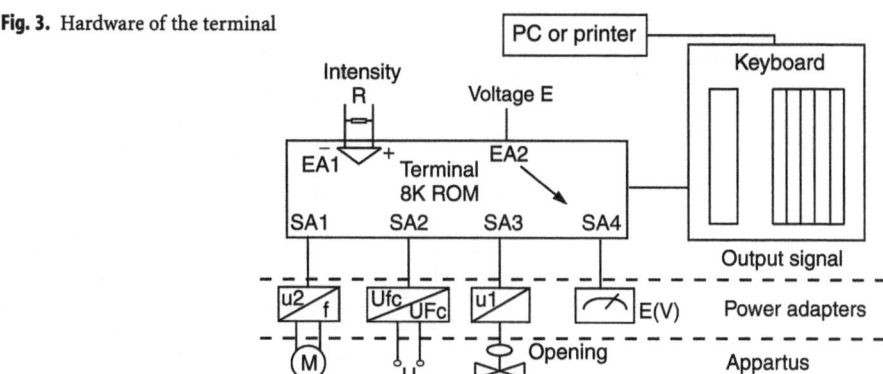

constant, but, if the particles are destabilized by the pulsation, the instantaneous intensity falls suddenly of the drop ΔI, as a function of the maximum velocity of the liquid. It is therefore necessary to limit the deepness of the drop intensity ΔI, characterizing that the bed is lifted up by the flow, at a value ΔI^* corresponding to the minimum flow velocity necessary to put the particles in suspension, by correcting the frequency of the pulsation.

The evolution of the pulsation parameters (notably the frequency because the amplitude stays fixed) and of the rising percolation is represented in Fig. 4. This figure restates the working region defined in Fig. 2.

The point A, where there is no pulsation (zero frequency) is chosen as the manual starting point. This regime is maintained in such a way that the deposed mass reaches a sufficient value to stick to the granules. When this condition is fulfilled, a pulsation of frequency f' is applied (point A'). The parameters must then evolve as presented in this figure. At first, only the frequency rises until it reaches a level such that it corresponds to a Strouhal number of 15. Later, the percolation velocity and the pulsation frequency increase in such a way that Strouhal number is maintained close to the value indicated above (B →C).

The calculation of the weight by program, using the expression of the minimum velocity of fluidization v_{mf}, given by equation (3), requires a difficult formulation using the characteristics of the fluid. A simple relation has been found from experiments, giving this expression of the velocity of destabilization v_{des}.

The velocity v_{des} of destabilization of particles at any time appears to be a function of the initial diameter d_{p0} and of the mass of metal deposited at this time Δm. Thus, relations as functions of d_{p0} connecting the velocity of destabilization to the mass has been found.

5.3.11
Conclusion

The study of pulsed porous electrodes has permitted researchers to determine their domain of application, which determines the range of variation of the functioning parameters of the reactor. It has shown the problems encountered in the continuous setting of pulsed electrochemical cells, which has led to the automatic control of the

Fig. 4. Evolution of the pulsation and percolation parameters

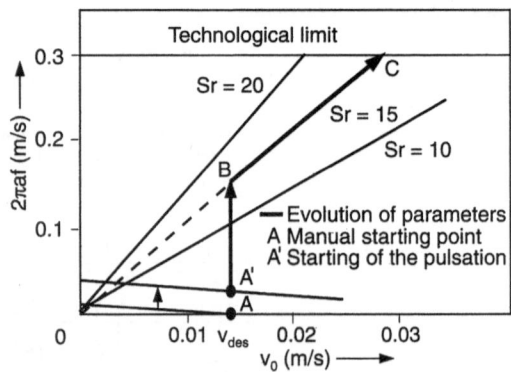

cell. Relationships have been drawn form experiments for the calculation by a computer of the electrical output values of the regulator. The hardware of the regulation system is then defined to ensure the evolution of both electrical and hydrodynamic behaviors.

The theoretical application field involves all electrolysable metals, whereas the present experimental knowledge is restricted, as the 3. P. E. process is quite new (Lacoste, 1986). It can be and must be modified, and adapted to specific industrial and environmental problems.

5.3.12
References

Duverneuil P, Lacoste G, (1987) Récente évolution dans la conception et la misc en euvre de réacteurs électrochimiques à cathode granulaire: Electrode Poreuse Percolée Pulsée. 1er congrès français de Génie des Procédés, Nancy

Duverneuil P, Lacoste G (1987) Elimination of metals form industrial wastes: a new recovery technic, 4ème congrès méditerranée de génie chimique, Barcelone

Fritze U Appareil à récupérer l'argent des bains de fixage. Brevet d'invention N° 1357 177

Krazuk JH, Smith JM (1983) A I Ch E J 10, 759

Lacoste G (1986) Procédé et installation d'électrolyse à travers une ou des électrodes volumiques poreuses. Brevet d'invention INP Toulouse

Olive H, Lacoste G (1980) Electrochim Acta 25, 1303

Ratel A, Duverneuil P, Lacoste G (1988) J Appl Electrochemistry 18, 394

Riba JP (1978) Etude des phénomènes de transfert en fluidisation homogène. Thèse de docteur ès sciences. ENSIGC Toulouse

Saint James R, Graham GP (1961) Génie Chimique 86, 1

Wragg A (1977) Chem Eng 49, 39

5.3.13
Further Readings

Title: **Electrochemical recovery of metals from effluent and process streams**
Author: Campbell, D. A.; Dalrymple, I. M.; Sunderland, J. G.; Tilston, D.
Corporate Source: EA Technology, Chester, Engl
Source: Resources, Conservation and Recycling v 10 n 1–2 Apr 1994. p 25–33
Publication Year: 1994
CODEN: RECREEW ISSN: 0921-3449
Abstract: The removal of copper and nickel from aqueous solutions using a simple tank cell, an improved mass transfer inert fluidised bed cell (Chemelec) and a three dimensional high surface area cell are examined. The high surface area cells which were examined closely consisted of a packed graphite particle bed cell using two different sizes of particle. The packed bed cells were found to be extremely effective in reducing metal concentrations below 1 ppm. The effect of current density and flow rate were studied for each type of cell. The most economic concentration range in which each cell should operate was determined. From this data suitable combinations of cells can be chosen to remove metal for high concentrations (20000 ppm) to very low concentrations (less than 1 ppm). (Author abstract) 7 Refs.

Title: **Kinetic studies of electrochemical generation of Ag(II) ion and catalytic oxidation of selected organics**
Author(s): Zawodzinski, C.; Smith, W. H.; Martinez, K. R.
Performing Organization: Los Alamos National Lab., NM.

Report No: LA-UR-93-2196; CONF-930571-24
Sponsoring Organization: Department of Energy, Washington, DC.
Contract No: W-7405-ENG-36
Notes: Electrochemical Society meeting (183rd), Honolulu, HI (United States), 16–21
May 1993. Sponsored by Department of Energy, Washington, DC
Date: 1993, Pages: 8 p NTIS Price Code: PC A02/MF A01

Abstract: The goal of this research is to develop a method to treat mixed hazardous
wastes containing selected organic compounds and heavy metals, including actinide
elements. One approach is to destroy the organic via electrochemical oxidation to carbon dioxide, then recover the metal contaminants through normally accepted procedures such as ion exchange, precipitation, etc. The authors have chosen to study the
electrochemical oxidation of a simple alcohol, isopropanol. Much of the recent work
reported involved the use of an electron transfer mediator, usually the silver(I)/(II)
redox couple. This involved direct electrochemical generation f the mediator at the
anode of a divided cell followed by homogeneous reaction of the mediator with the
organic compound. In this study the authors have sought to compare the mediated
reaction with direct electrochemical oxidation of the organic. In addition to
silver(I)/(II) they also looked at the cobalt(II)/(III) redox coupled. In the higher oxidation state both of these metal ions readily hydrolyze in aqueous solution to ultimately
form insoluble oxide. The study concluded that in a 6M nitric acid solution at room
temperature iso-propanol can be oxidized to carbon dioxide and acetic acid. Acetic
acid is a stable intermediate and resists further oxidation. The presence of Co(III)
enhances the rate or efficiency of the reaction.(ERA citation 18:029899)

Remediation of Metal Contaminated Soil by EDTA Incorporating Electrochemical Recovery of Metal and EDTA

Herbert E. Allen and Ping-Hsien Chen, University of Delaware, Newark;
Environ Progr, Nov 93, v 12, n 4, p 284 (10) research article

Abstract: A process has been developed by which ethylenediamine tetraacetic acid
(EDTA), which is used to extract heavy metals form contaminated soils, can be recovered and reused. The method involves the use of electrolysis of the metal chelates to
recover both the EDTA and the associated metals. Background data are presented on
metal-soil interactions, the release of metals form the soil matrix using EDTA, and the
electrochemistry of metal-EDTA complexes. Results are presented from a study of the
process using a soil contaminated with cadmium, copper, and lead. Polarography was
carried out at various pH values and at various metal-to-EDTA ratios. The recoveries
of Pb, Cd, Cu, and their EDTA-complex solutions were evaluated in the electromem-
brane cell. The use of a two-chamber electrolysis cell, in which the anode compartment
was separated from the cathode compartment by a cation-exchange membrane, prevented the EDTA from being oxidized at the anode during the electrolysis. Metal and
EDTA recoveries exceeded 90% in most cases. (2 diagrams, 10 graphs, 48 references,
4 tables)

5.4
An Overview of Plasma Arc Technology Applied Research Projects for the Vitrification of Hazardous Wastes

Hany Zaghloul · Ed Smith · Donald Freeman

Abstract

The U.S. Army Corps of Engineers (USACE) launched a study under the Construction Productivity Advancement Research (CPAR) program to determine the feasibility of using plasma arc technology to destroy asbestos contaminated material (ACM). The U.S. Army Construction Engineering Research Laboratories (USACERL) and the Construction Research Center (CRC) at Georgia Institute of Technology are the main partners in the ACM destruction projects. Phase I project results indicated that development of a mobile Plasma Asbestos Pyrolysis System (PAPS) for destroying asbestos and ACM on-site would have a high probability of success. CPAR is a cost-shared research and development (R & D) partnership between USACE laboratories and the U.S. construction industry (e.g., contractors, equipment and material suppliers) to improve construction productivity and competitiveness. The objective of CPAR is to facilitate productivity-improving research, development, and application of advanced technologies through cooperative R & D programs and field demonstrations as a means of technology transfer.

USACERL is exploring other uses for plasma arc technology in destroying hazardous wastes. DOD installations and depots have, for many years, accumulated hazardous wastes that resulted from their unique mission. In some cases, these wastes cannot be disposed in any type of landfill. Therefore, a separate group of plasma projects which apply to unique military hazardous wastes is being conducted for specific types of military wastes. One of these projects involves researching the disposal, gasification, vitrification, and destruction of waste materials and components in an environmentally safe manner using plasma arc technology. Vitrified waste candidates include: incinerator ash, polluted soils, obsolete thermal batteries and smoke signals, proximity fuses, and Army Materiel Command (AMC) production sludge. All the above candidates represent difficult wastes that cannot be treated or disposed of using conventional technologies.

In addition, USACERL is conducting a basic research project with the University of Illinois to study the fate of heavy metals in plasma arc processing. The volatility of some heavy metals (e.g., cadmium) still presents a problem for emission control during vitrification. Understanding and avoiding the potential for heavy metal volatilization with the emission will afford the plasma arc technology far reaching applications for processing other wastes. Finally, USACERL is representing DOD in a joint effort for plasma research between DOD and Department of Energy (DOE) common interest hazardous waste problems. The DOD\DOE project involves developing new technology applications, pushing the state-of-the-art equipment design, and transferring the technology to private sector potential users.

5.4.1
Introduction

Plasma arc technology offers many attractive features that could make it a feasible destruction method for a variety of hazardous waste candidates. A plasma is a gas that has been ionized by intense heat, such as that created by the electric arc of a plasma torch. Unlike non-ionized gases, plasma can respond to electrical and magnetic fields. The resistance of plasma to an electrical field converts the electricity into heat energy.

This technology was developed more than 25 years ago in the U.S. space program to simulate re-entry temperatures on heat shields. Only recently has it begun to emerge as a commercial tool in several industries, including steelmaking, precious metal recovery, and waste disposal. The heart of this technology is the plasma arc torch – essentially a steel cylinder several inches in diameter and several feet long (specific dimensions relate to torch power levels). Plasma torches operate in the 100 kilowatt to 10 megawatt power range, and can routinely produce controlled furnace temperatures that range from 3,000/C to more than 7,000/C. Thus, plasma torches can operate at much higher temperatures and at much greater efficiencies than fossil fuel burners. Further, plasma torches require only about 15 percent of the air necessary for fossil fuel burners. Therefore, effluent gases are greatly reduced compared to fossil fuel burners, allowing furnace systems to be built much more compactly than conventional furnaces and at lower capital costs.

This paper is based on the plasma arc work in-progress at USACERL with a number of partners. Figure 1 represents graphically the progress and evolution of different work units, or projects, aimed at turning plasma arc from a metallurgy industry tool into a hazardous waste destruction/disposal process. The arrow on Fig. 1 introduces historically the beginning of all plasma work at USACERL as a CPAR project with Georgia Tech for the destruction of asbestos, then an applied research project with AMC to apply the technology for a variety of military unique wastes, followed by a basic research project at the University of Illinois for the fate of heavy metals in plasma arc processing. The first three projects or blocks have been initiated and are in different stages of completion. Therefore, they are discussed in better detail in this paper. Next, both the two projects for process optimization research (conducted by a DOE contractor) and Army Environmental Center (AEC) technology transfer, as well as the proposed Fort Gordon Plasma Research Center are under development at the time of writing this paper, and therefore, will be briefly mentioned towards the end.

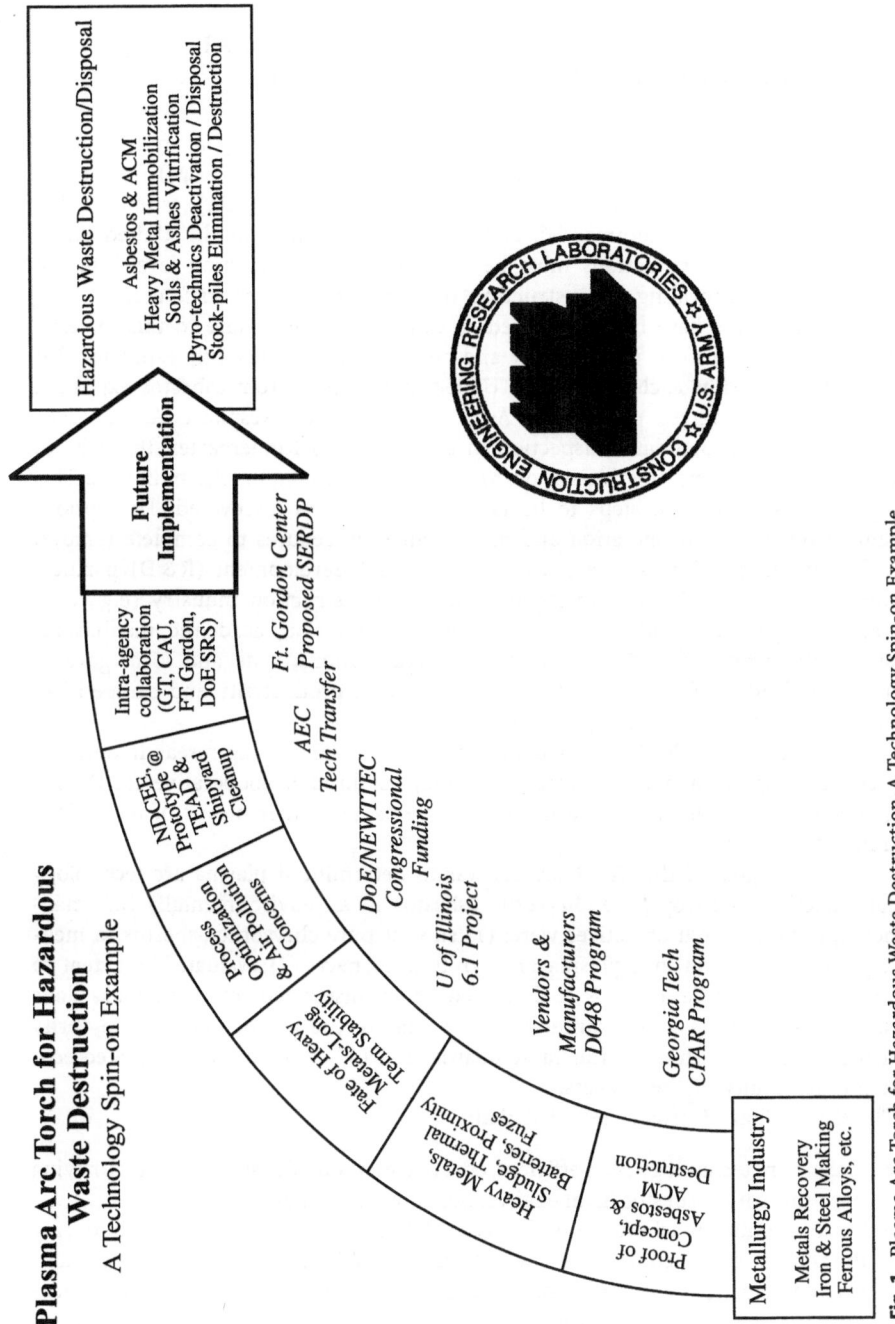

Fig. 1. Plasma Arc Torch for Hazardous Waste Destruction. A Technology Spin-on Example

5.4.2
Construction Productivity Advancement Research Program (CPAR) for the Destruction of Asbestos-Containing Material (ACM)

The U.S. construction industry currently faces multibillion-dollar rehabilitation costs for removal and disposal of asbestos-containing material (ACM) from existing buildings. The U.S. Environmental Protection Agency (EPA) estimates that about 750,000 public, commercial, and industrial buildings in the United States have ACM. Corrective actions must be taken when contamination levels are too high or when renovating, reconstructing, or demolishing these structures.

Asbestos fibers have been identified as carcinogenic. Exposure to this material creates a serious risk to public safety and health. The EPA currently estimates that up to 12,000 deaths each year in the United States result from asbestos exposure. As a result, Congress passed the Asbestos Emergency Response Act of 1986 (AHERA), which mandates inspection in school grades Kindergarten through 12. Congress may extend this legislation to all public and commercial buildings. This legislation specifies the steps to be taken when ACM is discovered. The options range from modified operation and maintenance procedures to complete removal and disposal. CPAR is a cost-shared research and development (R&D) partnership between USACE laboratories and the U.S. construction industry (e.g., contractors, equipment and material suppliers). In addition, academic institutions, public and private foundations, nonprofit organizations, state and local governments, and other entities interested in construction productivity and competitiveness also participate in this program.

The objective of CPAR is to facilitate productivity-improving research, development, and application of advanced technologies through cooperative R&D programs, field demonstrations, licensing agreements, and other means of technology transfer.

The main goal of the CPAR project was to determine if plasma arc technology can effectively destroy pure chrysotile asbestos in an environmentally safe manner. Specific research objectives were: (1) subject pure chrysotile asbestos to melting and vitrification in a plasma arc pyrolysis furnace; (2) evaluate the extent to which the chrysotile asbestos is destroyed and transformed into a nonhazardous material that meets EPA requirements; (3) analyze the gaseous effluent to verify that it complies with EPA effluent standards; and (4) evaluate the anticipated economic feasibility of the process.

The CPAR Phase I project revealed that:

- Plasma arc technology is an efficient, effective method of destroying and vitrifying pure chrysotile asbestos in an environmentally safe manner.
- Trace amounts of asbestos found in the solid residue and gaseous effluent during the analysis consisted of only a few scattered fibers. This amount of asbestos is considered negligible, and far below existing asbestos exposure standards and guidelines (i.e., less than 1% by volume in the solid vitrified material and a maximum airborne concentration of 0.2 fibers per cubic centimeter in the workplace). If necessary, a small increase in furnace temperatures or residence time should readily eliminate even these trace amounts.

- The capability of plasma arc technology to safely destroy asbestos, ACM, and any other contaminated building materials has a high probability of success.
- A mobile 7 ton-per-day PAPS could be expected to be commercially competitive right now in many regions of the nation where ACM landfill disposal costs are above average.

Currently, the CPAR Phase II research program is underway. The objective of Phase II is to conduct all the research necessary to test and evaluate the innovative concepts required to build a prototype mobile furnace system for the destruction of ACM. The on-going research program would consist of tasks which would modify and augment an existing plasma arc heating system to test and evaluate a wide variety of ACM taken from project sites. The results of the research program would be used to develop the design criteria for a prototype mobile Plasma Asbestos Pyrolysis System (PAPS). PAPS is a truck-mounted furnace for the on-site, safe and ultimate disposal of asbestos fibers and parasitic wastes. Any form of asbestos wastes may be fed into the PAPS and the wastes will be converted into a molten slag that hardens to a glassy solid when cooled. Disposal of the wastes will be continuous; feeding and tapping will be made at regular intervals. As a result, plasma arc vitrification of asbestos is likely to become increasingly competitive – using mobile PAPS units throughout the nation – as landfill disposal costs increase.

5.4.3
Army Materiel Command Unique Military Waste Experimental Program

The first stage of this study was the evaluation and selection of candidates for plasma arc pyrolysis. The main criteria for selection of a candidate was that its ultimate disposal would be placement in a hazardous waste landfill by virtue of its containing one or more hazardous substances, even after more conventional treatment methods. Four candidates were selected for initial feasibility tests at Ukiah: thermal batteries from Seneca Army Depot, Romulus, NY, contaminated soil from Picatinny Arsenal, Dover, NJ, incineration ash from Longhorn Army Ammunition Plant (LHAAP), Marshall, TX and reject pyrotechnic smoke assemblies from LHAAP. The first group of tests were conducted in a Retech (Ukiah, CA) plasma furnace. The laboratory scale batch unit furnace was originally built for use in research and development work, with a capacity of about sixty pounds of solid material. The furnace consists of a 1.5 foot diameter steel rotating drum with a 75 kW transferred arc torch. Tests for both of contaminated soil and incineration ash ran smoothly, and as both tests were complete, the resulting molten slag was satisfactorily vitrified.

Pyrolysis of the thermal batteries produced an unexpected phenomenon. Initially, batteries were fed one at a time. During this phase no smoke generation was observed and the flame cut through the steel case like a knife each time the battery passed through the flame. The combination of the plasma flame and molten bath caused the battery to disappear in a few minutes and another battery was fed to the furnace. After ten batteries were fed in this manner, the number was increased to two-at-a-time, then three-at-a time. Now, smoke appeared within ten seconds and lasted for about a minute, obscuring the flame during that time. This was very puzzling, as there are no organics in the batteries. Later observation of the scrubber water revealed a light rusty color. The most probable explanation was that the "smoke" was iron oxide particulate that sometimes forms when steel is subjected to a plasma flame with oxygen present. By comparison, the scrubber water from the Picatinny soil was completely clean. The scrubber water from the first ash was grey; from the second ash it was almost clear. This is in general consonance with the observed heterogeneity of the ash and the amount of smoke generation during the pyrolysis.

The smokes (and flares) caused some problems during pyrolysis. These items were live energetics and functioned when fed to the bath, releasing particulate smoke that obscured the plasma flame and continued to react with the oxygen from the lance. This caused the off-gas to overheat and burn out the off-gas thermocouple. The oxygen lance was turned off to prevent damage to the off-gas piping as a result of excessive heat. As a result, the scrubber water was black, with heavy sedimentation.

Next, a separate test series was scheduled using a larger Retech furnace (6 feet in diameter) at MSE in Butte, MT. The items destroyed in the MSE test were MK71 MOD 5 proximity fuzes. It was learned that these fuzes were undergoing demilitarization by incineration in an APE-2210 incinerator at Hawthorne Army Ammunition Plant (HWAAP) and were the cause of a processing problem. Because of a plastic windshield and potting compound, the fuzes exited the incinerator still burning and there was fear of possible bag-house fires. Second, these fuzes are classified "confidential" and the classified component was still recognizable after incineration, resulting in a classified waste. In addition, the fuze contains lead components and any contaminated incineration ash or scrap must be disposed of as a hazardous waste.

In the MSE test, the oxidizing atmosphere of the air plasma arc resulted in complete oxidation of the metal components of the fuzes, about 70% of the total weight. Some smoke generation was noted after each fuze hit the molten slag, but this quickly dissipated and caused no problem. This could have come from oxidation of organics, steel, or both. A total of 822 pounds of slag was poured into the slag mold. The 207 pounds of fuzes and 300 pounds of soil should have produced approximately 570 pounds of oxidized slag. This discrepancy illustrates the analytical problems caused by starting the test with a substantial skull remaining from prior tests. The slag was uniform throughout, with no indication of any remaining fuze components.

The total lead concentration for the fuze was calculated to be 1300 mg/L, not counting solder. The total lead concentration for the final slag was 2155 mg/L, indicating that the skull started with a higher lead concentration. In any event, the TCLP concentration for lead in the final slag was 0.062 mg/L, well below the regulatory limit (40 CFR 261.24) of 5.0 mg/L. As a result, this slag is a non-leachable, non-toxic waste that presents no security problems.

Finally, another group of tests was run for the evaluation of an improved small-scale plasma arc centrifugal treatment (PACT) unit – in Ukiah, CA – with a secondary combustion chamber (SCC) and a high-efficiency off-gas treatment baghouse to treat the IHAAP sludge. To underscore the ability of plasma arc pyrolysis to render TCLP metals non-leachable, 1–2% each of barium, chromium and lead compounds were added to one batch of sludge. In addition, 2 percent hexachlorobenzene, a semi-volatile TCLP organic used in pyrotechnic manufacture, was added in order to perform a modified DRE (destruction and removal efficiency) test to confirm organics destruction. Partitioning of the volatile metals, lead and chromium, between the slag and the off-gas particulate were also investigated as part of process development.

The problem with the candidate sludge was that it is a hazardous waste, both listed (KO44, wastewater treatment sludges from the manufacturing and processing of explosives) waste and toxicity characteristic (D005-barium, D007-chromium and D008-lead) waste. Hexachlorobenzene (D032) and chlorinated solvents are also present in small amounts. As a result, this sludge is drummed up and disposed of in a hazardous waste landfill. In order to delist this sludge, it must be rendered non-hazardous, thus the organics must be destroyed and the TCLP metals rendered non-leachable.

The sludge test series results showed that, in general, the slag products did meet TCLP requirements. Of particular concern is the barium leaching behavior, as there appears to be a history of the phenomenon. Organic components of the sludge feed were destroyed by the PACT. Final polishing in the SCC oxidized virtually all remaining total hydrocarbon and carbon monoxide to carbon dioxide, in spite of inadvertently operating the primary chamber under reducing conditions. The stoichiometry between oxygen and organics is a very critical performance determinant, affecting slag quality, complete destruction of organics and loss of volatile metals. Since controlling the oxygen lance flow by measuring the oxygen concentration in the primary chamber off-gas is unreliable because of inadequate mixing or residence time, or an incorrect meter reading, and alternate oxidation control procedure must be used. In future test, the required oxygen flow will be pre-determined, based on calculated feed oxygen demand. Both oxygen and carbon monoxide will be measured in the primary chamber off-gas, as carbon monoxide is a positive indicator of reducing conditions. The results also indicated that about 96% of input material was accounted for. Since it is known that a greater amount of material than this was unaccounted for, it is probable that some refractory fell into and added to the slag weight. The individual metal balances were poor, ranging from 1% recovery for chromium, to 10% for lead, to 72% for barium. Several thrusts are necessary to overcome this problem. First, all system elements must be examined and cleaned. Next, an Inductive Coupled Plasma (ICP) analysis of bulk oxides in the feed and slag would provide a check for mass balance calculation. Lead and chromium in the off-gas must also be determined. Additional work is necessary to determine the fate of barium, chromium and lead in sludge. For future work in this area, the cause and corrective action for barium analytical anomalies will be determined, evaluation appropriateness of approved test methods to the condition of the test samples has to be conducted, especially for the slag.

5.4.4
Plasma Arc Study for Fate of Heavy Metals During Vitrification

The suitability of thermal plasma for the immobilization of heavy metals into non-leachable slag is currently being investigated. USACERL and the University of Illinois Plasma Arc Facility have an on-going research project that provides critical information in the establishment of plasma arc technology as a key method for the treatment of heavy-metal-contaminated wastes. The research objective is to develop a quantitative and predictive understanding of plasma arc processing technology as a means for the safe conversion and destruction of heavy-metal-contaminated wastes.

The experimental system consists of a transferred-arc thermal plasma torch with 100 kW of power, which is capable of operating either in oxidizing or reducing atmospheres. Chamber pressure is automatically controlled, and the torch can be operated at any pressure between 400 Torr and atmospheric pressure (760 Torr). The installation of a residual gas analyzer (RGA), optical emission spectrometer (OES), automated controls of process gas flows, and computer monitoring of torch operating parameters and melt temperature has substantially enhanced the capabilities of the facility. Both the RGA and OES provide vital *in-situ* monitoring and characterization of the thermal plasma reaction, allowing a more precisely and more conveniently controlled procedure than in industrial scale plasma arc furnaces. As a result, a large data base documenting the effects of operating parameters and feed material flux on the composition of the product slag will be collected. Major process parameters are flux composition, process gas flow rates, and temperature of the melt. The experimental phase will obtain data on the physical and chemical processes that take place during thermal plasma processing, which will permit the optimization of processing parameters in order to minimize NO_x production, minimize the volatilization of high vapor pressure elements, and control the oxidation rate of heavy metals. In addition, analysis of dominant glass and ceramic phases in the slag will allow a determination of the suitability of the process for immobilization of heavy metals.

5.4.5
Plasma Arc Process Optimization Research Projects

The objective of the on-going efforts at NEWTTEC, a DoE contractor, is to assess the effectiveness and practicality of the Plasma Arc technology in the destruction and vitrification of hazardous waste components containing heavy metals in an environmentally safe and secure manner. The current goals are to evaluate the process capability for the ultimate destruction of the hazardous components, verify slag suitability for regular landfill disposal, identify potential hazards associated with the process emissions, and develop qualified cost estimate for the future utilization of the process on large scale operations.

Further USACERL R & D goals include the development of a quantitative and predictive understanding of plasma arc processing technology as a means for the safe conversion and destruction of heavy metal contaminated wastes. On-going experiments will obtain data on the physical and chemical processes that take place during thermal plasma processing: production of NO_x and other gases, and the volatilization and oxidations of heavy metals. Next, the research focus will shift to determine the dominant glass and ceramic metals by thermal plasma processing. In both phases of this research the data will be measured as a function of flux composition, process gas flows, and temperature of the melt. The leachability of the different slag phases will be documented.

5.4.6
Army Environmental Center (AEC) Technology Transfer Project

Currently, AEC is working on a joint plasma waste candidates study project with USACERL. The project will be managed by the National Defense Center for Environmental Excellence (NDCEE) at their contractor site - Retech, Ukiah, CA. The project objectives are to: evaluate the process capability of PAT for the ultimate destruction of hazardous item components, verify slag suitability for regular landfill disposal, identify potential hazards associated with the process emissions, and develop qualified cost estimates for the future utilization of the process on large scale operations. The project depicts, on a larger scale, the activities that were conducted throughout the earlier program for AMC military unique waste by USACERL. For AEC project testing, 27 test trials are currently planned, consisting of: three hazardous waste candidates, three different operating conditions (i.e., carrier gas type, feed rate, type of additive, etc.), and three replicates. A follow-up phase will be dependent on the findings and results of the first set of experiments.

5.4.7
References

Freeman DJ, Zaghloul HH. Evaluation of Plasma Arc Pyrolysis for the Destruction of Hazardous Military Wastes, presented and published in the 86th Annual Meeting and Exhibition of the Air & Waste Management Association, Denver, CO, June 1993

Freeman DJ, Zaghloul HH. Destruction of Pyrotechnics Manufacture Wastewater Sludge by Plasma Arc Pyrolysis, presented and published in the 1994 Joint Safety & Environmental Protection Subcommittee and Propulsion Systems Hazards Subcommittee Meeting, San Diego, CA, August 1994

Smith ED, Zaghloul HH. Applying Plasma Arc Technology to Asbestos Cleanup, published in Federal Facilities Environmental Journal/Spring 1994, pp 43–52

Zaghloul HH, Circeo LJ (1993) Destruction and Vitrification of Asbestos using Plasma Arc Technology, Technical Report CPAR-TR-EP-93/01, US Army Construction Engineering Research Laboratories, Champaign, IL

Zaghloul HH, Freeman DJ, Smith ED. Destruction of Proximity Fuzes Using Plasma Arc Pyrolysis, presented and published in the 17th Annual Army R&D Symposium and Third USACE Innovative Technology Transfer Workshop, Williamsburg, VA, June 1993

5.4.8
Further Readings

Applying Plasma Arc Technology to Asbestos Cleanup
Ed D. Smith and Hany H. Zaghloul, US Army Construction Engineering Research Lab, Champaign, IL; Fed Facil Environ J, Spring 94, v 5, n 1, p 43(10) research article. Plasma arc technology offers a solution for treatment of waste asbestos-con-

taining material (ACM) so that it can be placed in landfills. The high temperature attained with a plasma arc torch melts asbestos and ACM such that upon cooling, the waste vitrifies into a chemically inert residue that complies with all EPA environmental criteria. USACE research sought to determine the applicability of this technology to ACM destruction and addressed the destruction of pure chrysotile asbestos and economic factors. Plasma arc treatment was demonstrated an effective, efficient method for the destruction and vitrification of asbestos in an environmentally sound manner.

Plasma Pyrolysis of Medical Waste
T. A. Damberger, Kaiser Permanente; Assoc of Energy Eng World Energy Eng 14th Congr, Atlanta, GA, Oct 23–25, 91, p 75(6) conf paper
Plasma pyrolysis represents an environmentally benign solution to the medical-waste disposal problem. This electrotechnology converts organics into clean fuel gases and transforms inorganics into an inert, glassy slag. The vitreous slag can be exploited as concrete filler in roadbed construction or other applications, and the product gas can be used as boiler fuel or for methanol production. Types of plasma arc torches and operating prinicples of the technology are addressed, and results of demonstration applications to medical wastes are reported. (4 diagrams, 3 graphs, 1 photos)

DOD, DOE Clean-Up Efforts Promising
Diane Rose; Environ Prot, Jan 94, v 5, n 1, p 58(5) journal article
Over 7000 hazardous waste sites on DOD property require remediation in the near-term. New technologies are being applied to the restoration of soils tained with explosives, chemical weapons wastes, and radioactive wastes. Bioventing, a natural biodegradation process, has been demonstrated to effectively decontaminate soils polluted by hydrocarbon fuels. Composting offers potential to economically and effectively degrade explosives wastes. Hazardous and radiactive wastes at mixed waste sites are being separated for future treatment by DOE, which is also pursuing in situ vitrification and the plasma arc hearth process for mixed-waste treatment.

5.5
Permeable Barriers to Remove Cd and Cr from Groundwater

Jane Rael · Steve Shelton

Abstract

Laboratory investigations were conducted on permeable barrier media for in-situ removal of divalent cadmium (Cd) and hexavalent chromium (Cr^{+6}) from contaminated groundwater. Permeable barriers are in-situ treatment systems placed in the ground, down-gradient from a plume or source of groundwater contamination. They are designed to allow the groundwater to flow through them while retaining contaminants. This investigation sought to develop low cost, yet effective, barrier materials to expand the breadth of potential permeable barrier applications. The batch studies used a bottle point isotherm procedure to evaluate various combinations of powdered activated carbon (PAC), calcium carbonate ($CaCO_3$), and ferric oxide (Fe_2O_3). Silica sand was used as the inert media filler matrix. A 1:1:1 combination of PAC, agricultural limestone, and ferric oxide mixed at a 1:4 ratio with sand was selected for a column experiment based on data developed during the batch studies. Results of the column investigation indicated that the batch study sorption projections were low. This column investigation was conducted as the one-dimensional case for subsequent use of the media in a trench-based permeable barrier.

5.5.1
Introduction

Cadmium (Cd) and chromium (Cr) are two industrially significant heavy metals (Aylett 1975, Darrin 1956, "Chromium" 1974, Nriagu 1980, "Cadmium" 1975) that enhance the quality and performance of products into which they are incorporated. However, they are also potentially hazardous to the environment if allowed to escape from the industrial process without treatment, or if they are discarded with the product after use (Forstner and Wittmann 1979, "Chromium" 1974, "Cadmium" 1975, Sunderman 1977).

There are two fundamental objectives in groundwater remediation: (1) control or contain the pollutants and (2) remove or destroy the contaminant. Numerous technologies have been developed to restore groundwater quality; however, most are either labor- or energy-intensive. Since the permeable barrier considered in this investigation is neither labor- nor energy-intensive, it provides a new technology for groundwater remediation.

Permeable barriers are typically appropriate when the groundwater is located at shallow depths with a well-defined hydraulic gradient. Aquifer thickness is important for well-mixed contaminant plumes since the permeable barrier must extend to the base of the aquifer. Conversely, buoyant contaminant plumes do not require barriers that are sealed at the base of the aquifer. Permeable barriers are placed in a trench, similar to a grout curtain (Fig. 1); however, the barrier media is porous and designed to interact with the contaminants. While the barrier medium allows the flow of water, it also adsorbs, reacts, or otherwise precludes the contaminant from migrating beyond the barrier. Current slurry-trench technology can be used to place barriers as deep as 100 m; however, depths beyond approximately 30 m are normally cost prohibitive.

When the barrier reaches treatment capacity, it is replaced with fresh media. The contaminated media can either be recycled or managed as an industrial waste. Permeable barriers require less surface disruption and less energy than most other strategies. Beyond construction and renewal of the barrier, the only other requirements are groundwater monitoring and/or redundant barriers for highly contaminated systems (Thomson and Shelton, 1991). Permeable barriers also may be used as landfill liners, thereby providing in-situ treatment of leachate that may be generated.

5.5.2
Materials and Methods

Reagent grade calcium carbonate was used in the batch studies to ensure homogeneous material for small sample sizes. The reagent was crushed using a mortar and pestle. The fraction of material that passed a 200-mesh sieve, yielding particles less than 0.074 mm (0.0029 inch) in diameter, was used in the batch studies. Reagent grade calcium carbonate is a viable barrier material. Therefore, during the column study, un-sieved commercial grade agricultural limestone (53.94% $CaCO_3$ and 45.39% $MgCO_3$ from Bonham, Texas) was used as an inexpensive calcium carbonate source. Particle size characteristics of this material are presented in Table 1.

Fig. 1. Permeable Barrier Concept

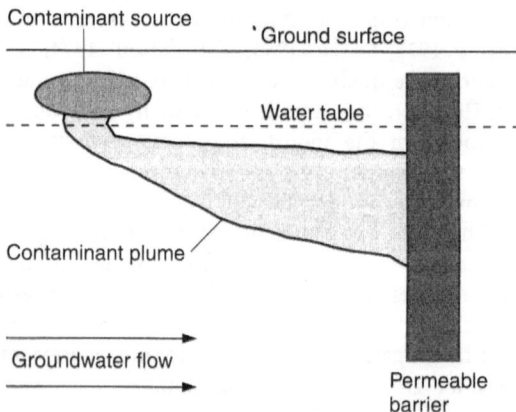

Table 1. Grain size distribution of agricultural limestone.

Sieve Size	Percent Passing
25	90
50	55
100	30
200	12

Ferric oxide is common rust taken from weathered steel. Ferric oxide used in both batch and column studies was ground using a mortar and pestle. The fraction of material passing a 200-mesh sieve, yielding particles less than 0.074 mm (0.0029 inch) in diameter, was used.

Powder activated carbon (PAC) used in both batch and column experiments (PAC Darco G-60) passed a 200-mesh sieve, yielding particles less than 0.074 mm (0.0029 inch) in diameter. This PAC was lignite based with a mean surface area of 750–800 m^2/g.

Washed silica sand (passing a 50-mesh sieve, but retained on a 100-mesh sieve) was used as the inert media component in both the batch study and the column study. These sand particles ranged from 0.15 to 0.3 mm (0.0059 to 0.012 inch) in diameter. Finer silica sand, passing a 100-mesh sieve but retained on a 200-mesh sieve, was used between the simulated aquifer and the permeable barrier in the column study. These sand particles ranged from 0.074 to 0.15 mm (0.0029 to 0.0059 inch) in diameter.

Cadmium stock solution was made by dissolving 1.000 g of mossy Cd metal in 50 mL hydrochloric acid (HCl) and then diluting the acid-cadmium solution to a volume of one liter with deionized water. This produces a 1 g/L Cd stock solution. Chromium stock solution was made by dissolving 3.735 g of potassium chromate (K_2CrO_4) in one liter of deionized water. This produces a 1 g/L Cr stock solution.

Batch Study

Two standard bottle point isotherm procedures exist. In one method, the amount of adsorbent is held constant while the initial concentration of adsorbate in solution is varied. In the second and more conservative approach, the amount of adsorbent is varied while the initial adsorbate concentration is held constant (Roy et al., 1992). This investigation used the second approach. Batch samples were mixed with an initial cadmium (adsorbate) of 10 mg/L and/or chromium (adsorbate) of 50 mg/L, 10 g of silica sand (the inert matrix), and various amounts of adsorbent.

Twenty milliliter batch samples, containing either 10 mg/L cadmium and/or 50 mg/L chromium, were made from the stock solutions. The adsorbent and silica sand were mixed and added to the 20 mL contaminant samples. The batch study initially considered single adsorbent tests, then progressed to multi-adsorbent tests. In the single adsorbent tests contaminated water, containing either cadmium or chromium, was mixed with the adsorbent (10 grams of silica sand and various amounts of one test medium – calcium carbonate, PAC, or ferric oxide) and agitated for two days. Atomic absorption was used to determine the concentration of cadmium or chromium liquid.

Using the data from these experiments, more complex tests were performed until both the cadmium and chromium were mixed with the silica sand and various amounts of all three test media. These data were used to calculate removal efficiency per gram of media. The most effective ratio of media components was used in the subsequent column study.

Column Study

This phase of the investigation used a one-dimensional (column) experiment to determine the effectiveness of the media. The experimental apparatus depicted in Fig. 2 was a column constructed of a 1.52 m (5 ft) length of Excelon 4000 series clear pipe with an interior diameter of 0.16 m (6.35 in). The media was placed in the center portion of the column over a length of 0.91 m (3 ft). The column diameter was greater than 50 times the mass-mean diameter of the largest particles comprising the media thereby reducing wall effects (Crittenden et al., 1985).

Active media for the column was prepared with a mixture of agricultural limestone, PAC, and ferric oxide in the ratio of 1:1:1. This was mixed with silica sand at a ratio of 1:4, resulting in a sand-active media mixture containing 80% sand and 20% active media. Sufficient mix was prepared to fill a 0.91 m (3 ft) section of the column. A 15 cm (6 in) layer of 200-mesh silica sand was placed above and a 6.4 cm (2.5 in) layer of 200-mesh silica sand was placed below the active media.

The column was flooded with deionized water and allowed to saturate the media/sand mixture for one week before initiating the experiment. A feed-stock solution containing approximately 1 mg/L cadmium and 1 mg/L chromium was prepared. The solution was pumped at 54 mL/min by a peristaltic digital drive pump into the base of the column, through the media, and out of the top of the column (an up-flow column).

Samples were collected in wide-mouth high-density polyethylene (HDPE) bottles at least once daily, from the influent sample port and the effluent stream. The samples were analyzed within three minutes of collection using an atomic absorp-

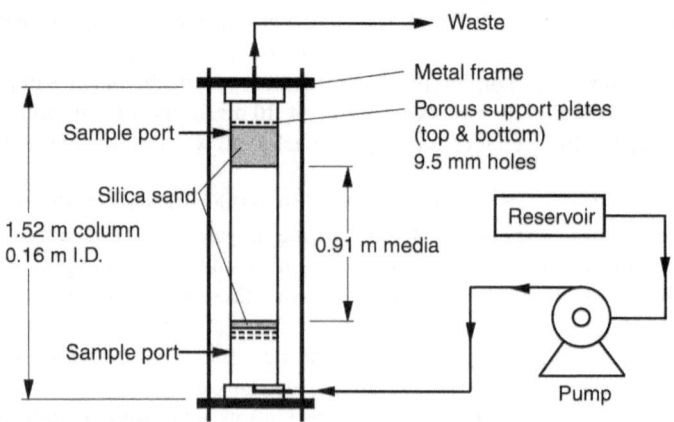

Fig. 2. Column Schematic

tion spectrometer to determine the concentrations of cadmium and chromium. The pH of each sample also was determined.

5.5.3
Results

Batch Study

Preliminary PAC data were evaluated using the Freundlich isotherm equation to estimate performance of candidate media as permeable barriers. The Freundlich model is an empirical equation of the form:

$$q = x/m = KC^{1/n}$$

where q is the amount of adsorbate x per mass of adsorbent m at equilibrium; C is the liquid-phase concentration; and K and n are the Freundlich isotherm parameters. Although empirical, the parameter K can be taken as a relative indicator of adsorption capacity, while $1/n$ describes the intensity of the reaction.

A plot of the ln q as a function of the ln C was prepared for each data set. Regression analysis was used to evaluate the linearity of the fit, with a slope of $1/n$ and intercept of ln K at ln C = 0 (C = 1). These data are presented in Figs. 3 and 4. A linear distribution was observed in all data sets.

Media mixtures were evaluated using their experimental rate reaction. Reactions typically exceeded 90% of their adsorption capacity within six hours. The most

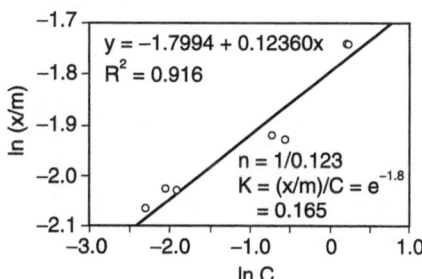

Fig. 3. Freundlich Adsorption Isotherm 10 mg/l Cd onto PAC

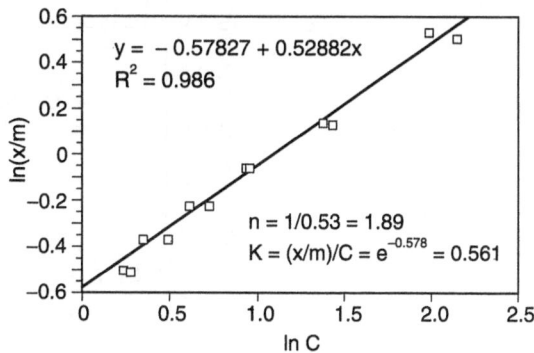

Fig. 4. Freundlich Adsorption Isotherm 50 mg/l Cr onto PAC

promising batch study test used a mixture of all three active media components. This system removed 0.10 mg of cadmium and 0.25 mg chromium per gram of media. The end point pH was 7.8.

Column Study

Data from the batch experiments were used to estimate time-to-breakthrough for the column study. Breakthrough was predicted to occur in eight days for an influent concentration of 1 mg/L Cd and 1 mg/L Cr in the previously described column. The first effluent sample, taken 18.2 hours after introducing the Cd and Cr, showed leakage of cadmium. The next effluent sample, taken 27.6 hours into the experiment showed cadmium leakage along with chromium leakage. Influent pH was 3.0, while effluent pH was 6.6 to 6.7 throughout the column study. The column breakthrough (and leakage) curves for Cd and Cr are presented in Figs. 5 and 6. From these data inferences can be made relative to media capacity and the rate at which capacity is approached.

Fig. 5. Cr Breakthrough Curve

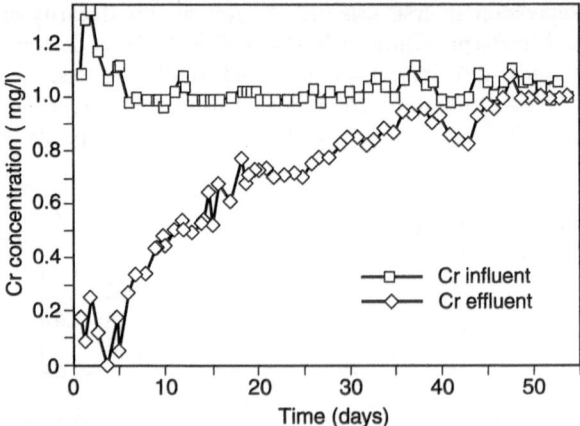

Fig. 6. Cd Breakthrough Curve

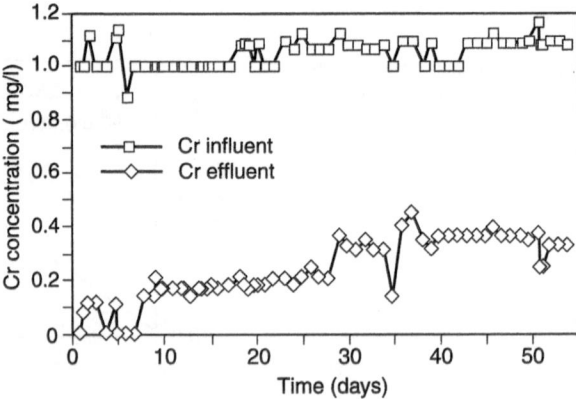

The length of the breakthrough curve is affected by contaminant concentration, media characteristics, temperature, and contact time (hydraulic retention time). The contact time for this column, based on experimental measurements of the media, was 2.3 hours. The voids ratio, e, for the media was measured at e = 0.704. This produced a contact time of 4.0 hours in the column. At higher flow rates (lower contact time) the breakthrough curve is drawn out, while at low flow rates (higher contact time) the breakthrough curve is better defined.

The total mass of the media material in the column was 24 kg – 4.8 kg of the active media (1.6 kg each of PAC, ferric oxide, and agricultural limestone) plus 19.2 kg inert silica sand. The media was designed to react with up to 1.25 g of cadmium and/or chromium. Figure 7 presents a mass diagram of cadmium and chromium precipitated or adsorbed during the column experiment. These data suggest that 1.3 g of cadmium and 3.4 g chromium were removed by the 4.8 kg of active media in the column (0.27 mg of cadmium and 0.71 mg of chromium precipitated or adsorbed per gram of media). Thus, the media capacity measured in the column study was approximately three times the batch study estimate of 0.10 mg of cadmium and 0.25 mg of chromium precipitated or adsorbed per gram of media.

5.5.4
Discussion

Batch Study

Treatment efficiencies as high as 98.7% for cadmium and 97% for chromium were achieved during batch studies. These data indicate that:

- PAC adsorbs chromium better than cadmium.
- Agricultural limestone increases pH that may facilitate precipitation in the presence of ferric oxide.
- More limestone went into solution during the batch study than the column study. This was evidenced by the effluent pH values in each experiment – 7.8 for the batch study and 6.6 for the column study.

Fig. 7. Mass Diagram

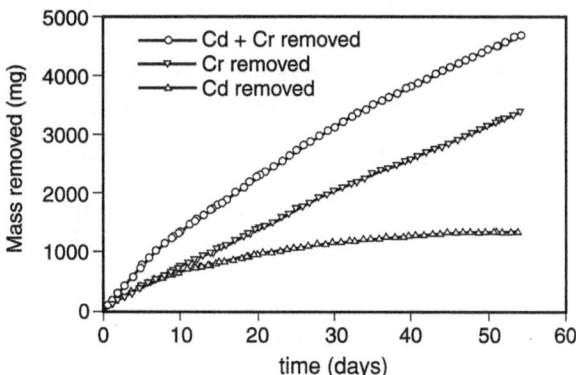

Column Study

The column study was designed for an eight day breakthrough. This estimate was based on calculations using data from the batch study. Both cadmium and chromium were detected in low concentrations almost immediately; however, classical breakthrough took much longer than indicated by the batch study data. These variations may be attributed to the following:

- A conservative batch study method.
- Leakage along the walls of the column.
- Agricultural limestone has less surface area per gram than calcium carbonate. Therefore, it is more difficult to dissolve. This could account for the difference between the batch study pH (7.8) and the column study effluent pH (6.6).

5.5.5
Permeable Barriers

The trench-based permeable barrier provides in-situ treatment by allowing water to migrate through the barrier while contaminants are retained by the barrier. As the contaminated groundwater flows through the barrier (PAC, ferric oxide, agricultural limestone, and sand media placed in the trench) the active media affects in-situ treatment. As the active media (PAC, ferric oxide, agricultural limestone, and sand mix) nears removal capacity, it can be replaced with fresh media. Contaminated media may require special handling as a hazardous waste; however in many instances the media can be regenerated using physical/chemical treatment techniques. The permeable barrier may offer the following advantages over conventional groundwater remediation techniques:

- Low operating cost in contrast to pump and treat technology.
- Fast containment of contaminant plumes.
- Low technology construction and installation techniques.
- Low technology recovery and replacement techniques for barrier materials.
- Low operation and maintenance costs (monitoring and barrier replacement).
- Less surface disruption, less labor, and less energy than most other processes.
- Can be applied horizontally as a liner for landfills.
- Can be used as a prophylactic in unpaved work areas that may be subject to spills.

The permeable barrier may be subject to the following disadvantages over conventional groundwater remediation technologies:

- Many groundwater remediation efforts require technologies that are responsive to contaminants in deeper aquifers.
- Not all groundwater contaminants are well suited to treatment using reactants and adsorbents that can be incorporated into a trench based permeable barrier media.
- It is unlikely that permeable barriers would be cost competitive at total depths over 30 meters.
- Efficient media design may be inhibited by the uncertainties associated with the hydrogeology of the site.

The permeable barrier may be subject to the following special considerations, not normally evaluated in conventional groundwater remediation technologies:

- Redundant barriers, as shown in Fig. 8, with both intermediate and tertiary monitoring may be required. In this mode the contaminant plume migrates through both barriers. When the first barrier experiences breakthrough the media is replaced. The second barrier serves as the treatment system during breakthrough and media replacement.
- Contaminant plumes may vary in concentrations as a function of depth. When this occurs the permeable barrier may experience breakthrough in some areas while other areas have significant remaining capacity. This case can be addressed by increasing the media PAC, ferric oxide, and agricultural limestone concentration in areas of higher contaminant concentration. The problem could also be addressed by increasing the physical dimensions of the permeable barrier in areas of higher contaminant concentration.

5.5.6
Conclusion

The batch studies provided conservative estimates of media capacity. Considering data from the column investigation, a mixture of agricultural limestone, PAC, ferric oxide, and sand may be a reasonable media for use in a trench-based permeable barrier to treat cadmium and chromium groundwater contamination. Advantages of this media include:

- Remains structurally sound and permeable throughout life cycle.
- Can be tailored to meet site-specific needs.
- Will remove a broad spectrum of inorganic and organic contaminants.
- Has the structural strength to maintain permeability when used horizontally as a prophylactic measure in unpaved work areas that may be subject to chemical spills.

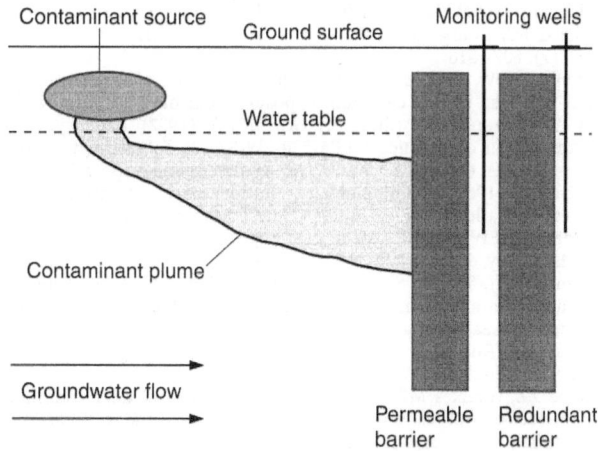

Fig. 8. Dual Permeable Barrier

Further investigations should be conducted to better define the impacts of:

- Hydraulic retention time.
- A separate agricultural limestone barrier to adjust the pH prior to a PAC and ferric oxide media barrier.

The objective of any groundwater remediation technique is to control and/or contain the contaminant of concern. We believe that the permeable barrier is another tool to meet this objective in selected applications. This investigation demonstrated that a trench-based permeable barrier using a PAC, ferric oxide, agricultural limestone, and sand media should be effective in removing the heavy metal contaminants.

5.5.7
References

Artiola J, Fuller WH (1979) Effect of crushed limestone barriers on chromium attenuation in soils. J Envir Quality 8(4), 503–510

Aylett BJ (1975) The chemistry of Zinc, cadmium, and mercury, Pergamon Press

Bartlett RJ, Kimble JM (1976a) Behavior of chromium in soils: I: trivalent forms. J Envir Quality 5(4), 379–383

Bartlett RJ, Kimble JM (1976b) Behavior of chromium in soils: II. hexavalent forms. J Envir Quality 5(4), 383–386

Cadmium and the environment: Toxicity, economy, control (1975) Organization for Economic Co-Operation and Development. Paris

Chromium (1974) National Academy of Sciences. Washington, DC

Crittenden JC, Luft P, Hand DW, Oravitz JL, Loper SC, Ari M (1985) Prediction of multicomponent adsorption equilibria using ideal adsorbed solution theory. Environmental Science & Technology 19(11), 1037–1043

Darrin M (1956) Chromium chemicals – their industrial use. Chromium: Vol. I, chemistry of chromium and its compounds. MJ Udy, Reinhold Publishing Corp, New York

Federal Register (1990) 55 (No 61), 11803–11815

Forstner U, Wittmann GTW (1979) Metal pollution in the aquatic environment, Springer-Verlag

Gleason MN, Gosselin RS, Hodge HC, Smith RP (1969) Clinical toxicology of commercial products: Acute poisoning (home and farm). Williams and Wilkins, Baltimore

Kido T, Shaikh ZA, Kito H, Honda R, Nogawa K (1991) Dose-response relationship between dietary cadmium intake and metallothioneinuria in a population from a cadmium-polluted area of Japan. Toxicology 66, 271–278

Mentch RL, Lansche AM (1958) Cadmium, a materials survey. National Technical Information Service, Rep AD 680–443, Springfield, Va

Nriagu JO (1980) Cadmium in the environment, part I: Ecological Cycling, Wiley & Sons, Inc

Nriagu JO (1981) Cadmium in the environment, part II: Health effects, Wiley & Sons, Inc

Richard FC, Bourg AC (1991) Aqueous geochemistry of chromium: A review. Water Resources 25(7), 807–816

Rosenberg DW, Kappas A (1991) Induction of heme oxygenase in the small intestinal epithelium: A response to oral cadmium exposure. Toxicology 67, 199–210

Roy WR, Krapac IG, Chou SF, Griffin RA (1992) Batch-type procedures for estimating soil adsorption of chemicals. EPA/530-SW-87-006-F, Washington DC

Schroeder HA (1970) Chromium. Air quality monograph #70-15. American Petroleum Institute

Sullivan RJ (1969) Preliminary air pollution survey of chromium and its compounds, a literature review. National Air Pollution Control Administration

Sunderman FW Jr (1977) Metal carcinogenesis. Toxicology of trace elements, Goyer RA, Mehlman MA, Wiley & Sons, Inc, New York, NY

Thomson BM, Shelton SP (1991) Permeable barriers: A new alternative for treatment of contaminated ground water, 45th Purdue Industrial Waste Conference Proceedings, Lewis Publishers, Inc, Chelsea, Michigan, 73–80

Turner MA, Rust RH (1971) Effect of chromium on growth and mineral nutrition of soybeans. Soil Sci Soc Am Proc 35, 755

Wahba ZZ, Waalkes MP (1990) Effect of in vivo low-dose cadmium pretreatment on the in vitro interactions of cadmium with isolated interstitial cells of the rat testes. Fundamental and Applied Toxicology. 15, 641–650

Notation

The following symbols are used in this paper:

C = Liquid-phase concentration (mg/L);
e = Void ratio;
K = Freundlich capacity parameter (L/g)
ln = Natural log
m = mass of adsorbent (g);
n = Freundlich isotherm empirical coefficient, dimensionless;
q = Amount of adsorbate per mass of adsorbent at equilibrium;
V = Volume of solution (mL); and
x = mass of adsorbate (mg).

5.5.8
Further Readings

Title: **Removal of cadmium (II) from saturated kaolinite by the application of electrical current**
Author: Acar, Y. B.; Hamed, J. T.; Alshawabkeh, A. N.; Gale, R. J.
Corporate Source: Louisiana State Univ, LA, USA
Source: Geotechnique v 44 n 2 Jun 1994. p 239–254
Publication Year: 1994
CODEN: GTNQA8 ISSN: 0016-8505
Abstract: The fundamentals of electrokinetic remediation (the removal of contaminants from soils by the application of electrical current across electrodes inserted in a soil mass) are presented and the findings of recent studies are reviewed. An improved theoretical formalism is presented for conduction phenomena under electrical currents. Predictions of effluent pH using this model compared excellently with the results of one-dimensional experimental studies. The results of four one-dimensional tests conducted to assess the removal of 99–114 µg/g of Cd(II) loaded on kaolinite specimens are presented. The flow conditions, chemistry, efficiency of removal and energy expenditure in these tests are evaluated. Cd(II) was removed from the kaolinite specimens by 90–95%. It was deposited close to the cathode and on the cathode due to the basic environment and electrodeposition. The energy expenditure in these tests was 50–106 kW h/m³ of soil processed. The test results also showed that when the initial pH in the specimen is low, high removal rates are achieved by electrical migration rather than electro-osmotic flow. (Author abstract) 51 Refs.

Title: **Feasibility of extracting toxic metals from soil using anhydrous ammonia**
Author: Clifford, Dennis A.; Yang, Ming; Nedwed, Tim
Corporate Source: Univ of Houston, Houston, TX, USA
Source: Waste Management v 13 n 3 1993. p 207–219
Publication Year: 1993
CODEN: WAMAE2 ISSN: 0956-053X

Abstract: It was the purpose of this research to establish the feasibility of using anhydrous ammonia alone and with enhancing ligands to extract common metal contaminants-lead (Pb), cadmium (Cd), mercury (Hg) copper (Cu), and zinc (Zn)-from soil. Lead and mercury removals were low (less than 15%) with pure ammonia. Lead removal increased dramatically, however, when ethylenediaminetetraacetate (EDTA) was added to the ammonia and when aqueous EDTA alone was used as an extractant. When the metal-nitrate spiking concentration was lowered to 5000 mg/kg, the percent removals for all the metals were generally less than 30%. The presence of water in the soil at the start of extraction significantly reduced the removals of all the metals except cadmium. Overall, ammonia extraction of toxic metals does not appear promising for complete site cleanup. However, to determine the practical effectiveness of ammonia extraction as a pretreatment for soil remediation, additional tests need to be performed, and the criterion for success should be a Toxicity Characteristic Leachate Procedure (TCLP) test on the extracted soil. (Edited author abstract) 16 Refs.

Title: Removal of chromium from groundwater using permeable barriers: an aquifer simulation study
Author: Schmidt, Mark D.; Shelton, Stephen P.
Corporate Source: Univ of New Mexico, Albuquerque, NM, USA
Conference Title: Proceedings of the 1994 National Conference on Environmental Engineering
Conference Location: Boulder, CO, USA
Conference Sponsor: ASCE
Source: Critical Issues in Water and Wastewater Treatment National Conference on Environmental Engineering 1994. Publ by ASCE, New York, NY, USA. p 792–803
Publication Year: 1994
CODEN: NCEEDO ISSN: 0731-1516 ISBN: 0-7844-0031-8
Abstract: Previous efforts to remediate groundwater contaminated by chromium-bearing industrial wastes have involved post-extraction methods, whereby groundwater is pumped to the surface treated and returned to the aquifer. This practice has proven effective for removing soluble pollutants. However, it is often costly and labor intensive and requires treating large volumes of water. Also, institutional obstacles such as ground and surface water discharge permits and groundwater rights must be considered. An alternative to conventional remediation methods is the in situ permeable barrier process, which intercepts soluble contaminants from solution but allows groundwater to flow through. Trench-based barriers, backfilled with reactive media, result in the direct adsorption of chemical species or the oxidation or reduction of chemical species followed by precipitation. Laboratory studies were conducted to determine the technical feasibility of using trench-based media to remove chromium from groundwater. In batch tests various doses of candidate media and 10 g of silica sand were added to glass vials containing a chromium solution. Candidate media included powder-activated carbon, ferric oxide and agricultural limestone. Adsorption isotherms were plotted from batch test results. In aquifer simulation tests a chromium-containing solution was passed through an aquifer simulation model containing silica sand and a vertical barrier of candidate media. Removal of soluble hexavalent chromium

(Cr(VI)) to concentrations less than the maximum contaminant level (MCL) for total chromium in drinking water (0.05 mg/l) was demonstrated with all candidate media in the aquifer simulation model. Adsorption appeared to be the principle mechanism of removal for all candidate media considered. Information gained from experience with the physical model was used in developing a computer generated, 1-D solute transport model to predict the movement of hexavalent chromium in an aquifer system. (Author abstract) 16 Refs.

6 Soil and Groundwater Contamination

6.1
How Technology is Improving Decision Making for Environmental Restoration

JOHN DITMARS

Abstract

Environmental restoration, or the cleanup of contaminants from past activities, at its core depends on a series of decisions about the nature and extent of contamination, the risk to human health and the environment, and the potential effectiveness of remediation techniques and technologies to reduce the risk to acceptable levels. The effectiveness with which these decisions are made has significant impacts on the cost and duration of the cleanup efforts. The decisions must often be made on the basis of incomplete and uncertain data. Emerging environmental information and data acquisition technologies together with appropriate strategies to support decision making are beginning to change the way environmental restoration occurs in the United States.

Past environmental restoration activities too often relied on prescriptive data collection activities to generate the information upon which decisions were to be made. Retrospective studies of such activities have shown that, while often data were gathered for the purpose of reducing the risk in consumed. Recent examination of the failures in the United States to achieve many complete cleanups, despite the investment of large sums and time, points to the inability to have decisions made efficiently. The solution to the problem involves both regulatory change to allow more flexibility in decision-making and the introduction of technology to improve decision making.

This paper reviews the recent assessments made of the cleanup process and application of strategies and technologies to enhance decision-making for cleanup. It provides examples of the new decision approaches and the technologies that have been employed to speed up characterization and to optimize the implementation of remediation.

6.1.1
Introduction

The cleanup of contaminants from past activities, environmental restoration, has been a major environmental activity in the United States for the past fifteen years. The Comprehensive Environmental Response, Compensation, and Liability Act of 1980 (CERCLA or Superfund) and the Superfund Amendments and Reorganization Act of 1986 (SARA) provided the U.S. Environmental Protection Agency (EPA) with authority and funding to establish cleanup requirements and to cleanup the nation's worst hazardous waste sites. The Corrective Action portions of the Resource Conservation and Recovery Act (RCRA) likewise require the closure and cleanup of waste facilities.

The EPA's National Priorities List (NPL) now contains over 1200 seriously contaminated sites and is projected to grow to about 2100 sites by the year 2000 (GAO, 1992). In addition there are over 300 sites at federally-owned facilities that have been added to this total. There have been few completed cleanups and less than 50 sites have been removed from form the NPL. Estimates of the costs of cleaning up the remaining sites vary widely, but they are all in the hundreds of billions of dollars. Large federal facilities, particularly those related to the defense establishment, have received increasing attention in recent years and now represent a significant portion of the cleanup effort.

At its core environmental restoration of contaminated sites depends on a series of decisions about the nature and extent of contamination, the risk to human health and the environment, and the potential effectiveness of remediation techniques and technologies to reduce the risk to acceptable levels. The Superfund program in the United States established a series of regulatory steps leading from the discovery of a potential contamination problem to its cleanup. They are designed to provide information for progressive decision making. In simple terms, the steps and their purpose are as follows:

- preliminary assessment

  review of available information and reconnaissance visit to the site to determine if a release of hazardous substances requires further investigation or action.

- site inspection

  collection of samples to describe known contaminants, the surrounding area, and potential human and environmental receptors of contamination.

- remedial investigation

  characterization, through sampling and analyses, of the nature and extent of contamination. It affords sufficient information to support the evaluation of remedial alternatives and remedy selection.

- baseline risk assessment

  qualitative and quantitative evaluation to define the risk posed to human health and/or the environment by the presence of specific contaminants.

- feasibility study

  evaluation of alternative remedial actions screened against several technical, cost, and regulatory criteria, leading to selection of a preferred remedy.

- public comment

  community commentary on the selected remedy.

- record of decision

  documentation of the final remedy selection.

- remedial design

  development of engineering plans and specifications for implementation of remedy.

- remedial action

  action construction and implementation of the cleanup.

As logical as the above sequence of activities appears, practice has shown that the "feed forward" requirements for information to support decisions do not work effectively. It is common in practice to find that the data brought forward for decisions at a particular step are inadequate or inappropriate or both. The subsequent return to earlier steps to acquire the necessary data becomes a significant consumer of time and resources.

The progressive steps in the cleanup process are geared to be initially open to multiple possibilities and only later to focus by screening out alternative remedies. Practically, it has proved difficult to capture data in early investigations appropriate for all potential forthcoming remedies. "Feedback" mechanisms are needed to focus the information and data acquisition activities of the process – particularly during the site characterization phase of the remedial investigation step.

A consequence of the "feed forward" nature of information flow in the Superfund process was need to establish data quality objectives (DQOs). These were meant to assure that the data were appropriate for decisions at the next step. While early guidelines for the establishment of DQOs were rational and linked to decision making, they were often reduced in practice to simply require that the highest level of laboratory chemical analysis quality assurance (following EPA's Contract Laboratory Program procedures) be followed. This was not always a technical expedient but was driven in large measure by the potential of legally imposed penalties on those thought to be responsible for creating the contamination. Thus, "highest quality" chemical analyses were sought not to support technical decisions affecting cleanup but to be acceptable in the legal setting.

There are many reasons why the cleanup process has not been effective. Principal among them is the fact that activities are more focused on the processes and related data acquisition steps than the decisions that they are to support. The drive to gather "enough data" to characterize contamination has not been tempered by the realization that it is usually practically impossible to obtain sufficient data to characterize with complete certainty the nature and extent of contamination. Decision making has, in many cases, been subordinated to data gathering. Consequently, many cleanups have been bogged down in the early steps of the process and have expended cons.derable resources on problem assessment.

Recognition that these shortcomings have impeded cleanup has led to the exploration of new approaches to improve the effectiveness and efficiency of cleanup activities in the United States. Part of the impetus has come from federal agencies. They have estimated the potential cost of cleanup of their many installations, and it has become apparent that current practices are very expensive. And, in the case of facility closure actions, present cleanup practices appeared extremely slow and likely to impede the transfer of these facilities for other uses. In addition, for federal facilities the issue of legal responsibility for the cleanup was generally clear – the federal government would pay for cleanup. Thus, extensive data gathering for the purpose of establishing responsibility for the site cleanup was deemed unnecessary. These considerations have fostered a significant refocusing of strategies for cleanup on those decision process that have been barriers to actual restoration.

The emphasis on decision making is changing the approach to cleanup. The essential features of these changes are addressed below. They include requirements that data be generated in a timely manner and of adequate quality to support the decisions. Such requirements have led to the introduction of new technologies into the process to:

- generate data rapidly and often at the sites of cleanup studies,
- manage, display, and analyze data efficiently, and
- integrate information to support the decision process for both technical and nontechnical decision makers.

6.1.2
Re-examination of the Decision Process for Environmental Restoration

Several different approaches have been undertaken to streamline the cleanup process in the United States in recent years. They range from modifications in regulatory practice that encourage rapid removals and cleanups where problems are well-established and additional process-driven studies are unnecessary (EPA's Superfund Accelerated Cleanup Model) to early inclusion of stakeholders (regulatory and community members) in the cleanup process. However, re-examinations of the basic relationship between the decisions governing the cleanup process and the information upon which the decisions are made have had a profound effect on streamlining efforts.

The re-examination has occurred at several levels of the cleanup process. It can be characterized by two related, but different, approaches – the "observational approach" and "data quality objectives," – aspects of which appear separately, or integrated, in streamlining plans. The observational approach draws upon the well-known principle in geotechnical engineering that regardless of the degree of site exploration. Prudent engineering therefore requires that some level of uncertainty be accounted for in designs, that contingency plans be prepared for potential deviations from "normal" conditions, that observations be made during construction to detect changes form expectations, and that contingency plans be invoked should deviations from "normal" conditions appear. The subsurface nature of many environmental restoration problems makes the transfer of this principle to restoration problems obviously attractive (Smyth et al., 1992; Wallace and Lincoln, 1989).

More generally, however, the observational approach confronts a fundamental decision issue in the cleanup process – that the information available is almost always incomplete. It postulates that acknowledgment of those certainties and preparations to deal with unlikely consequences, should they be observed to occur, establish a path forward to cleanup that rational decision makers can follow. It holds out the promise of breaking the loop of never ending data acquisition that plagued past efforts because it removes the goal of eliminating all uncertainty in the nature and extent of contamination.

Concomitant with the application of observational approach principles to the cleanup decision process were recent attempts to clarify the concept of data quality objectives as they applied to the cleanup process (Neptune et al., 1990; Ryti and Neptune, 1991; EPA, 1993). While often viewed as simply a means of defining the level of quality control for chemical analyses of samples of environmental media, DQOs in fact referred to a logical process for planning and designing data acquisition activities that addresses the types, quantity, and quality of data required to support a specific decision. Couched in "total quality" terminology and following standard scientific logic, the DQO approach requires the decision-maker to:

- determine what decision addresses a restoration problem,
- identify the inputs affecting the decision,
- develop a logic statement (quantitative, if possible) relating the inputs to the decision,
- establish the constraints on uncertainty acceptable to stakeholders, and
- optimize the design for obtaining the required data (i. e., minimize resources while satisfying uncertainty constraints).

For example, in application to a contaminated soil problem, the DQO approach can lead to the identification of specific contaminants that are of primary concern because of their contribution to a particular level of human health risk. Depending on the quantitative level of certainty a decision-maker may demand in assessing that risk, there are different requirements for determining the extent of contaminant. Data acquisition schemes for sampling and analyzing soil can be developed to obtain the appropriate quality and quantity of data to meet these requirements – and no more. The important contribution of the DQO approach to the restoration decision-making process is the rigorous linkage that it establishes between the planning for data acquisition and the decisions that are to be made on the basis of the data. This a priori analysis of the potential value of the data is in striking contrast to open-ended data collection that vaguely promises to provide the information needed.

The bias for action in the face of uncertainty of the observational approach and logic appraisal of uncertainty in quantitative terms of the DQO approach were combined to form a strategy for expedient cleanup decision making by the Department of Energy known as the Streamlined Approach For Environmental Restoration (SAFER) (Gianti et al., 1993). Variants of these approaches are appearing in environmental restoration projects in the Untied States but most retain these features:

- acknowledgment of the inability of data to provide complete characterization of problems,
- accounting for uncertainty,
- specific linkage of data quantity and quality to the decisions to be made.

An excellent application of these concepts to hydrogeologic decision analysis is given by the work of Freeze and others. (Freeze et al., 1992). They demonstrate the approach in the design of landfill systems to meet regulatory and other risk constraints in the face of uncertainties regarding subsurface conditions and hydrogeologic transport. Tradeoffs among design costs, monitoring well costs, and the costs of "failure" illustrate clearly the tension between the decision process and the information needed to support decision making.

The author participated recently in the development of a decision strategy to address the problem of the characterization of over 2000 potential release sites at a large federal facility. Many of the sites were thought not to contain serious contamination, but a strategy was necessary to make that determination to the satisfaction of the stakeholders. Resource constraints dictated not undertaking an open-ended characterization at each site. A decision logic was developed that proposed screening action levels of all expected contaminants. These concentrations related to very low values of potential human health risks under conservative exposure scenarios appropriate to the facility. These screening action levels and some existing data form the sites provided data acquisition planners the rules for designing data collection schemes that would permit a rapid initial determination of whether a particular potential release site could be put aside for "no further action" or whether additional characterization was warranted. The data collection designs were sensitive to uncertainty constraints particularly those associated with false negatives. This relatively simple decision strategy holds the possibility of saving significant costs and time by not "over-characterizing" non-problem sites and allows using resources for true problem areas.

The decision-driven data acquisition activities resulting from the re-examination of the cleanup process have highlighted the contributions that technology can make in the implementation of the process.

6.1.3
Application of the Decision Process in the Field

Execution of accelerated or streamline approaches to cleanup in the field have taken many forms. A streamlined pilot project at two wood treating sites contaminated with creosote and pentachlorophenol (Sieminski et al., 1993) began by streamlining the planning documents and evaluation procedures. The risk assessments and remedy selection activities were then directed to obvious potential problems as were the field data acquisition activities. Global positioning systems and 48-hour turnaround of analytical results (delivered electronically to the site) reduced the duration of data validation from months to weeks.

Similarly, a study in Canada faced with a large number of sites with potential mercury contamination in soil developed a decision methodology to guide sampling and to focus cleanup actions (Roffman et al., 1993). Plans were dynamically adjusted to account for screening results. A screening measurement device to estimate mercury fluxes form soils was developed and employed to drive the characterization. The application of a rigorous decision logic, driven by risk assessments, together with dynamic field screening rapidly focused the study on 15 of 190 possible sites. The field investigation effort was reduced significantly without compromising the overall study objective.

Potential groundwater contamination at landfill sits in New Mexico was investigated efficiently through a characterization program directed toward investigations that would decrease the need for sampling and drilling (Burton, 1992). Following a strategy similar to that successfully used in pretroleum and minerals exploration programs, geological and site history models were generated to guide the selection of multiple surface geophysical techniques. A dynamic, multidisciplinary field approach using these technologies and data interpretation by experienced technical staff on site resulted in significant reductions in time, cost, and intrusive sampling.

While each of the applications summarized above was different with regard to streamlining or expediting some portion of the cleanup process, they and other recent activities in this area share the following characteristics:

- an objective exists to minimize the measurements required to make a defined decision,
- a decision logic, explicit or implicit, is developed before measurements begin to permit dynamic changes in the sampling as the results become available,
- technology is employed that provides data appropriate to the required decision – often rapid turnaround and in-field screening technologies, and
- data integration or information fusion supports the decision making.

6.1.4
Technologies that Provide Data for Decisions

The regulatory-driven volume of environmental restoration activity in the United States has grown sufficiently large to have spawned the development of technologies to assist in the rapid and cost-effective acquisition of data. While such technological advances have had positive impacts on portions of the cleanup process, they represent a significant opportunity for improving the decision process. When coupled with decision-based characterization schemes, they have been effective in minimizing oversampling and in limiting investigations to locations where contaminant levels pose real risks to human health or the environment.

The need to understand subsurface conditions has pushed the development of technologies toward rapidly gathering samples form the subsurface, or measuring there, so as to delineate the extent of contamination in soil or groundwater in a relatively short period of time. Direct push, or cone penetrometer, technology has been used successfully in this regard, and holds the potential for other advances. The small diameter penetrations into soil are minimally intrusive and have been used to sample aquifers – often reducing the need to establish monitoring wells. The flexibility afforded by rapidly sampling at multiple ovations in an aquifer permitted characterization of a hexavalent chromium plume that had eluded earlier well sampling (Cherry et al., 1992). Combining the groundwater sampling capabilities of these devices with field gas chromatograph technologies streamlined the delineation of a 40-acre chlorinated solvent plume to within 6 weeks (Prochaska et al., 1991). Coupling sensors to cone penetrators is an area of continuing development – with fiber optics and laser-induced fluorescence used to detect pretroleum contaminants (Bratton et al., 1993) is just one example of such developments. Multilevel drive point samples (Widdowson et al., 1994) have been used to sample soil gas and groundwater as well as to determine aquifer permeability.

Several screening methods for chemical analyses of soil and water have been developed for field use with detection levels and turnaround times that make them appropriate for supporting dynamic decision plans. Immunoassay techniques allow detection of PCBs in soil samples at concentrations below action levels and have contributed to speeding on-site decision making (Mapes, 1991). Colorometric methods for field screening explosives residues such as TNT, RDX, and 2,4-DNT, in soil and water yield results that compare favorably with those from standard laboratory methods (Jenkins and Walsh, 1992). These techniques have proved effective in directing more traditional sampling and analysis activities on large installations where potential contamination sites are many. Likewise, hand-held instruments such as x-ray fluorescence devices have been used to locate hot spots of metals contamination as part of a screening decision strategy.

The chemical analysis of samples of environmental media, particularly for organic compounds, has resulted in an industry of commercial laboratories operating under prescribed regulatory quality control and assurance procedures. The time and cost involved in packing and shipping samples to these laboratories and waiting for the results of the analyses to become available for decision making have been identified often as a significant impediment to process. Consequently, substantial effort has gone into the development of field analytical methods and field, or transportable, gas chromatography/mass spectrometer instruments. A recent critique of the state of such methods and instruments (McDonald et al., 1994) concludes, among other things, that improvements in instrumentation are removing earlier concerns that only expert mass spectrometrists could employ the techniques successfully in the field. It suggests that early integration of the technology into the planning and decision process is essential for the realization of the benefits of on-site data acquisition.

While the development of new sensor and analysis technologies to assist characterization (Wang et al., 1994) enhances the possibilities supporting the decision process even further, there is a cautionary note regarding the infusion of new measurement technologies into the environmental restoration process in the United States. Regulatory agencies have not uniformly accepted the use of filed analytical instrumentation. In recent analysis of the adoption of such instrumentation in practice (Moore, 1994), regulatory legislative, and institutional barriers were shown to have slowed the implementation of field instrumentation. Strategies to streamline the decision making process have required stakeholder approval, and the means of expediting the availability of data appears to require acceptance as well.

6.1.5
Technologies That Provide Data Integration for Decision Making

Advances in technologies associated with data acquisition and personal computing systems have made the electronic collection, storage, display, and manipulation of data in the field a practical reality. Systems that acquire and process data from multiple sources are routinely integrated with laptop computers to guide decision making in the field as well as the office.

Most data supporting environmental restoration decision have a spatial, or geographic, reference, and a variety of geographic information systems (GISs) are used to display and integrate information. Traditional GIS database and map layer systems have served to integrate data by overlaying information on topography, contamination contours, groundwater levels, and the like to support remedial decision making (Woodside and Otis, 1991). More recently, GISs with object-oriented databases and vertical dimension capabilities have provided powerful displays data related to subsurface conditions. Some of these systems integrate data of multiple types in rapidly displayed easily queried modes so "what if" questions can be posed of complex spatial data sets (McGrath and Brown, 1991). These systems provide the means to support "observational approach" decision making. Multimedia applications for GIS hold the promise of even more effective displays of decision-related information (Huber, 1994).

While the data integration and display capabilities of GISs are powerful aids to decision making, the analytical capabilities of these systems are generally limited. Several recent efforts have added the capabilities of models or other analytical tools to GIS configurations. For example, simple groundwater transport models have been interfaced with the GIS, so as too exploit its display capabilities (Rafai, 1994). In a more active decision supporting mode, an object-oriented database/GIS has been combined with Bayesian and geostatistical models to assist in sampling decisions (Johnson, 1993a; Johnson, 1993b). This adaptive sampling model seeks to fuse both "hard" sampling data with related "soft" information on subsurface soil contamination to indicate where additional sampling would have the greatest impact on the uncertainty on plume extent. Retrospective analyses suggest that the model guidance produced significant reductions in the numbers of soil bores and samples required to reach a prescribed level of certainty in the volume of contaminated soil.

Modeling of contaminant transport and fate has historically been a component of the environmental restoration process. However, in many cases the modeling has been carried out by specialists somewhat removed from the data acquisition and decision processes. This role has changed as several factors seem to be bringing modeling more actively into the decision process:

- increased capabilities of modelers to use field data to determine model parameters, particularly in the face of uncertainty and large spatial variability,
- models that predict the performance of remedial technologies, and
- improved visualization capabilities that rapidly reduce complex model results to displays readily grasped by non-technical decision makers and stakeholders.

The coupling of seismic and tracer test data for aquifer property characterization (Hyndman et al., 1994) and stochastic approaches to characterizing groundwater contamination (McLaughlin et al., 1993) provide examples of how modelers have become integral pats of the data acquisition activities. This is a welcome sign given the importance of the relationship between acceptable levels of uncertainty in model predictions and decisions regarding the extent of data acquisition activities (Ditmars, 1988).

Models of remedial actions are becoming essential elements in the decision process. Hydraulic control of contaminated groundwater throughout pump-and-treat systems has been modeled creatively in an attempt to optimize the cleanup (Ahlfeld, 1988) and to demonstrate the effects of alternative solutions to the stakeholders. Advanced visualization systems now allow modelers to provide easily generated multiple three-dimensional views of groundwater contaminant transport during the model application to remedial decision makers (Williams and Durham, 1992). These visualization tools were used to demonstrate the efficacy of soul vapor extraction systems (Shikaze et al., 1993). Advanced graphics also permit rapid observation of whether predictions will satisfy remedial goals (Sepehr and Samani, 1993).

6.1.6
Conclusions

The slow pace of cleanup and the realities of the resource and time expenditures have caused a re-examination of the process that leads to cleanup. New decision strategies have been developed that attempt to link closely the decisions and the information on which they depend. The realizations that uncertainty cannot be eliminated by sampling and that a finite risk is associated with all decisions are embedded in these strategies. Technologies for sampling, chemical analysis, and data integration are essential to implementing this new approach. While several such technologies are presently employed, the challenge remains to develop and utilize technologies that will continue to improve the decision making that leads to cleanup.

Acknowledgments

This work was supported by the U.S. Department of Energy, Assistant Secretary for Environmental Management, under contract W-31-109-Eng-38.

6.1.7
References

Ahlfeld DP, Mulvey JM, Pinder, GF, Wood EF (1988) Contaminated Groundwater Remediation Design Using Simulation, Optimization, and Sensitivity Theory, 1. Model Development. Water Resources Research 24(3): 431-441

Bratton JL, Bratton WL, Shinn JD (1993) In-Situ Laser Induced Fluorescence. In Proceedings on 19th Environmental Symposium & Exhibition, American Defense preparedness Association. Albuquerque, NM, March 22-25, pp 26-81

Burton JC (1992) Prioritization to Limit Sampling and Drilling in Site Investigations. In Proceedings Federal Environmental Restoration Conference & Exhibition. Vienna, VA, April 15-17 pp 242-251

Cherry A, Walsh JJ, Towson GD, Handley JP (1992) Rapid Aquifer Characterization of a Hexavalent Chromium Plume Using Direct Push Technology and Field Chemical Analysis. In Proceedings of Hazardous Materials Control/Superfund '92. Washington, DC, December 1-3, pp 207-213

Ditmars JD (1988) A Method for Evaluating the Effectiveness of Site Characterization Measurements. In Land Disposal of Hazardous Waste: Engineering and Environmental Issues. JR Gronow, AN Schofield, RK Jain, eds John Wiley & Sons, New York, pp 59-68

EPA, Environmental Protection Agency (1993) Data Quality Objectives Process for Superfund. Interim Final, EPA-540-G-93-071, Washington, DC, September

Freeze RA, Massmann J, Smith L, Sperling T, James B (1992) Hydrogeological Decision Analysis, National Ground Water Association, Dublin, OH

GAO, United States General Accounting Office (1992) Superfund: Problems with the Completeness and Consistency of Site Cleanup Plans. GAO/RCED-92-138. Washington, DC, May

Gianti S, Dailey R, Hull K, Smyth J (1993) The streamlined Approach for Environmental Restoration. In Proceedings of Waste Management '93. Tucson, AZ, February 28-March 4, pp 585-587

Huber M (1994) Multimedia Enhances GIS Applications. GIS World 7(8):51-52

Hyndman DW, Harris JM, Gorelick SM (1994) Coupled Seismic and Tracer Test Inversion for Aquifer Property Characterization. Water Resources Research 30(7):1965-1977

Jenkins TF, Walsh ME (1992) Development of Field Screening Methods for TNT, RDX and 2,4-DNT in Soil. Talenta 39(4):419-428

Johnson RL (1993a) A Bayesian Approach to Contaminant Plume Delineation. In Proceedings 1993 ground Water Modeling Conference. Colorado School of Mines. Golden, CO, June 9-12 pp P-87-P-95

Johnson RL (1993b) Adaptive Sampling Program Support for Expedited Site Characterization. In Proceedings ER '93 Meeting the Challenge, Environmental Remediation Conference. Augusta, GA, October 24-28, pp 781-787

Mapes JP, Stewart TN, McKenzie KD, McClelland L, Mudd RL, Manning WB, Studabaker WB, Friedman SB (1991) PCB-RISc - An On-Site Immunoassay for Detecting PCBs in Soil. In Proceedings Hazardous Materials Control/Superfund '91. Washington, DC, December 3-5, pp 437-440

McDonald WC, Erickson MD, Abraham BM, Robbat AB Jr (1994) Developments and Applications of Field Mass Spectrometers. Environmental Science and Technology 28/7):336A-343A

McGrath LA, Brown S (1991) The Virtual Site as a Tool for Remediation Using the Observational Method. In Proceedings Hazardous Materials Control/Superfund '91. Washington, DC, December 3-5, pp 93-96

McLaughlin D, Reid LB, Li S-G, Hyman J (1993) A Stochastic Method for Characterizing Ground-Water Contamination. Ground Water 31(2):237-249

Moore J Jr (1994) Barriers to Technology Adoption Using Field Analytical Instruments as an Example. Environmental Science and Technology 28(4):193A-195A

Neptune MD, Brantly EP, Messner MJ, Michael DI (1990) Quantitative Decision Making in Superfund: A Data Quality Objectives Case Study. Hazardous Materials Control 3(3):19-27

Prochaska K, Hartness J, Lanis L (1991) Rapid Delineation of a Chlorinated Solvent Plume Using a Hydro Punch and Field GC. In Proceedings Hazardous Material Control/Superfund '91. Washington, DC, December 3-5, pp 54-57

Rafai HS (1994) Decision support Systems in the Public Domain: Issues and Considerations. In Proceedings EPA Decision Support Tools Workshop. Seattle, WA, June 28-29, pp 15-17

Roffman A, Verbanic EJ, Sherrill RP (1993) A Dynamic Methodology for Estimating Cleanup Efforts for Mercury in Soil at Gas Utility Gate Stations. Remediation 3(3):413-424

Ryti RT, Neptune D (1991) Planning Issues for Superfund Site Remediation. Hazardous Materials Control 4(1):47-53

Sepehr M, Samani ZA (1993) In Situ Soil Remediation Using Vapor Extraction Wells, Development and Testing of a Three-Dimensional Finite-Difference Model. Ground Water 31(3):425-436

Shikaze SG, Sudicky EA, Mendoza CA (1994) Simulation of Dense vapor Migration in Discretely Fractured Geologic Media. Water Resources Research 30(7):1993-2009

Sieminski PE, Griswold RM, Koski WA, Mosher JM (1993) Accelerated Cleanup: A Case Study of the Streamlined Remedial Project. In Proceedings on 19th Environmental Symposium & Exhibition, American Defense Preparedness Association. Albuquerque, NM, March 22–25, pp 64–69

Smyth JD, Amaya JP, Peffers M (1992) Observational Approach Implementation at DOE Facilities. Federal Facilities Environmental Journal, 3(Autumn):345–355

Wallace WA, Lincoln DR (1989) How Scientists Make Decisions About Ground Water and Soil Remediation. In Proceedings National Research Council Water Science and Technology Board Colloquium. Washington, DC, April 20–21, pp 151–165

Wang PW, Purdy CB, Tardiff AN, Lightner EM, Stallings EA (1994) Characterization, Monitoring and Sensor Technology Integrated Program: An Overview of Emerging Technologies in Site/Waste Characterization and Waste Treatment Monitoring. In Proceedings of Waste Management '94. Tucson, AZ, February 27-March 3, pp 1085–1088

Widdowson MA, Suddeth CK, Scaturo DM (1994) Cost-Effective Contaminant Plume Delineation Using Multilevel Drive Point Samplers. Remediation 4(1):59–76

Williams G, Durham L (1992) Efficient Analysis Using Custom Interactive visualization Tools at a Superfund Site. In Proceedings Hazardous Materials Control/Superfund '92. Washington, DC, December 1–3, pp 1060–1072

Woodside GD, Otis L (1991) Use of a GIS Database in the Whitter Narrows Operable Unit Feasibility Study. In Proceedings Hazardous Materials Control/Superfund '91. Washington, DC, December 3–5, pp 131–135

6.1.8
Further Readings

Innovative Remedial Cost Estimating for Multiple Sites with Limited Environmental Site Characterization Data

Hasan Cirpili, Environ Strategies Corp, Reston, VA;
Environ Claims J, Summer 94, v 6, n 4, p 591(13) journal article

Abstract: The transfer of multiple sites during corporate mergers and acquisitions involves the determination of remedial costs where environmental media may have been impacted by former industrial activities. To facilitate the process in cases where little analytical data are available on the environmental conditions at each site, a model has been developed that combines historical information, knowledge of typical operations at sites in the same industry, degrees of contamination at these related sites, and remedial costs. Probability modeling can be used to develop reasonable-case remedial scenarios, which rely on taking best-, medium-, and worst-case cost and environmental scenarios and applying a random distribution to the costs and effects. A specific model framework is presented, involving an inventory of potential processes and the determination of site variables, site information and tabulations, and volume calculations and remediation modeling.

Case Study: Application at a Superfund Cleanup

Stanley Blacker, MAC Technical Services, Germantown, MD and Daniel Goodman, Montana State University, Bozeman;
Environ Sci Technol, Oct 94, v 28, n 11, p 471A(7) journal article

Abstract: A new approach toward Superfund site cleanup has been developed at EPA, which involves the incorporation of the Superfund Accelerated Cleanup Model and Data Quality Objectives. The approach, which focuses on standardizing the level of residual risk and data acquisition and separating risk-goal policy from negotiable technical factors, is described as it was applied to a Superfund site near Times Beach, MO, which had been contaminated by waste oil used as a dust suppressant on dirt roads. The methodology is described in terms of the separation between policy decisions about risk-goals and technical discussions of alternative for implementation, the translation of regulatory policy decisions into measurable criteria, the documentation of the agreements reached between stakeholders on the critical requirements for cleanup, and the use of all available technical optimization methods to develop the most efficient design for the site remediation. The use of the approach effectively achieved at least 50% of the initial site-cleanup costs. (4 diagrams, 4 references)

Risk and the New Rules of Decisionmaking: the Need for a Single Risk Target

Doulgas J. Sarno, Phoenix Environ Corp, Alexandria, VA;
Environ Law Report, Jul 94, v 24, n 7, p 10402(4) journal article
Abstract: Three issues have emerged to push the government toward goals of increased rationality, consistency, and accountability in the Superfund program. Concerns about environmental justice have moved to the forefront, and restrictions on future land use have become widely accepted tools for improving the rationality of decision making. The decision making role of the public has been enlarged, as DOE and EPA have given full support to increasing citizen participation in decision making. In light of these changes, the government must establish a single, nationwide risk target to apply at each site to foster effective public participation in cleanup decisions and to achieve a consistent level of protection, regardless of the socioeconomic status of affected communities. (21 references)

Site characterization by artificial neural networks

Author: Rizzo, Donna M.; Dougherty, David E.; Lillys, Theodore P.
Corporate Source: Univ of Vermont, Burlington, VT, USA
Conference Title: Proceedings of the 21st Annual Conference on Water Policy and Management: Solving the Problems
Conference Location: Denver, CO, USA
Conference Sponsor: ASCE; American Water Resources Association; Center for Advanced Decision Support for Water and Environmental Systems; Colorado State University; Denver Water Board; et al
Source: Proceedings of the 21st Annual Conference on Water Policy and Proc 21 Annu Cof Water Policy Manage Solving Probl 1994. Publ by ASCE, New York, NY, USA. p 250–253
Publication Year: 1994
ISBN: 0-7844-0020-2
Abstract: Recently, an optimal groundwater management model has been developed to treat groundwater remediation problems at Lawrence Livermore National Laboratory (LLNL). The objective of the model is to identify the best remediation strategies (well site selection and pumping rates) so that water quality standards are met at a specified reliability level within a given time frame. A thorough understanding of the hydrodynamic behavior of aquifer systems requires a complete and accurate determination of the physical parameters of the groundwater system. An example, the one we will examine here, is the identification of three-dimensional hydraulic conductivity fields for LLNL's Main site. 9 Refs.

6.2
Soil Decontamination Using Electrokinetics, with Application to Urban Residual Sludges

MICHEL ASTRUC · SYLVAINE TELLIER · ISABELLE LE HÉCHO · JÉSUS LARRAÑAGA · MARIE-CHRISTINE FOURCADE

Abstract

The electrokinetic soil decontamination process has been applied, at laboratory scale, to a clay model (kaolinite) and organic rich solids (sewage sludges) to study transport of anionic (nitrate) or cationic (heavy metals) contaminants.

Heavy metal decontamination rats in kaolinite are essentially controlled by liquid-solid interactions. These processes are also predominant in sludges. However, it is possible to extract metals by this process in conditions where they are only slightly solubilized. In well defined experimental conditions, the removal of cadmium, zinc and nickel form sewage sludges was quantitative. It remained satisfying for copper (87%), whereas chromium and lead were only partially mobilized (44 and 40% respectively).

6.2.1
Introduction

Soil remediation, with regard to heavy metals, is a palliative aimed at preventing dissemination of the pollution; this is the case for confinement, solidification or vitrification methods. The long term efficiency of isolation technologies has recently been questioned (Diependaal, 1993). Washing techniques are only concerned with sandy soils and do not avoid the dumping of fine particulates such as clays and silts carrying the better part of metals.

During the last few years a new electrokinetic treatment process allowing the decontamination of fine particulates, possibly in-situ, has been developed by several research teams, including ourselves. This process, relying on ionic transport in an electric potential gradient, had been proposed very early for the desalinization of alkaline soils (Puri, 1936) and direct extraction of copper in mines (Drinkard, 1976). A first attempt on copper, nickel and manganese (Hamnet, 1980) was unsuccessful. More recently the method was developed in theoretical and experimental laboratory studies of the conditions of elimination of model pollutants such as acetic acid or phenol (Renaud, 1987; Shapiro, 1989a, 1989b, 1993; Probstein, 1993).

Phenol elimination was also studied by (Acar, 1992). Transport of hydrocarbons by an electroosmotic process as well as electrokinetic nutrient transport (NO_3^-, PO_4^-) was also investigated (Segall, 1992a, 1992b).

As regards heavy metals the first attempts were realized by Lageman, (1989) with lead, copper and arsenic whereas Pamucku (1990, 1992, 1993) studied zinc, strontium, nickel, cobalt, cadmium, cesium and uranium.

Working with kaolinite as a soil model, Hamed (1991) and Acar (1994) demonstrated that lead and cadmium could be eliminated by this process (respectively 75–95% and 90–95% elimination). Our own research demonstrated that copper, cadmium, zinc and nickel could be efficiently extracted from various model soils (Le Hécho, 1992) and sewage sludge (Larranñaga, 1994).

Field experiments have been performed in the Netherlands (Lageman, 1989, 1993) or are in progress in the USA (Trombly, 1994)

In this paper the principles of the electrokinetic treatment process will be illustrated by experiments using kaolinite as a model soil and as a possible application to the decontamination of sewage sludge presented.

6.2.2
Principle of the Electrokinetic Soil Decontamination Process

The global migration process of ionic substances is ruled by several phenomena, the most important being:

- the electric field which induces purely electrokinetic effects such as electroosmosis and electromigration
- the electrolysis of water on the electrodes
- the physico-chemical processes at the solid/solution boundary such as adsorption, complexation, precipitation of oxydes, hydroxydes, carbonates etc…
- the gradient of chemical potential inducing diffusion and osmosis phenomena which have however, usually much smaller effects than electrokinetic processes
- complex formation in solution (including colloids)
- hydraulic pressure (mainly in field studies).

While combined fluxes of water, ions and electricity lessen the effect of the electric potential, hydraulic and chemical gradients have been made clear by (Yeung, 1990) using phenomenologic coefficients. Alshawaken (1992) modeled the process, including physico-chemical phenomena in solution and at the solid surface.

The influence of some of these phenomena may be evidenced using a clay such as kaolinite for model soil.

6.2.3
Processes Involved in Electrokinetic Soil Decontamination

Physico-chemical phenomena involved are governed mainly by pH gradients appearing in the interelectrode space which are susceptible to the modification of electroosmosis, chemical equilibria or adsorption.

pH Gradients and Their Consequences

Electrolysis of water on the electrodes, generating H^+ and OH^- ions on the anode and cathode respectively, creates pH gradients. An example is given in Fig. 1 in the case of kaolinite wetted by a slightly conducting water.

Initial pH is quite homogeneous (5 < pH < 6) in this unbuffered medium. Acidification and alcalinization of the anodic and cathodic area, respectively, appear. Concerned areas extend with times towards cell center (Fig. 1).

In these areas the ionic content of water is deeply modified, inducing drastic local changes of conductivity and, therefore, of the potential gradient (Fig. 2).

The homogeneous potential gradient of $1 \, V \cdot cm^{-1}$, approximately verified at the beginning of the experiment, is rapidly modified. After 48 hours, very low gradients exist in both anodic and cathodic areas, whereas a very steep potential gradient prevails in a central area.

These pH gradients may be used to modify chemical conditions prevailing in the soil and to favour desorption phenomena and free metallic cations, as will be shown later (cf §3) in the case of sewage sludges. This principle has also been used by Acar (1994) who modelized the shift of the acidic pH front.

On the other hand, an experimental approach to the rate of the electrokinetic process, which depends on the electrical field, can only be easily realized when the potential gradient is homogenous, therefore when pH profile is under control.

Electrokinetic Ion Transport Rate in Soil – Example of Nitrate in Kaolinite

Experimental set-up
Electrodes are contained in compartments separated from test soil by inert porous membranes. This set up allows the injection of an aqueous solution of an acid (cathodic

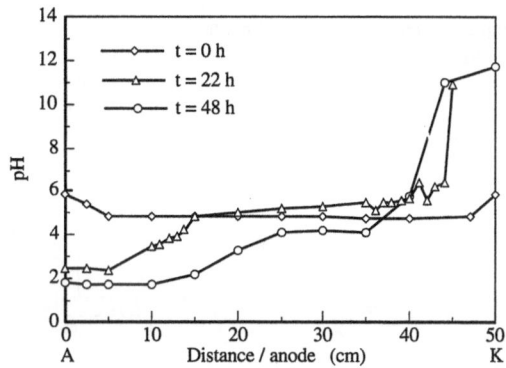

Fig. 1. Evolution of the pH gradient in the interelectrode domain on a 48 hours period. Kaolinite. Cell length: 50 cm between electrodes. Cell voltage: 50 V

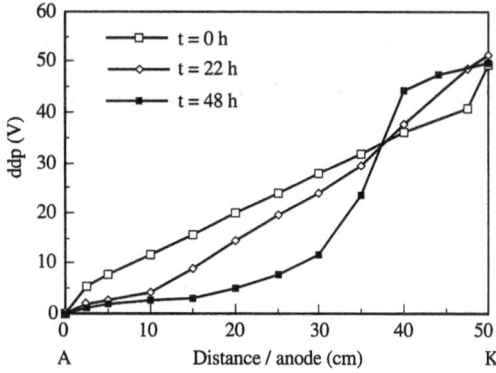

Fig. 2. Evolution of potential profiles with time on a 48 hours period. Same conditions as in Fig. 1

side) or a base (anode) in each electrode compartment. Controlled concentrations and flow rates are chosen so that pH and potential gradient are maintained constant throughout the interelectrode space (Fig. 3).

Purely electrokinetic process

A purely electrokinetic control of ion migration rate may be obtained only in the case of non adsorbed ionic species, such as nitrate in the presence of kaolinite.

When nitrate is injected in the cathodic compartment it is possible to follow its migration towards the anode by analysis of interstitial waters along the cell. An example is given in Fig. 4.

Experimental migration rates ($V_{EC(exp)}$) have been thus determined in several experiments (Table 1). They are in reasonable agreement with theoretical previsious calculated from the simplified equation (1)

$$V_{Ec(theo)} = V_{Em} - V_{Eo} \tag{1}$$

where V_{Es} is the total electrokinetic speed, V_{Eo} the electroosmotic speed and V_{Em} the electromigration speed. This last parameter may be calculated using an ionic mobility $u = 6.46 \ 10^{-4} \ cm^2 \ V^{-1} \ s^{-1}$, a porosity of 0.619 and a tortuosity factor of 1.27 close to that

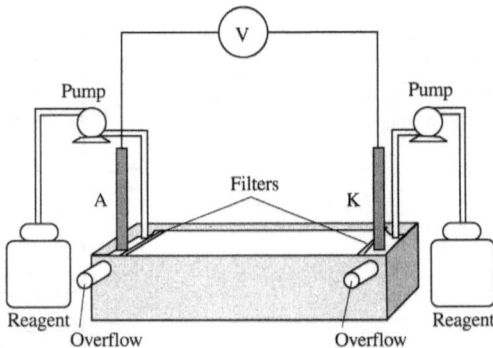

Fig. 3. Scheme of the experimental set-up

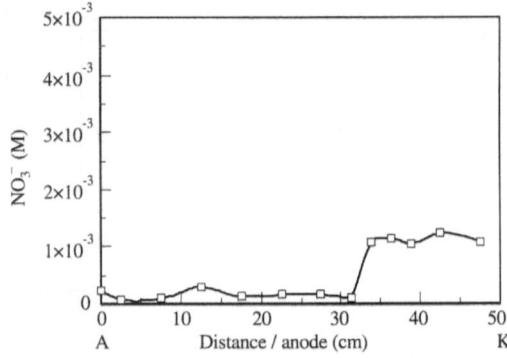

Fig. 4. Concentration profile of nitrate ions – pH = 8.4 Injection concentration of NO_3^-: 0.01 mol · l^{-1}

Table 1. Electrokinetic speed of nitrate ions at different pH values

pH	Potential gradient $(V \cdot cm^{-1})$	$V_{Ec}(exp)$ $(cm\ h^{-1})$	$V_{Eo}(exp)$ $(cm\ h^{-1})$	$V_{Ec}(theo)$ $(cm\ h^{-1})$
4.3	0.97	0.83	0.36	0.77
8.4	1.1	0.42	0.84	0.43

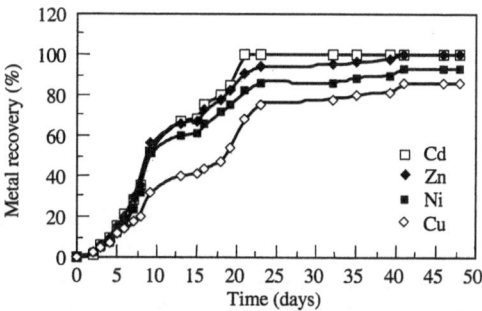

Fig. 5. Cumulated elimination in the cathodic effluent of heavy metals (Cd, Zn, Cu, Ni) introduced at $0 \le t \le 1$ day at the anode – kaolinite (45 cm cell) – pH = 5

determined by Shapiro (1993). The electroosmotic speed, V_{Eo}, is experimentally measured because it is dependent on solution composition (Lockhart, 1983). Results in Table 1 demonstrate the influence of this parameter on the movement of nitrate ions.

Nitrate ions are a simple model for which transport is essentially controlled by electrokinetic speed. Transport of heavy metal ions, on the contrary, is affected by liquid-solid interactions.

Effect of Adsorption: Case of Heavy Metal in Kaolinite

Experimental data presented in Fig. 5 have been obtained in experiments where kaolinite was the soil model and a pH of 5 was maintained throughout the 45 cm long cell. Heavy metals (Cd, Zn, Cu, Ni) were injected in the anodic compartment during 24 hours. Their elimination was followed by analysis of the cathodic effluent of the cell.

Theoretical electrokinetic speed in these conditions may be calculated as close to 31 cm day^{-1}. It is evident from Fig. 5 that liquid-solid interactions considerably reduced transport rates of all the metallic cations studied, copper being the more retained substance and cadmium the more mobile. Ion exchange is probably responsible for this retention.

After 23 days cadmium was completely recovered, together with 94, 86 and 76% of zinc, nickel and copper respectively.

6.2.4
Decontamination of Sewage Sludge Polluted by Heavy Metals

Progressive acidification induced by the anodic reaction may be used to promote the solubilization of heavy metals present at undesirable concentrations in most sewage

Fig. 6. Chemical extraction yields of various heavy metals from primary sludge as a function of final pH. Acidification by nitric acid

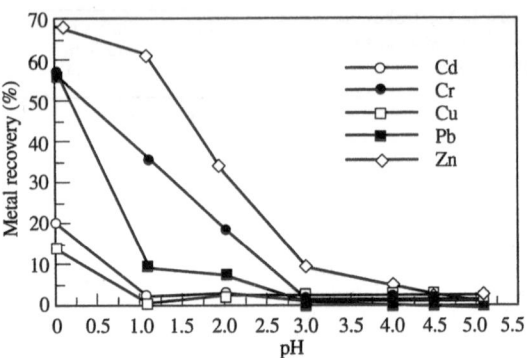

sludges, following a process described by Wozniak (1982). Figure 6 presents chemical extraction yields of various heavy metals from one of the primary sludge samples used in this study at various pH values obtained by nitric acid additions. One may note in Fig. 6 that zinc solubilization begins at pH < 3, that of lead and chromium at pH < 2, while pH < 1 is necessary for copper and cadmium. Extraction yields for cadmium are noticeably lower than those published by Wozniak (1982) and Logan (1985).

Electrokinetic decontamination of sewage sludge with acidification has been addressed in two experimental methods:

- leaving both acidic and alkaline fronts to develop freely under the influence of electrode reactions
- controlling cathodic pH by suitable addition of acid to prevent the formation of an alkaline area.

6.2.5
Electrokinetic Decontamination of Sewage Sludge without pH Control

This method has been applied essentially to primary sludge samples. A pH front rapidly appears in the vicinity of the cathode (Fig. 7), and fluctuates little after the 7th day. The normalized distance of this pH front to the anode (ratio of its distance to the anode to the total cell length) has been found close to 0.85 in several experiments. This value is appreciably different from what was predicted by a simple calculation involving only ionic mobilities of H^+ and OH^- ions and the assumption that these species are produced in equal amounts by electrode reaction (0.64).

The difference may, for the better part, be explained by the consumption of hydroxyl ions in the precipitation of hydroxydes and carbonates of calcium and magnesium, elements quite abundant in the sludge and which migrate towards the cathode (Fig. 8).

In these conditions of uncontrolled pH, the most mobile metals accumulate behind the pH front. Normalized concentrations (i.e. the ratio of local measured concentration to the initial homogenous concentration) of cadmium, zinc and nickel are presented on Fig. 9. These data show that an important concentration decrease of these elements has taken place in 90% of the sludge volume (80, 74 and 58% respectively for

Fig. 7. Evolution of pH as a function of normalized distance to the anode and time 1 V · cm⁻¹. Primary sludge

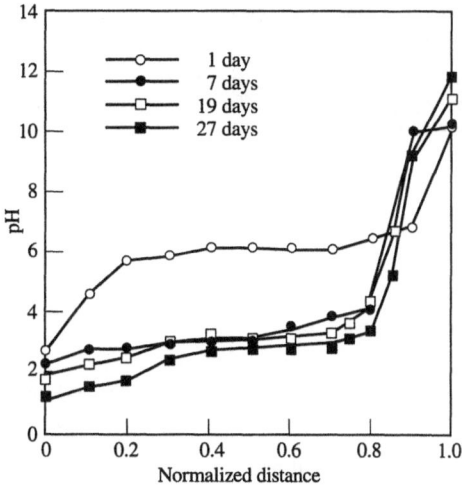

Fig. 8. Normalized concentration profiles as a function of normalized distance to the anode. Conditions as in Fig. (–): pH profile at the end of the experiment

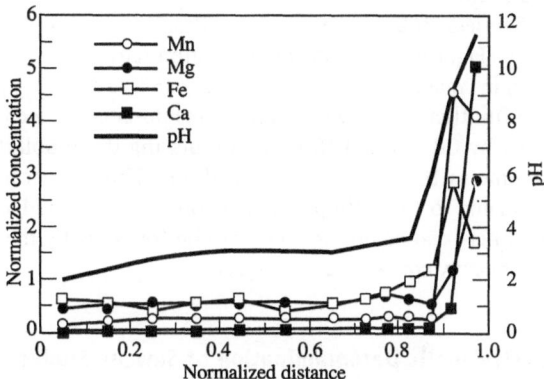

Fig. 9. Normalized concentration profiles of heavy metals as a function of the normalized distance to the anode. Conditions as in Fig. 7 (–): pH profile at the end of the experiment

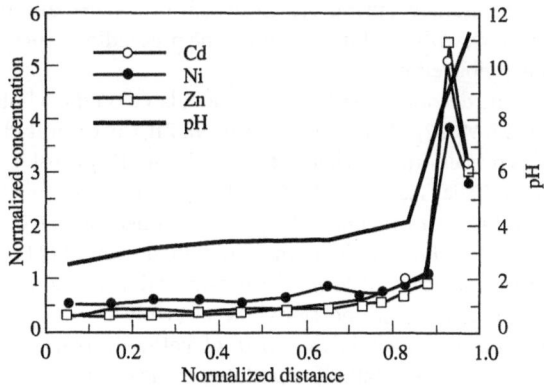

Fig. 10. Normalized concentration profiles of Pb, Cr, Cd as a function of the normalized distance to the anode. Conditions as in Fig. 7 (–): pH profile at the end of the experiment

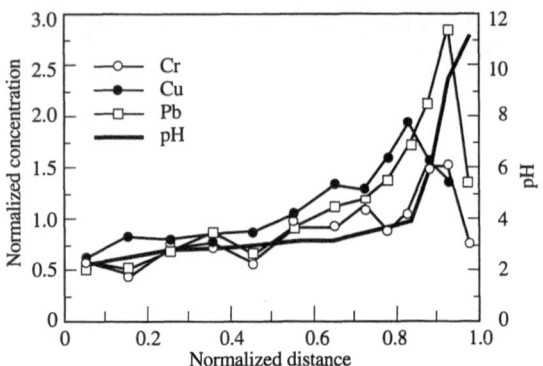

Cd, Zn and Ni). It may be noted also that this electrokinetic migration has taken place in a $2 < pH < 3$ range, whereas chemical extraction data presented in Fig. 6 indicated the necessity of much lower pH values ($pH < 1$). This effect is particularly important for cadmium.

The less mobile elements, lead, chromium and copper, are accumulated before the pH jump (Fig. 10) and the decontamination rates are much lower (35, 31 and 27% respectively) in smaller portions of the sludge (60, 70 and 50% of the interelectrode space).

With this simple method it is thus possible to decontaminate a sewage sludge polluted by Cd, Zn and Ni by accumulating these pollutants in c.a. 10% of the initial volume in the vicinity of the cathode. The electrical power consumption has been evaluated to 165 kWh per metric ton of treated sludge. It remains however necessary to either treat again the 10% residue (possibly by the same process) or to dispose of this heavily contaminated residue.

6.2.6
Electrokinetic Decontamination of Sewage Sludge with pH Control

By control of the pH in the cathodic compartment it is possible to avoid the appearance of the pH front and collect metals in the cathodic effluent. Figure 11 presents the data obtained during the treatment of a thermally treated sewage sludge (48% dry matter); with this kind of sludge the preceeding process without pH control resulted in negligible efficiency.

The decontamination yields calculated on the whole sludge sample are respectively 99, 97, 98, 87, 44 and 40% for Cd, Ni, Zn, Cu, Cr and Pb after 40 days. The efficiency of this process appears better than without pH control. The electrical power consumption after 40 days is higher (500 kWh per metric ton) for this sludge with a high solid content (48% dry matter). The power cost decreases when dealing with primary sludges, down to values similar to those obtained without pH control. It may be further reduced, even drastically, if lower decontamination levels are accepted, the power cost being approximately proportional to electrolysis time. However, economical evaluation should also include the costs of cathodic reactants and secondary treatment (or disposal) of the catholyte polluted by heavy metals.

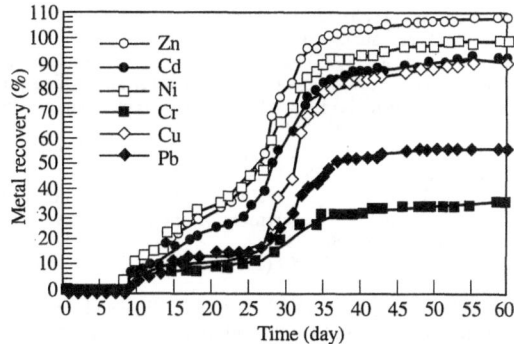

Fig. 11. Cumulated fractions of metals collected in the catholyte as a function of time. Thermally treated sludge Cell length: 50 cm – Voltage: 50 V

6.2.7
Conclusion

Electrokinetic decontamination is a process which may be applied to anionic or cationic species, to simple solids such as clays but also, soils, or even sludges.

In all situations it has been possible to obtain very good efficiencies for the removal of cadmium, zinc and nickel. Copper is not as well eliminated, but yields of 85–87% have been attained.

Two processes are proposed and may find applications; without pH control, metallic pollution may be concentrated in a small fraction of the solid treated; with a control of cathodic pH, metallic pollution is collected in the catholyte. The last process appears more efficient but somewhat more costly.

6.2.8
References

Acar YB, Gale RJ (1992) Phenol removal from kaolinite by electrokinetics. Journal of Geotechnical Engineering 118 (11), 1837–1852

Acar YB, Gale RJ, Putnam GA, Hamed J, Wong RL (1990) Electrochemical processing of soils: Theory of pH gradient development by diffusion, migration and linear convection. J Environ Sci Health A25, 687–714

Acar YB, Hamed JT, Alshawabkeh AN, Gale RJ (1994) Removal of Cadmium (II) from saturated kaolinite by the application of electrical current. Geotechnique 44 (2), 239–254

Alshawabkeh AN, Acar YC (1992) Removal of contaminants form soils by electrokinetics: a theoretical treatise. Environ Sci Health A27 (7), 1835–1861

Diependaal MJ, Klein AE, Oudee boerrigter PBJM, Van der Meij JL, De Walle FB (1993) Waste Management & Research 11, 481–9. Long term effectiveness of isolation techniques for contaminated soils.

Drinkard WF (1976) Electrochemical mining of copper. US Patent 956–087

Hamed J, Acar YB, Gale RJ (1991) Pb (II) removal from kaolinite by electrokinetics. Journal of Geotechnical Engineering 117, 241–271

Hamnet C (1980) A study of processes involved in the electroreclamation of contaminated soils. M Sc Thesis University of Manchester – Angleterre

Lageman R, Pool W, Seffinga G (1989) Electro-reclamation: theory and practice. Chemistry and Industry, 585–590

Larranaga J, Tellier S, Astruc M (1994) Décontamination électrocinétique de métaux lourds dans les boues urbaines in Recueil de Conférences 11e colloque Journées Information Eauc, Poitiers, 28–30 septembre, pp 43-1–43-9

Le Hécho I, Larranaga J, Eyrolles F, Tellier S, Astruc M (1992) Protection des eaux naturelles par décontamination in situ de sols pollués par des espéces ioniques à l'aide d'une méthode d'électromigration. Récents progrés en génie des procédés 6, 167–172

Lockhart NC (1983) Electroosmotic dewatering of clays II Influence of salt acid and flocculants. Colloids and surface 6, 239–51

Logan TJ, Feltz RE (1985) Effect of aeration cadmium concentration and solids content on acid extraction of cadmium from a municipal wastewater sludge. Journal WPCF 57 (5), 406–12

Pamukcu S, Wittle JK (1992) Electrokinetic removal of selected heavy metals from soil. Environmental Progress 11 (3), 241–250

Pamukcu S, Khan LI, Fang HY (1990) Zinc detoxification of soils by electroosmosis. Transportation Research Record 1288, 41–46

Pamukcu S, Wittle JK (1993) Electrokinetics for removal of low-level radioactivity form soil. Fourteenth Annual U.S. Department of Energy Low-Level Radioactive Waste Management Conference, Proceedings, pp 256–78

Probstein RF, Edwin Hick SR (1993) Removal of contaminants from soils by electric fields. Science 260, 489–503

Puri AN, Anand B (1936) Reclamation of alkali soils by electrodialysis. Soil Science 42, 25–27

Renaud PC, Probstein RF (1987) Electroosmotic control of hazardous wastes. Physicochemical Hydrodynamics 9, 345–360

Segall BA, Bruell CJ (1992b) Electroosmotic contaminant-removal processes. J of Environmental Engineering 118 (1), 84–100

Segall BA, Bruell CJ, Walsh T (1992a) Electroosmotic removal of gasoline hydrocarbons and TCE from clay. J of Environmental Engineering 118 (1), 68–82

Shapiro AP, Probstein RF (1993) Removal of contaminants from satured clay by electroosmosis. Environ Sci Technol 27, 283–291

Shapiro AP, Renaud PC, Probstein RF (1989b) Preliminary studies on the removal of chemical species from saturated porous media by electroosmosis. Physicochemical Hydrodynamics 11, 785–802

Shapiro AP, Renaud PC, Probstein RF (1989a) In-situ extraction of contaminants from hazardous waste sites by electroosmosis, Waste management and productivity enhancement, International Symposium (Muralidhara HS, Ed), Batelle Press, 346–353

Trombly J (1994) Electrochemical remediation take to the field. Environ Sci Technol 28 (6), 289A–291A

Wozniak DJ, Huang JYC (1982) Variables affecting metal removal from sludge. Journal WPCF, 54 (12), 1574–80

Yeung AT (1990) Coupled flow equations for water, electricity and ionic contaminants through clayey soils under hydraulic, electrical and chemical gradients. J non-equilib Thermodyn 15, 247–267

6.2.9
Further Readings

Impact of system chemistry on electroosmosis in contaminated soil

Author: Eykholt, Gerald R.; Daniel, David E.
Corporate Source: GE Co, Schenectady, NY, USA
Source: Journal of Geotechnical Engineering v 120, n 5, May 1994. p 797–815
Publication Year: 1994
CODEN: JGENDZ ISSN: 0733-9410
Language: English
Document Type: JA; (Journal Article) Treatment Code: X; (Experimental)

Abstract: Electroosmosis in a copper-contaminated koalinite was highly sensitive to chemical treatment schemes to remove the contamination. Nonuniform profiles of electric field intensity and pH as well as negative pore-water pressure develop during sustained electrokinetic treatment of clays. These nonlinearities and nonuniform pore-water pressures cannot be adequately described by classical analysis. Classical analysis is based on assumptions of a uniform and constant electroosmotic permeability coefficient, for instance. An extended capillary model which includes nonuniform contributions to electroosmosis and pore pressures that vary with space and time, is developed and compared with experimental findings. Subtle changes in initial and boundary conditions of the system chemistry have a very large effect on electroosmosis of soils. (Edited author abstract) 40 Refs.

Principles of Electrokinetic Remediation

Yalcin B. Acar and Akram N. Alshawabkeh, Louisiana State Univ, Baton Rouge;
Environ Sci Technol, Dec 93, v 27, n 13. p 2638(10) journal article

Abstract: Electrokinetic remediation uses low-level DC to remove chemical contaminants from soils. The principles of species removal and metals removal in soils under an electric field are explained, highlighting electrolysis, diffusion, icon migration, electroosmosis, dissolution, and precipitation processes. Ionic migration is the most important component of mass transport in electrokinetic remediation in most soils. The efficacy of species transport is linked to species transference number, which in turn depends on its ionic mobility and concentration. Recent research on the electrokinetic remediation process is surveyed. (1 diagram, 8 graphs, 1 photo, 38 references, 1 table).

Fundamentals of extracting species from soil by electrokinetics

Author: Acar, Yalcin B.; Alshawabkeh, Akram N.; Gale, Robert J.
Corporate Source: Louisiana State Univ, Baton Rouge, LA, USA
Source: Water Management v 13, n 2, 1993. p 141 – 151
Publication Year: 1993
CODEN: WAMAE2 ISSN: 0956-053X
Language: English
Document Type: JA; (Journal Article) Treatment Code: A; (Applications); G; (General Review)
Abstract: Bench-scale and pilot-scale studies demonstrate that ionic contaminant species, some organic contaminants, and certain radionuclides can be removed efficiently from fine-grained deposits by application of electrical currents across a soil mass, through electrodes that allow egress and ingress of a pore fluid. This technique (electrokinetic remediation or electrochemical soil processing) results in soil acidification, contaminant description, transport, accumulation, and removal. A review of the fundamentals of the process and the theoretical development, together with a review of considerations and limitations for full-scale application of the technique for site remediation, are presented. (Author abstract) 39 Refs.

6.3
A Systematic Approach to Groundwater Management

ROBERT HAUSLER · PATRICK BÉRON · ANDRÉ HADE

Abstract

A multidisciplinary approach is recommended in the search for long lasting solutions to environmental problems, and the working team addressing any given problem must be able to interact successfully. The principle of total quality is a philosophy that applies very well to environmental engineering. However, tools must be made available to engineers at their training and working stages in order to ease decision making. In order to choose the best technology to treat lixiviated water, one must analyze how valuable the project is by evaluating the quality-price ratio. This simple and rigorous method allows all contributors to understand and easily accept the chosen option. Its implementation in the applied research field allows young researchers to make decisions and to convince themselves of their accuracy. This tool helps the personnel to become self-sufficient and responsible and can be taught in "the learning through problems" form.

6.3.1
Introduction

Total quality is a philosophy developed in the 1950's which aimed to increase the productivity of businesses by making the employees feel responsible. The industrial world has rediscovered this philosophy during the 1990's. This revival of interest is due on one hand to the two global economic crises and recessions, and on the other hand, to the worldwide application of the markets, which raises the competition factor. In this context, management based on total quality may destabilize the employees who are not prepared to assume this responsibility. However, total quality is an interesting philosophy for environmental protection and for a global approach envisioning risk management. Indeed, it allows the reconciliation of the economic development and the environmental safeguard, two concepts which have been opposed for too long.

The population of Quebec is so concerned about the environment that a change in the decision making of business and government should be implemented. Large firms, for example, are creating environmental divisions within their companies even though their line of work is not directly related to environmental issues. On the other hand, the government is growing more strict with regard to industrial environmental impacts, invoking sanctions which can lead to closing firms. This change in the mentality of Quebec's society calls for the training and role of engineers to be reviewed and modified. Indeed, a non-repressive approach would be to train executives as well as employees to anticipate the sources of pollution instead of having the industries treating them. The population potentially affected by the environmental problems must also be integrated into the mitigation procedures in order for it to suggest solutions instead of slandering the problems.

This dialogue is one of the three fundamental principles of the total quality philosophy. The second is the proaction and retroaction process. This process entails evaluation of the consequences of the decisions made in time or in space. When this spatiotemporal vision is applied to engineering, it not only requires the evaluation of the protection means, but visualization and global knowledge of the client's problems. The client aspect brings us to the third principle. The latter consists of putting oneself at the disposal or the service of the community. In other words, the engineer must fuse himself with the company as if it were his own and must develop a sense of belonging. None of the above notions are new and they have always been at the root of education and savoir-faire. These notions are even part of the code of ethics of the "L'ordre des ingénieurs du Québec". Nevertheless, these traditional notions might have lost some significance.

These reflections lead us to conclude that the role and the relations universities have with the industrial world and the general public should change. Indeed, within the context of total quality, the universities may fill out a new mandate in accordance with their training and research vocation. This mandate consists in opening oneself to the private companies and making their infrastructures and their experts available to the community. Universities can then work hand in hand when searching for solutions and carrying out concepts. However, the research and philosophy of the researchers must be brought into this mandate. The fundamental and applied research can no longer clash in global research, they must become integrated into it. On the other hand, tools needed to measure the quality of a project must be perfected in order to allow for an unbiased judgment. One of these tools is the analysis of the value of a project. One example of treating lixiviated water of a landfill site is addressed here in order to clearly visualize the total quality approach to the problem.

6.3.2
Methodology

The work program of the value analysis leads the contributors to question and investigate some of the fundamental aspects of the subject. It ensures that all needs are identified and understood and that the search is directed to reach optimal solutions. The solutions can be classified in ascending order from the level of highest satisfaction to the lowest cost. The stages in the value analysis range from identifying the needs to performing a functional analysis and identifying its characteristics and, finally, evaluating the suggested solutions.

The functional analysis consists in identifying the functions required to satisfy the need. This need is naturally issued by the project's promoter. However, it is preferable that all actual or future contributors step in as soon as the need is identified. The functions are expressed with a verb and a noun in order to communicate clearly and to bring the participants back to the root of the problem. With this fundamental need, one must elaborate the identification of the basic and secondary functions from their criteria of importance and their interactions. In the same way, the solutions are brought forward and are viewed according to the knowledge of the contributors, the experts in the field or the bibliographical journals. A multidisciplinary team is also preferable at this stage. This analysis of the different functions allows a functional graph to be determined. This graphic solution shows overall, the hierarchical organization of the functions between the need and the suggested solutions.

The characterization of the functional analysis is a stage that consists in explaining the need by deciding the criteria of achievement and the measurable element used to satisfy it. Therefore, the characterization consists in setting forth the levels of criteria for assessment, and specifying the flexibility of those levels. For each function, the characterization of the functional analysis expresses their level of tolerance and flexibility by using criteria. These levels are shown generally with numbers ranging from 0 to 10. The flexibility "0" means that the function has no possible tolerance where the flexibility "10" expresses much tolerance, that is to say the criteria can have many levels of flexibility.

Finally, the evaluation of the proposed solutions is also based on criteria which are expressed by a multidisciplinary team, including all concerned contributors. In order to determine the level of an evaluation criterion, a 1 to 10 factor is attributed to it by the multidisciplinary team (10 being the most important factor and 1 the least important). Even if this weighting is arbitrary, it is relative to the best technology and is based on current knowledge and theories. This weighting is also obtained through consensus among the contributors. The level of satisfaction of a criterion is obtained by multiplying the weight of the criteria by the given profit and compiling the data in a chart. The sum of the results obtained for each of the solutions allows their rapid classification in ascending order of interest.

A trial program must be created for the most interesting solutions in order to determine that which is the most favorable and economically most feasible. Thus, the quality-price ratio must be assessed according to the updated estimated cost of building and operating in order to finalize the value analysis.

6.3.3
Implementation (Example)

Context

Urbanization and economical development have created the need for solid waste disposal. It is foreseeable that solid residues will still have to be landfilled even if the three "R" principles (reprocessing, revaluation and recycling) are used and the quantity of waste is reduced to a minimum. Moreover, one must foresee systems to treat the lixiviated water of the old disposal sites in order to preserve ground water. The variety of buried waste and the diversity of dumping areas are characteristics of landfill sites. Under those circumstances, the use of a unique or universal technique is unrealistic. The creation of a treatment system requires that the best applicable techniques be chosen and an experimental validation be executed. Nevertheless, those best applicable techniques have often been restricted to process domestic liquid waste in a conventional way and therefore didn't necessarily respond to the owner's and the environment's real needs.

The given example has been based on the City of Montreal's true situation which manages the disposal site of "the sorting and waste disposal center" (CTED). For political reasons, this municipality has envisaged different closing scenarios for the site in order to satisfy the surrounding population and to protect the environment. These closing scenarios recommend choosing a treatment for lixiviated water and its flows to the sewer. Besides having the obligation to respect the City of Montreal's "flows to the sewer" standards which manages the liquid waste treating center, the process had to consider certain particular constraints. The site's geographical and social situation at the heart of an urban center, as well as the site's ultimate calling, were among the important parameters. However, the site's future isn't as clear. A rigorous strategy is then required. The value analysis which searches for the best quality-price ratio of a project by using a dialogue process, is ideal in solving this kind of problem.

Results

The first step was to determine the indispensable condition in order to satisfy the essential requirements for the development and the future closure of the disposal site. This need is defined as follows:

To Protect the Underground Water
Identifying the basic and secondary functions of this fundamental need has enabled one to work out carefully their criteria of importance and their interactions. The graphical and functional representation is partially illustrated in Fig. 1. In Fig. 1, each function is characterized by criteria linked with a tolerance and flexibility level. The characterization of the functional analysis in order to dispose lixiviated water and to choose a treatment for it is generated in Table 1.

The average flow has no flexibility since this parameter is one of the main criteria when creating the treatment. The maximum flow through must be treated by the installations, but it is temporary. The mitigation measures for reducing the water that needs to be treated (reduce the lixiviat level, eliminate the water inflow, etc.) should minimize these peaks. The flexibility criterion is, in this case, higher. Even if the total characterization of the lixiviated water exceeds the flow norms in natural environment for many chemical parameters (DCO, DBO$_5$, phosphorus, certain heavy metals, etc.), the phenol composites and the sulphides alone exceed the "flow to sewer" criteria.

The toxicity isn't presently constraining since there isn't any norm to regulate it. Nevertheless, the St. Laurent Center, a governmental organization which monitors the quality of the St. Laurence, evaluates among other things, the performance of the industrial and municipal purification treatments with ecotoxicologic tests. It is sensible to project that one day, the municipalities will use this criterion to evaluate the quality of the flows to the sewer. In the case of the flexibility of con-

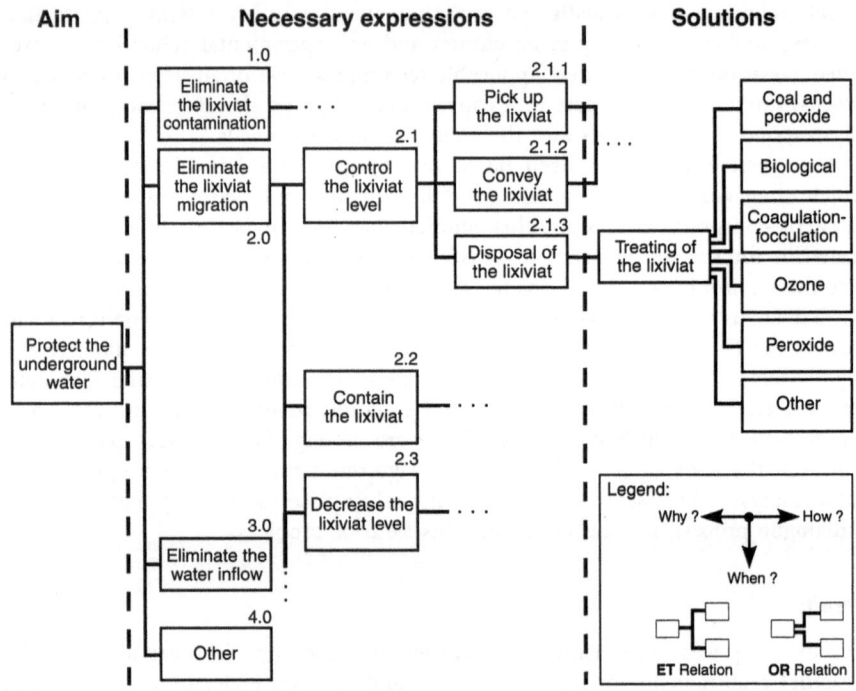

Fig. 1. Schematization of the functional analysis

Table 1. Characterization of the functional analysis

Number	Function	Criterion	Level	Flexibil.	Remarks
2.1.3	Disposal of lixiviated water (treatment)	Flow	2660 m³/d	0	Average
			6400 m³/d	5	Maximum
		Characteri-zation	Phenol composites	3	Up to 5 mg/L
			Sulphides		Up to 50 mg/L
			Toxicity		Potentially restricted
		Flow Norms	Phenol composites	0	Present: 1 mg/L Future: 0,5 mg/L
			Sulphides	0	Present: 5 mg/L Goal: 2,5 mg/L

centrations in the untreated water, it is inadequate to weight the sulphides and the toxicity since the concentrations are uncontrollable and they vary in time. However, a low flexibility exists in the phenol composites by controlling the sheet of water's level on the site. Finally, the norms don't offer any flexibility since they must be followed at all times.

In order to evaluate the treatment technology, a first selection was performed on the basis of a literary review. The technologies which weren't retained during the first selection, aren't excluded and could be reintegrated and reevaluated at a later date. The retained technologies are shown in Table 2. They include the classic technologies (coagulation-flocculation, biological treatment, etc.), costly technologies, (ozone, activated coal, etc.) and innovative technologies (peat and peroxide).

Consequently, the multidisciplinary team chose the criteria and weighted them. Certain criteria are traditional (efficiency, reliability, etc.) but others aren't. The latter can trouble the society (odor pollution, transportation of chemical products or sediments, etc.) or generate future treating administrators (easy access to development, execution delay, available expertise, etc.). Table 2 summarizes the values obtained for each of the solution-criteria couples as well as the results for the quality of each solution.

In order to better understand the weighting system, the multidisciplinary team's discussion could be summarized and could accompany the evaluation chart.

6.3.4
Discussion

The result analysis shows that the ozonization of the lixiviated water is the solution that offers the best quality according to the needs elaborated earlier. However, the best technology must be chosen on the basis of quality-price ratio. The main difficulty of a value analysis, especially when choosing a treatment technology, is assessing the cost. Indeed, the economical analysis, based on previous achievements, can involve a certain amount of risks, since the characteristics of the water which needs to be treated vary constantly and greatly influence the criteria of creation. Moreover, the principle of the safety factors existing in the traditional engineering field can result in efficiency losses and, consequently, cause the results to be falsely interpreted. Therefore, a technical demonstration program is essential to complement the value analysis. In order to minimize the investments in such a program, it is preferable to perform an efficiency preliminary validation of the technologies used in the laboratory and to carry out the pilot trials on site to get the creation's exact parameters. This program can be performed on the technologies in descending order according to their quality. In the case of the City of Montreal's CTED, the trials were based on the ozonization and on joint treatment of coal-peroxide. The trials performed in the laboratory confirmed the ozonization's efficiency and, surprisingly, they concluded that the ozonization was one of the most economical solutions. Since the interpretation of the value analysis was approved, CTED set the study out on the site.

Table 2. Evaluation of the proposed treating solutions

N°	Criteria	Evaluation (degree)	Oxidation		Adsorption		Biological	Coagulation-flocculation
			Ozone	Peroxide	Coal and peroxide	Peat and peroxide		
1	Phenolics	10	9/90	0/0	9/90	7/70	9/90	5/50
2	Sulphides	10	9/90	9/90	8/80	8/80	5/50	5/50
3	Reliability	8	9/72	9/72	8/64	8/64	3/24	5/40
4	Safety	10	7/70	5/50	5/50	5/50	9/90	8/80
5	Flexibility	9	10/90	5/45	8/72	8/72	5/45	9/81
6	Approval	10	9/90	0/0	8/80	8/80	8/80	8/80
7	Integration to director plan	6	9/54	8/48	8/48	7/42	5/30	5/30
8	Ability to operate	7	9/63	8/56	5/35	5/35	2/14	8/56
9	Implementation time	8	8/64	3/24	8/64	8/64	2/16	9/81
10	Expertise availability	7	9/63	6/42	8/56	7/49	9/63	9/63
11	Tranquillity	8	8/64	8/64	5/40	5/40	6/48	3/24
12	Odor elimination	6	9/54	9/54	8/48	9/54	6/36	6/36
Quality			864	545	727	700	586	671

In this context, it is important to emphasize that the university was involved in the project right from the beginning. It created the bibliographic journal, planned and carried out the trials performed in the laboratory, as well as the ones performed onsite. This integration is very positive for the students; they better understand the concrete problems. It justifies the financing of university research and leads fundamental research towards the applied field. This kind of service is offered more and more by Quebec's universities. Selective associations between the public and the private sector, as well as the universities, are interesting and advantageous for the engineering firms since the associations allow them to increase their overall knowledge at a low cost. They needn't manage the research laboratories; the idea providers (undergraduate and graduate students) are constantly renewed. However, this interweaving between the private sector and the universities requires for the work tasks among professors to be reorganized and for their ethical code to be redefined in order to retain their impartiality in the public's eye.

This choice of technology can also be used for educational purposes. Defining the problems and analyzing the value are part of the present trends that exist in the offering of applied courses. Students can, individually or in group, put their knowledge to practice and learn methods that will increase self-sufficiency. This "learning through problems" process must complement their theoretical knowledge. Self-sufficiency is the cornerstone of the training requested by the industrial world. It will allow the engineers to increase their efficiency and their productivity. The productivity notion must not be directly related to the quantity notion in a total quality situation; but it must be linked to high work performance. Under these conditions, the client's savings and trust will increase the productivity and the efficiency on a long-term basis. Moreover, the long-lasting operating notions will be greatly understood by the whole society. University training will last longer and will become more efficient.

6.3.5
Conclusion

The three-part experience for the choice of a treatment technology for lixiviated water of the City of Montreal's CTED has led to uncommon and interesting effects. The first effect is to zero in, more concretely, on the relations between the private sector (engineering firms and others), the public sector and the universities. The second is to permit the validation of a tool which can be easily used in order to advance the total quality idea and lasting development. Furthermore, this tool permits an understanding of the reasoning mechanism of experts in various fields (biologists, chemists, ecologists, sociologists, theologians, etc.), particular to the environment. When adapted and applied to environmental engineering, it will help develop student autonomy. This autonomy will later permit students to collaborate more efficiently within multidisciplinary teams. An engineer specializing in the environment can integrate ideas of purification at source, of revaluation, of reprocessing, of recycling and finally of restoration. Their mandate will permit the consideration of waste or pollution as a source of raw materials. This new type of engineer will be recruited by industries as he corresponds to the development and the orientation adopted by our society. Finally, the reevaluation of the role and code of ethics of professors in environmental specialization of the University du Quebec B Montreal, is perhaps the most fundamental effect of this experience.

6.3.6
Further Readings

Challenge for Sustainable Development
Chesapeake Experience: NPS Chesapeake
Bauereis E. I.
Baltimore Gas and Electric Co., MD. Environmental Programs.
Water Science and Technology WSTED4, Vol. 26, No. 12, p 2723–2725, 1992, Journal Announcement: SWRA2608
Abstract: Without quantification of compliance concepts as applied to non-point sources (NPS) in areas such as the Chesapeake Bay (Maryland), it is impossible to design control strategies that are effective, implementable, attainable, measurable, and flexible. The quantities required for each watershed are the existing loadings and the loading standard (or carrying capacity) of that watershed. The Chesapeake Experience has been reasonably successful with measurable reductions in phosphate and some other contaminants but recent analyses have identified NPS pollution as the culprit in potential failure to attain nitrogen reductions. The NPS loading

could also be implicated as a major source for some heavy metals, organics, and sediments to the Chesapeake Bay. These NPS loadings may become a major impediment to attaining designated uses of water bodies, and there is a need for focus on the loading quantification of NPS. There are three areas which will impact the Chesapeake NPS program positively in the future: the Toxics Research Program, the Clean Air Act of 1990, and the implementation of a cultural change embracing a philosophy of total quality management. The need for cost-effective controls and innovative methods to accomplish NPS goals is obvious. Land use issues are local jurisdiction issues as well as state and federal issues, which leads to confusion and even conflict over program objectives and overlapping authority. There is a need to provide understanding, greater choice, and more individual responsibility to attain better environmental stewardship.

If innovation is important to your organization's survival, how will TQEM help?
Johannson, Lynn.
Total Quality Environmental Management 1995, v 4 n 3, Spring p. 135–141 (7 pages) TQE 1055–7571.
Abstract: TQEM is a systematic approach to managing environmental issues within an organization's business process. Inherent to both TQEM and innovation is change. An organization must be flexible, and its people must have an attitude and willingness not only to accept change, but also to relentlessly seek out and grab onto change. TQEM philosophy, like its quality base, is founded on key commitments, some of which are: 1. a dedication to meeting or exceeding customer needs, which includes the environment, 2. an obligation to solving the core problem, and 3. an adherence to continuous improvement. It is adherence to continuous improvement that ties to innovation. Solving environmental problems requires new ideas, and a fresh look at resource management, business practices, and industry's traditional relationship with the planet. In the realm of TQEM, it is necessary to improve the management of resources in an entirely new way. An organization successful in TQEM will be the competition.

Economic incentives for total quality environmental management
Bhushan, Abhay K.
Corporate Source: Xerox Corp, Palo Alto, CA, USA
Conference Title: IEEE International Symposium on Electronics and the Environment
Conference Location: Arlington, VA, USA
Conference Sponsor: IEEE
Source: IEEE International Symposium on Electronics and the Environment IEEE Int Symp Electron Environ 1993. Publ by IEEE, IEEE Service Center, Piscataway, NJ, USA. p 173–177
ISBN: 0-7803-0830-1
Abstract: We show how the TQEM approach combines quality with environmental responsibility, and has powerful economic incentives that can improve profitability for any company. How the economic incentives have worked at Xerox Environmental Leadership program is also discussed. 4 Refs.

7 Environmental Trends and Policy Perspectives

5 Environmental Trends and Policy Perspectives

7
Environmental Trends and Policy Perspectives

Ravi K. Jain · Stephen P. Shelton

Environmental quality represents the basic life support systems of society and their relationships to one another and, as such, it is one of society's most important assets. Thus, understanding complex scientific and technical issues related to emerging environmental technologies and future trends based upon research and engineering practice is important. Environmental technologies are applied over a wide variety of public services such as water supply, water and wastewater treatment, air pollution control, solid waste management, and hazardous waste treatment and disposal. A contributing factor to the failure of many ancient societies was the inability to deal with environmental issues which were the by-products of their growth; thus, societies have risen and fallen in direct relationship to the intelligent use of their resources and management of their environment.

Technological innovation needs to lead to improved environmental quality, reduced costs and, thus, sustainable development. Investment in research and development can lead to technological innovation. There are often considerable impediments to implementing new technologies and to reaping benefits from technological innovation. This needs to be understood and policies developed to foster the *technology transfer* and *utilization* process.

Several issues relate to *environmental technologies and regulations*. Development of a clear, technical understanding of technologies is needed to effectively drive regulatory policy. Traditionally, environmental regulations and treatment technologies are viewed in a piecemeal manner focusing on a single media (e.g., water, air, soil, etc.). Arguably, a *holistic approach* to addressing environmental problems, i.e., a multimedia approach, is needed. With the considerable investment nations are making in the environmental area, *environmental technologies* and related *trends* and policy considerations become significant issues.

Increasingly, more cost effective means of building and maintaining environmental infrastructure are needed. *Privatization of the environmental infrastructure* is one approach that is becoming increasingly popular. In implementing environmental regulations and in order to achieve certain levels of environmental protection, increased use of *economic instruments* is being proposed as a prudent policy. Because industry produces a great majority of the goods and services that a society consumes, *industry trends* related to environmental compliance are of import.

Pollution prevention has become one of the important goals of the US Environmental Protection agency. Now issues going *beyond pollution prevention* such as *industrial ecology* are being postulated. Industrial ecology concepts try to mimic ecological systems where resources are used efficiently and waste products of one system become food or fuel for the follow-on system. All of the resulting information is then analyzed in a broader economic context.

These overarching issues related to environmental trends and policy are further elaborated on in the following sections.

7.1
Technology Transfer and Utilization

Numerous impediments exist in the path from technology development to its transfer and utilization. Fundamentally, the ability to transfer a new technology is limited by its utility. Utility encompasses such items as relative advantage, marketability, economic feasibility, and user acceptability. Trying to push a new technology that is marginally utilitarian is likely to result in failure in the end (Jain and Traindis, 1990). In addition to the characteristics of the technology, other impediments to technology utilization relate to characteristics of the organization, resource requirements and general behavior/related items. In the case of environmental technologies, the way regulations are formulated and implemented also plays an important role in the technology diffusion process.

7.2
Environmental Technologies and Regulations

The industrial world is currently faced with a massive need to plan, develop, and construct new environmental infrastructure and to repair and maintain existing facilities. The increasing needs of a growing society, coupled with an historical "out of sight, out of mind" attitude, have created a demand for efficient environmental facilities. As environmental regulations evolve, policy makers, scientists, and engineers will be challenged by the task of simultaneously maintaining existing environmental infrastructure and developing new and innovative technical strategies to address impending challenges.

Technical issues encompass many media: air, water, soil, and groundwater; and many contaminants that may be contained in those media. Development of a technical understanding of the issues to drive regulatory policy is a prerequisite to formulating cost efficient environmental management scenarios.

7.3
Holistic Approach to Environmental Problems

Many environmental scientists and engineers propose a holistic approach to understanding and solving environmental problems. The traditional piecemeal approach treats each environmental problem as an isolated entity, in most cases involving a single medium (water, air, soil, etc.). Solutions are often proposed to address these isolated problems, without further examination of the entire context of the problem. It would be prudent to view environmental problems in their larger context and solutions would then involve a multimedia approach. This would require policy makers to take a broad based holistic approach while moving away from incremental, uncoordinated, proscriptive, single-media regulations. A coordinated outcome-based approach to regulations, which recognizes the interrelatedness of both the causes of environmental problems and the effects of environmental solutions, will be needed to effectively coordinate the environmental policy of the future. Gibson (1994) discusses this trend and provides examples of the ways in which companies and agencies are beginning to use these new approaches. In addition, risk criteria is gaining momentum as a common platform form which the entire context of an environmental problem (e.g. human health, ecological, economical, and political concerns) may be referenced and it can provide a solid foundation from which solutions may be proposed.

In addition, risk assessment and risk management are currently gaining momentum as an effective holistic approach. Risk management possesses the unique ability to address human health, ecological, economical, and political concerns utilizing both the quantitative, technical tools of scientists and economists and also the qualitative, amorphous concerns of the public in general.

7.4
Environmental Forecasting and Technology Trends

Another important trend in environmental policy is the attempt to anticipate technology trends and future environmental developments, taking steps to avoid problems rather than just responding to them after the fact. This trend is discussed in *Beyond the Horizon; Using Foresight to Protect the Environmental Future*, published by the Environmental Futures Committee of the EPA in January 1995. After-the-fact responses are not effective enough and do not take effect quickly enough to protect important economic and environmental resources. The very creation within the EPA of this Environmental Futures Committee is a strong indicator of this trend toward anticipating rather than simply reacting to environmental problems.

As an example, a summary of the *Report of the Indoor Air Quality and Total Human Exposure Committee* (EPA-SAB-IAQ-95-005) is as follows: The Indoor Air Quality and Total Human Exposure Committee studied opportunities for advances in the science and art of human exposure assessment, and the opportunities that such advances could offer EPA and the nation for improving risk assessment and management. The committee recognized that significant advances could be made in three critical areas:

- Microsensor and microprocessor technologies;
- Biomarkers and exposure; and
- Database resources.

As a result of its study, the committee made five specific recommendations to EPA:

1) Develop a mechanism to support the research, validation, and application of:
   - more sensitive and specific microsensors, biomarkers, and other monitoring technologies and approaches for measuring exposures; and
   - validated data on associated exposure determinants, including demographic characteristics, time-activity patterns, locations of activities, and behavioral and lifestyle factors.
2) Establish a mechanism to develop, validate with field data, an iteratively improve models that integrate:
   - measurements of total exposure and their determinants;
   - a better knowledge of exposure distributions across different populations; and
   - the most current understanding of exposure-dose relationships.

3) Develop, in cooperation with other agencies and stakeholders, a robust database that reflects the status and trends in national exposure to environmental contaminants.
4) Develop sustained mechanisms and incentives to ensure a greater degree of inter-disciplinary collaboration in exposure assessment and, by extension, in risk assessment and risk management activities.
5) Take advantage of improving capabilities in exposure assessment technology, electronic handling of data, and electronic communications to establish and disseminate early warnings of emerging environmental stressors.

Other authors have discussed related issues. Lindsey (1994) suggests technological innovation to achieve pollution prevention (not producing waste and emission streams to start with) as a more cost-effective method than the current end-of-the-pipe treatment technologies being used. In *The Future of the Environment*, Duchin and Lange (1994) construct and analyze scenarios of possible futures, based on different economic and technological assumptions. They then evaluate the economic and environmental consequences of following a particular path over several decades. This attempt to provide a policy framework for building and evaluating alternative visions of the future exemplify the trend toward anticipation of future environmental problems.

7.5
Privatization of the Environmental Infrastructure

The concept of privatization of environmental utilities, such as water and wastewater treatment facilitates, is already well-accepted in Europe. Now that the federal government has begun to encourage privatization in the US, it is likely to become a major trend for the future. Many instances of partial privatization, in which public utilities hire private firms to manage selected plants or aspects of their operations, can be seen in the US today. Advantages of privatization include more efficient use of labor as well as reduced building and operating costs. This trend, as well as the remaining financial and legal obstacles to its full development, are discussed in Powers (1994) and Lindsey (1994).

7.6
Increased Use of Economic Instruments in Environmental Policy

Another trend identified by some experts is the move toward utilizing economic instruments such as pollution taxes, tradable permits, bubble concept, etc., as part of an overall economic policy to encourage environmentally responsible behavior. Economic instruments can be developed within the context of an overall policy and act as an incentive to encourage compliance. Barde (1994) traces the development of these incentives, discusses their present features and outlines obstacles and remaining questions regarding their future application. A corollary trend to the use of economic instruments is the removal of incentives such as subsidies that currently regard environmentally irresponsible behavior. This issue is discussed in Pope (1994). Considerable pioneering work related to designing market-based environmental strategies has been done under the sponsorship and direction of Wirth, Heinz, and Stavins (1988, 1991).

7.7
Industry Trends

In a recent survey by Price Waterhouse an increasing number of US corporations has placed the responsibility for environmental issues in the hands of their boards of directors. Companies are becoming more proactive on environmental issues, as evidenced by establishing policies and programs as well as the above-cited assignment of environmental responsibility to the board level. Other findings reported by Drake (1995) include:

- increasing number of companies are conducting environmental audits;
- public environmental annual reports are now issued by some companies;
- 38% of the Price Waterhouse respondents now factor environmental performance into incentive compensation for executives and management; and
- companies recognize liabilities for environmental responsibilities sooner than in the past.

Gibson (1994) points out that companies are also attempting to go beyond simple compliance with regulations, stressing instead an integration of environmental concerns in all decision-making and planning activities. Incorporating environmental goals into the company's mission has led to innovative solutions at companies such as Monsanto.

7.8
Industrial Ecology-Going Beyond Pollution Prevention

Under the Pollution Prevention Act of 1990, Congress established as a national policy that:

- pollution should be *prevented or reduced at the source* whenever feasible;
- pollution that cannot be prevented should be *recycled* in an environmentally safe manner whenever feasible;
- pollution that cannot be prevented or recycled should be *treated* in an environmentally safe manner whenever feasible; and
- *disposal* or other release into the environment should be employed only as a last resort and should be conducted in an *environmentally safe* manner.

Pollution prevention means *"source reduction"*, as defined under the Pollution Prevention Act and other practices that reduce or eliminate the creation of pollutants. This is a sound concept. This, however, focuses primarily on technology issues and does not adequately encompass economic and industrial productivity issues.

The concept of industrial ecology incorporates technology, industrial productivity and economic issues. This provides for an analysis of energy and material flow in technological and economic contexts and attempts to minimize wastes produced by the total system (Allenby, 1994; Graedel, 1995).

Industrial ecology mimics ecological systems where resources are used efficiently and waste products of an organism become food or fuel for another, thus creating an efficient natural equilibrium.

In a similar manner, in industrial processes, wastes from one industrial process become raw material or fuel for another and the wastes produced by the larger system as a whole are minimized. Further, all of this is accomplished in an economic context, i.e., keeping economic efficiency, utility and equity in mind. Since economic considerations would differ in different industrial processes and in different nations, this concept can provide a framework for environmental policy responsive to industrial processes, organizations, and nations at different stages of development.

7.9
Summary

Improvements in the quality of life, protection from disease, and preserving the long-term viability of the life support system are the benefits of effective environmental policy. Political systems are often the vehicle used to apply institutional pressure to further enhance environmental quality. Unfortunately, many of these pressures have been misdirected by those who write and enforce environmental regulations. At the expense of maintaining and renovating viable existing systems, socially correct but technically deficient regulations have often mislocated resources by funding program with exotic solutions to simple problems. Furthermore, many environmental laws and regulations generate institutional disincentives for progress. The laws typically require a public agency to promulgate regulations as a framework for compliance. In so doing, regulations have become increasingly more technology-specific, thereby forcing use of existing technologies rather than encouraging development of new, more innovative solutions. Issues discussed here may provide some new policy perspectives.

References

Allenby BR, Richards DJ (1994) The Greening of Industrial Ecosystems. National Academy of Engineering: National Academy Press
Barde J, Opschoor JB (1994) From Stick to Carrot in the Environment. The OECD Observer, 1994 (186), 23–27
Drake T (1995) US Companies Put Environmental Responsibilities on Top of Priority List. Water Engineering & Management 142(3), 16
Duchin F, Lange G (1994) The Future of the Environment: Ecological Economics and Technological Change, New York: Oxford University Press.
Environmental Futures Committee (1995) Beyond the Horizon: Using Foresight to Protect the Environmental Future, Washington, DC: Report Prepared for the US Environmental Protection Agency
Gibson WD (1994) Waste Management '94: The Big Picture. Chemical Marketing Reporter 246 (23), SR8
Graedel TE, Allenby BR (1995) Industrial Ecology, Englewood Cliffs, NJ: Prentice-Hall
Jain RK, Triandis HC (1990) Management of Research and Development Organizations, New York: John Wiley & Sons.
Lindsey AW (1994) Environmental Technology Innovation: A Twenty-first Century Imperative. Environmental Progress 13(4), N2–N3
Pope C (1994) Looking for Gold in All the Wrong Places. Vital Speeches of the Day, 60(24), 759–61
Powers MB (1994) Bureaucrats Begin Leaning Toward Privatization of Infrastructure. Engineering News Record (ENR) 233(21), 46–50
Wirth TE, Heinz J, Stavins RN (1988) Project 88: Harnessing Market Forces To Protect Our Environment: Initiatives For the New President, Washington, DC
Wirth, TE, Heinz J, Stavins RN (1991) Project 88 – Round II: Incentives for Action: Designing Market-Based Environmental Strategies, Washington, DC

Editors

Ravi K. Jain, Ph. Dr., P. E., DEE. Associate Dean for Research and International Engineering, College of Engineering, University of Cincinnati, Mail Location 0018, Cincinnati OH 45221–0018, USA

Yves Aurelle, Prof. Dr., INSA Toulouse, Complexe Scientifique de Rangueil, Directeur du Departement G. P. I., 31077 Toulouse Cedex, France

Corinne Cabassud, Dr., INSA Toulouse, Department Genie des Proced Industriels, Complexe Scientifique de Rangueil, 31077 Toulouse Cedex, France

Michel Roustan, Ph. Dr., INSA Toulouse, Departement Genie des Proced Industriels, Complexe Scientifique de Rangueil, 31077 Toulouse Cedex, France

Stephen Shelton, Ph. D., Department of Civil Engineering, University of New Mexico, NM 87131 Albuquerque, USA

Principal Authors

Gérard Antonini, Professor, Université de Technologie de Compiègne, Division Génie des Transferts et Energétique, Département Génie Chimique, B. P. 64206 Compiègne Cédex, France

Philippe Aptel, Université Paul Sabatier, Laboratoire de Génie Chimique et electrochimie – URA CNRS – 192, 118, Route de Narbonne, 31062 Toulouse Cédex, France

Michel Astruc, Professor, Université de Pau et Pays de l'Adour, Laboratoire de Chimie Analytique, Avenue de l'Université, 64000 Pau, France

Jean-Marc Audic, Senior Research Engineer, Centre International de Recherche sur l'Bau et l'Environment (Research Center of Lyonnaise des Eaux) 38, Rue du Président Wilson, 78 230 Le Pecq, France

Dominique Bastoul, Associate Professor, Institut National des Sciences Appliquées de Toulouse (INSA), Laboratoire d'Ingéniérie des Procédés de l'Environnement, Complexe Scientifique de Rangueil, 31077 Toulouse Cédex, France

Nouredine Bel Hadj Tahar, Ph. D. Student, Groupe Génie des Procédés de Séparation et Membranes, Laboratoire de Génie Chimique et Electrochimie – URA CNRS – 192, Université Paul Sabatier, 118, Route de Narbonne, 31062 Toulouse Cédex, France

Vincent Boisdon, Research Engineer (Research Center of Compagnie Generale des Eaux), Anjou Recherche, Chemin de la Digue, BP 76, 78850 Maisons Laffitte, France

Christophe Bonnin, Gorup Manager, Anjou Rcherche (Research Center of Compagnie Générale des Eaux), Chemin de La Dique, BP 76, 78603 Maisons Laffitte, France

Richard C. Brenner, Environmental Engineer, U.S. Environmental Protection Agency, 26 W. Martin Luther King Drive, Mail Stop 420, Cincinnati, OH 45268, USA

Bernard Capdeville, Professor, Institute National des Sciences Appliquées de Toulouse (INSA), Laboratoire d'Ingéniérie des Procédés de l'Environnement, Complexe Scientifique de Rangueil, 31077 Toulouse Cédex, France

Michael Clifton, Senior Scientist, Groupe Génie des Procédés de Séparation et Membranes, Laboratoire des Génie Chimique et Electrochimie – URA CNRS – 192, Université Paul Sabatier, 118, Route de Narbonne, 31062 Toulouse Cédex, France

Hubert Debellefontaine, Professor, Institut National des Sciences Appliquées de Toulouse (INSA), Laboratoire d'Ingéniérie des Procédés de l'Environnement, Complexe Scientifique de Rangueil, 31077 Toulouse Cédex, France

Géraldine Deiber, Ph. D. Student, Institut National des Sciences Appliquées de Toulouse (INSA), Laboratoire d'Ingéniérie des Procédés de l'Environnement, Complexe Scientifique de Rangueil, 31077 Toulouse Cédex, France

Bernard Delanghe, Associate Professor, Université de Pau et Pays de l'Adour, Laboratoire de Chimie Analytique, Avenue de l'Université, 64000 Pau, France

John D. Ditmars, Manager, Environmental Sciences and Engineering Group, Environmental Assessment Division, Argonne National Laboratory, 9700 South Cass Ave.-EAD/900, Argonne, IL, 60439, USA

Christian Fonade, Professor, Institut National des Sciences Appliquées de Toulouse (INSA) Département GBA – URA CNRS 544, Complexe Scientifique des Ragueil, 31077 Toulouse Cédex, France

Jean-Noel Foussard, Associate Professor, Institut National des Sciences Appliquées de Toulouse (INSA), Laboratoire d'Ingéniérie des Procédés de l'Environnement, Complexe Scientifique de Rangueil, 31077 Toulouse Cédex, France

Donald J. Freeman, P. E., Environmental Engineer, US Army Armament Research Development and Engineering Center (ARDEC), Picatinny Arsenal, New Jersey 07806-5000, USA

Dimitre Hadjiev, Associate Professor, Institute Universitaire de Technologie, Université de Caen, 14000 Caen Cedex, France

Robert Hausler, Professor, Université du Québec à Montréal, Case postale 8888, succursale Centre-Ville, Montréal (Québec), Canada, H3C 3P8

Edmond Julien, Associate Professor, Institut National des Sciences Appliquées de Toulouse (INSA), Laboratoire d'Ingéniérie des Procédés de l'Environnement, Complexe Scientifique de Rangueil, 31077 Toulouse Cédex, France

Germain Lacoste, Professor, ENSIGC, 18, Chemin de la Loge, 31078 Toulouse Cédex, France

Alain Laplanche, Professor, Ecole National Supérieure de Chimie de Rennes, Laboratoire Chimie des Nuisances et Génie de l'Encironnement, Avenue du Général Leclere, 35700 Rennes, France

Pierre Le Cloirec, Professor, Ecole National Supérieure des Techniques Industrielles et des Miens d'Alés, 6, Avenue de Claviéres, 30319 Alés Cédex, France

Xavier Lefebvre, Research Engineer, Institut National des Sciences Appliquées de Toulouse (INSA), Laboratoire d'Ingéniérie des Procédés de l'Environnement, Complexe Scientifique de Rangueil, 31 077 Toulouse Cédex, France

Marie-Hélène Manero, Associate Professor, IUT Génie Chimique, Chemin de la Loge, 31078 Toulouse Cédex, France and Institut National des Sciences Appliquées de Toulouse (INSA), Laboratoire d'Ingéniérie des Procédés de l'Environnement, Complexe Scientifique de Rangueil, 31077 Toulouse Cédex, France

Hugo Matamoros, Ph. D. Student, Institut National des Sciences Appliquées de Toulouse (INSA), Laboratoire d'Ingéniérie des Procédés de l'Environnement, Complexe Scientifique de Rangueil, 31077 Toulouse Cédex, France

Michel Mauret, Research Engineer, Institut National des Sciences Appliquées de Toulouse (INSA), Laboratoire d'Ingéniérie des Procédés de l'Environnement, Complexe Scientifique de Rangueil, 31077 Toulouse Cédex, France

Pierre Mocho, Ph. D. Student, Ecole National Supérieure des Techniques Industrielles et des Mines d'Alés, 6, Avenue de Claviéres, 30319 Alés Cédex, France

Philippe Moulin, Ph. D., Groupe Génie des Procédés de Séparation et Membranes, Laboratoire de Génie Chimique et Electrochimie – URA CNRS – 192, Univesité Paul Sabatier, 118, Route de Narbonne, 31062 Toulouse Cédex, France

Etienne Paul, Lecturer, Institut National des Sciences Appliquées de Toulouse (INSA), Laboratoire d'Ingéniérie des Procédés de l'Environnement, Complexe Scientifique de Rangueil, 31077 Toulouse Cédex, France

Jane E. Rael, Ph. D. Candidate, Department of Civil Engineering, University of New Mexico, Albuquerque, NM 87131, USA

Jean-Christophe Rouch, Research Engineer, Groupe Génie des Procédés de Séparation et Membranes, Laboratoire de Génie Chimique et Electrochimie – URA CNRS – 192, Université Paul Sabatier, 118, Route de Narbonne, 31062 Toulouse Cédex, France

André Savall, Professor, Université Paul Sabatier, Laboratoire de Génie Chimique et Electrochimie – URA CNRS – 192, 118, Route de Narbonne, 31062 Toulouse Cédex, France

Christophe Serra, Ph. D. Student, Groupe Génie des Procédés de Séparation et Membranes, Laboratoire de Génie Chimique et Electrochimie – URA CNRS – 192, Université Paul Sabatier, 118, Route de Narbonne, 31062 Toulouse Cédex, France

Ed Dean Smith, Chief, Environmental Engineering Division, US Army Construction Engineering Research Laboratory, 2902 Newmark Drive, Champaign, IL 61821, USA.

Francis L. Smith, Graduate Assistant, University of Cincinnati, Civil & Environmental Engineering, P. O. Box 210071, Cincinnati, OH 45221-0071, USA

Ian Smith, Senior Research Fellow, Department of Mechanical Engineering, University of Southampton (UK)

George A. Sorial, Senior Research Associate, University of Cincinnati, Civil & Environmental Engineering, P. O. Box 210071, Cincinnati, OH 45221-0071, USA

Makram Suidan, Schneider Professor and Department Head, University of Cincinnati Civil & Environmental Engineering, P.O. Box 210071, Cincinnati, OH 45221-0071, USA

Martin Thew, Professor of Fluid Technology, Mechanical Engineering, University of Southampton (UK)

Hany Zaghloul, Environmental Engineer, US Army Construction Engineering Research Laboratory (USACERL), P. O. Box 9005, Champaign, IL 61826-9005

Other Contributors:

Susan Bill, Coordinator, Professional and Executive Development, Research and International Engineering, University of Cincinnati, P. O. Box 210018, Cincinnati, OH 45221-0018, USA

Todd Bragdon, Graduate Student, Civil and Environmental Engineering, University of Cincinnati, P. O. Box 210071, Cincinnati, OH 45221-0071, USA

Shafiq Islam, Assistant Professor, Civil and Environmental Engineering, University of Cincinnati, P. O. Box 210071, Cincinnati, OH 45221–0071, USA

Andrew Martyniuk, Research Associate, Research and International Engineering, University of Cincinnati, P.O. Box 210018, Cincinnati, OH 45221–0018, USA

Jean-Pierre Soula, Professor, Centre de Communication et de Gestion, INSA Toulouse, Complexe Scientifique de Rangueil, 31077 Toulouse Cedex, France

Donna Vitucci, Consulting Editor, Research and International Engineering, University of Cincinnati, P.O. Box 210018, Cincinnati, OH 45221–0018, USA

Francoise Voillot, Directeur, Service des Relations Exterieures et Interantionales, INSA Toulouse, Complexe Scientifique de Rangueil, 31077 Toulouse Cedex, France

Subject Index